INTRODUCTION TO CHEMICAL PROCESS TECHNOLOGY

Introduction to Chemical Process Technology

P. J. VAN DEN BERG and W. A. DE JONG

Delft University of Technology,
Laboratory of Chemical Technology, Delft, Holland

Delft University Press

D. Reidel Publishing Company

Library of Congress Cataloging in Publication Data

Berg, P. J. van den.

Introduction to chemical process technology.

1. Chemical processes. I. Jong, W. A. de, joint author. II. Title.
TP155.7.B46 1979 660.2 80–13300
ISBN 90–277–1099–6
ISBN 90–277–1100–3 pbk.

Published by Delft University Press, Mijnbouwplein 11, 2628 RT Delft, Holland
and by D. Reidel Publishing Company, P.O. Box 17, 3300 AA Dordrecht, Holland

Sold and distributed in the U.S.A. and Canada
by Kluwer Boston Inc., Lincoln Building,
160 Old Derby Street, Hingham, MA 02043, U.S.A.

In all other countries, sold and distributed
by Kluwer Academic Publishers Group,
P.O. Box 322, 3300 AH Dordrecht, Holland

D. Reidel Publishing Company is a member of the Kluwer Group

Printed in The Netherlands

TABLE OF CONTENTS

PREFACE

This book has developed out of a lecture course given at Delft University of Technology to undergraduate students of chemical engineering, chemical technology and applied chemistry. Starting from the chemical reactions, it analyzes the process steps required to produce chemicals from various raw materials and shows how the choice of a process depends not only on technological factors but also on external effects such as social and economic developments, political factors affecting the availability of raw materials, environmental legislation and many others. As a result, several solutions to a given problem have often been developed and are practised side by side. In short, the chemical and allied process industries are viewed as a subsystem of a dynamic system of much higher order.

The book demonstrates analogies between processes for different products or based on different raw materials and aims at getting a feeling for process development strategy. The approach is purely qualitative and does not cover the entire industry; rather, examples are given to demonstrate the principles of process technology.

When teaching a course on the subject matter starting from this book it is obviously necessary to add recent material from the literature because the field covered is in rapid development. The lecturer may also find it necessary to add problems for students or to use this book in conjunction with preliminary design studies by students.

The material presented has been reviewed by many colleagues and friends working in the chemical and oil industries. Their constructive criticism and help in finalizing the text is gratefully recorded here. Many thanks are also due to Mrs. Monna, Wisgerhof and Van der Loo-Vreeburg, who helped in typing and preparing the manuscript.

Delft, May 1980.

This book provides a qualitative analysis of chemical processes by showing the factors determining process structures. These include the underlying chemistry but also feed and product specifications, methods of mass and heat transfer, reactor type and many others. External factors are also taken into account, such as economic developments, availability of raw materials and environmental aspects.

Quantitative engineering aspects are not treated. Rather, the analogies between different processes and operations are emphasized demonstrating the logic in the structure of processes as well as the main technological principles. Although many industrial processes are analyzed the entire chemical industry is not covered.

The matter presented here has been reviewed by many colleagues and friends working in the chemical and oil industries.

ABOUT THE AUTHORS

Pieter J. van den Berg is chairman of the Institute of Chemical Technology, Delft University of Technology, the Netherlands. Willem A. de Jong is professor in chemical technology at the same Institute.

Both authors lecture in process design and development as well as in chemical reactor engineering. Their research interest is in the field of conversion of hydrocarbons, urea and other nitrogen technology, gas purification processes and generally in the field of developing laboratory chemistry to industrial processes.

Professor De Jong is also chairman of the Board of T.N.O. (the Netherlands Organization of Applied Scientific Research).

Both professors are author or co-author of more than 100 papers on chemical technology.

I. INTRODUCTION

Four periods in chemical technology

1. Period of the crafts

The discovery of fire and its use in cooking food and forging meant the emergence of 'technique' in history. (A technique is an often repeated working method. The man who masters such a technique is called a technician or craftsman.) The rise of different crafts also brought about the birth of chemistry. In early history chemistry played a role in the making of pottery for the kitchen and in beautifying the ceramics by glazing. The Egyptians used lead oxide, cobalt compounds and iron oxide to make enamels. Also stemming from them is the discovery of fermentation which first produces alcohol and, with continued reaction, acetic acid. The latter enabled the first preparation of a color pigment:

lead oxide + acetic acid → white lead

The apparatus used in this craft consisted of colander, filter-cloth, sieve, oven, mortar and pestle etc. A strong impulse to the development of empirical chemistry was also induced by the refining of gold and copper.

Primitive chemistry, however, was not concerned exclusively with such empirical crafts, it was also important in the defense against unattractive climatic conditions. Very soon people knew how to burn limestone and gypsum and how to use these materials in building. Chemistry made it possible to tan hides and to dye wool for clothing, using natural products as the basic materials (for example, madder or cochineal extracted from wood lice, and indigo dye, which was extracted by fermentation of the indigo plant). The dye craft has been a great stimulus for the development of chemistry: textiles were treated by Fuller's earth and soda solution and then dyed. Later, soap was produced from alkali and fatty oils. In the textile industry aluminum salts were used to mordant the fibers. Tin salts produced scarlet. A yellow dye was made from safflower, giving a green color when used mixed with indigo. A blue dye was produced from Campeachy-wood.

At the same time the visual arts were developing, using pigments such as iron oxide, red-lead, arsenic sulfide and yellow lead oxide. Many chemical conversions resulting in these pigments had been discovered by the Egyptians who named their land 'chem'. This is also true for another application of colors: cosmetics. Cleopatra accentuated her natural beauty by these chemicals.

Such crafts were associated with a craftsman (or -woman). Any understanding of the conversion of matter was non-existent. The knowledge was 'know-how' and far removed from 'know-why'.

A better insight into chemistry originated in the seventeenth century when Boyle defined exactly what is an element ('Sceptical Chymist', 1661), Lavoisier discovered the law of conservation of mass and Dalton attributed typical properties to atoms. This better understanding of chemistry was very important in the eighteenth century, for in that period an untenable situation had risen. To meet the ever increasing demands for potash, enormous amounts of wood had to be burned and the ash extracted by water. In Canada alone 4 million tons of trees were felled. The Académie Française tried to find a solution and offered in 1775 a reward of 2400 livres for the manufacture of soda ash to replace potash. On the basis of research elsewhere, rock salt appeared to be the most attractive basic material. The personal physician of the Duke of Orleans, Leblanc, succeeded in converting salt to soda by reaction with sulfuric acid, reducing the resulting sodium sulfate with hot coke to give sodium sulfide which, when treated with carbon dioxide from limestone, reacts to give sodium carbonate. This process was a great success. No limits were set to the manufacture of glass and soap. Unfortunately, Leblanc did not receive his prize. He was swept away by the Revolution and committed suicide in 1806 after the government annexed his patent.

2. Period of industrial development

With the Leblanc process a period of fast development started - the true birth of the chemical industry. The crafts are now pushed back; much human labor is taken over by machines (the spinning jenny, the steam engine); the production of goods takes place in 'manufactories'. Thanks to this Industrial Revolution the industrial products benefit a fast growing population. The role of agriculture diminishes.

The Leblanc process was the first process to be carried out on an industrial scale. Charles Tennant's factory, where more than 1000 people worked, covered an area of 100 acres in 1830. The process was not without its problems, one of the largest being the liberation of hydrochloric acid gas. Tennant solved this first by building a stack, 140 meters high. The population was not very happy with this solution and the chemists regarded it as a loss. Under these circumstances there was an impulse for Deacon to look for an alternative solution, which he found in the oxidation of the hydrochloric acid to chlorine using copper catalyst supported on pumice-stone. This became a process of great importance for the cotton industry which started to use bleaching powder produced by reacting slaked lime and chlorine. The Leblanc process appeared on the way to becoming a nucleus for further industrial development. Soda ash manufacture required sulfuric acid, made initially by oxidation of sulfur by nitric acid. Muspratt, however, developed a continuous process for sulfuric acid production - the lead chamber process.

The change from discontinuous to continuous processes was an important event. Many developing techniques show such a transformation: e.g., the roller displaces the discontinuous pace, the plough supersedes the spade, and the propeller ousts the paddle. It is the high efficiency associated with the continuous mode of operation which makes it possible to reach high production. It is an important technological principle.

Such a change also manifested itself in the soda ash industry. Solvay invented a process, starting from salt, which used ammonia and carbon dioxide to convert it to sodium bicarbonate which was heated to produce soda ash. The main feature is that it can be carried out in a continuous manner. Economically the Solvay process has surpassed the Leblanc process. In fact, it is so efficient that nowadays the same process is still in use. Nevertheless, the first Solvay plants have been cause for anxiety. Wisdom was gained only by bitter experience, technique being more advanced than the underlying scientific basis (= technology).

The rapid increase in chemical knowledge also produced, in the nineteenth century, the synthesis of indigo by Perkin, achieved by using an, at that time, almost worthless by-product of coal-gas manufacture: tar. In Germany the first dyestuffs plants were constructed.

3. Period of scientific development

A third and completely new era was ushered in about sixty years ago by the ammonia synthesis of Haber and Bosch. Haber's laboratory studies of the formation of ammonia from hydrogen and nitrogen were taken out of the University by Bosch, who was the engineer responsible for translating the laboratory data to an industrial process. Bosch's discoveries make it possible to carry out the ammonia synthesis at high temperatures and very high pressures. The ammonia produced on a large scale is nevertheless relatively cheap. This is an important advantage because Crookes forecasts, around the turn of the century, that the increasing world population will suffer hunger because of the short supply of natural fertilizers.

The development of the Haber-Bosch process was therefore tremendously important, based as it was on an industrial application grounded on pure scientific research. A systematic investigation was also necessary to develop efficient catalysts for the process. It is justified to call the impact of science in this way a scientific revolution.

Already at the start of this new era it was felt by some that different chemical industries have a common basis in their engineering. In 1885 Armstrong tried for the first time to combine an education in physics with the training of a works engineer. Around 1900 Davies, in Manchester, described for the first time the operations which different chemical industries have in common. He classified these in the same manner as turning, gear-cutting or constructing in other industries. The accent was on handling matter, expecially two-phase (or multiphase) mixtures. One objective was 'mixing', the achievement of the intensive contact between the phases necessary for chemical reactions to occur. Techniques employed were: crushing, grinding, blending, dispersion etc. An inevitable second objective was 'separation', necessary in order to isolate desired products. Such techniques as sieving, filtration, distillation etc. were used to this end.

Slowly a general chemical-technical discipline emerged from these skills. In chemistry, also, a more general knowledge was developed, in addition to specific knowledge pertaining to single compounds. Simultaneously with the scientific background of techniques (= technology), for which a combination of different disciplines like physics and mechanics is essential, chemical knowledge was developed in thermodynamics, chemical kinetics, catalysis and the structure and properties of matter. Such a scientific background remarkably accelerates the development of techniques. Inventions are no longer a process of haphazard discovery but are the results of systematic investigation.

The character of the chemical conversion associated with a change in material properties brings about an immense diversity. Possibilities have now become available not only to produce a large number of products from one basic material, but also one special product from many different starting materials. This leads to integration and conglomeration, and even pipeline systems are introduced connecting the various production centers in a vast network. In this way maximal use is made of production on a large scale. It necessitates cooperation.

4. Period of electronic revolution

It is now clear that we are in a fourth period of development. Another type of labor is made superfluous, comparable with the replacement of muscle labor by the steam engine in the last century: mental repetitive work is taken over by computers and other electronic instruments. They also collect, compile, compare and process more information than people can ever do. Therefore, new insight is gained into actual processes, both in production and laboratory equipment. Many processes can now be better controlled and automated. In this way not only a maximal production is guaranteed but also quality and safety can be improved. Although we are only at the beginning of this electronic revolution, its effects have become noticeable in production, research and development as well as in plant design.

5. Further outlook

With increasing industrialization it is now clear that the chemical industry in developed countries feels the impact of scarcity of raw materials and energy because it has grown faster than the rate at which sources of raw materials can be found and exploited. Much of modern industry is based on oil and natural gas. It is indeed a challenge to use coal as a raw material. This, however, may not be necessary if a timely switch can be made from part of the oil to coal as an energy carrier (power stations, large ships etc). Today only 7% of all the oil produced world-wide is used as a raw material for the chemical industry. The higher costs of materials and energy will inevitably lead to more complicated processes because of the tendency to recover heat more completely and to prevent thermal losses. A major impact on industrial development could come from nuclear energy, but its acceptance for large-scale use is as yet uncertain.

Another area of problems is the increased burden on air, water and soil. This will undoubtedly lead to more research on higher efficiency, the cleaning of stack gases and effluents etc. Moreover, further attention must be given to safety in and around plants. For optimal safety engineering many more data are required about reliability of equipment and instruments, and more studies must be made on non-stationary states of processes.

Finally, the economic development of non-industrial countries will make it necessary to design processes and plants under new constraints. The development of oil-producing countries, in the Middle-East, North Africa and Asia, will be particularly stimulating in this respect.

II. STRUCTURE OF PROCESS SYSTEMS

Definition

Chemical technology, in the preceding chapter described as the scientific background of techniques used in the chemical industry, is defined as the science of those operations which convert raw materials into desired products on an industrial scale, applying one or more chemical conversions. (τεχνη = skill; λογος = theory)

Process structure

The route along which a raw material is converted to products is a logical coupling of interconnected operations: the process. Such a coordinated set is a system containing subsystems. At least one of the subsystems (process units) is the chemical reactor in which chemical conversion takes place. Process units preceding the reactor prepare the feed and succeeding units treat the converted mass, often including separation steps.
The first steps in a process, such as the mechanical operations of crushing or grinding, are often followed by physical treatments such as mixing, heating, evaporating. Such process steps are described by theories based on conservation of mass, energy and impulse and are generally called 'unit operations'. Unit operations succeeding the reactor are also physical in nature. Final treatment is often mechanical (e.g. agglomeration, packing). In practice the system is more complex. Raw materials generally are impure or even consist of mixtures of different compounds.

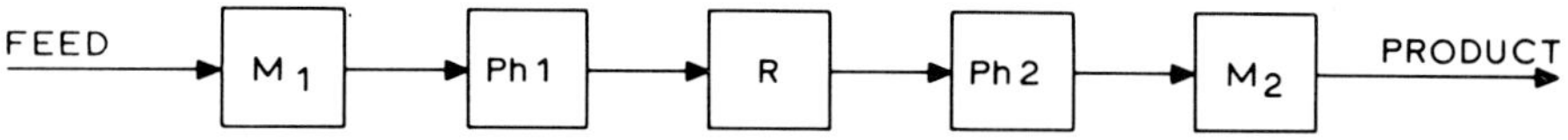

Figure II.1. Process structure

Every chemical reaction shows in principle incomplete conversion and often the formation of by-products. In this way the complexity of the mixtures increases.

Auxiliary materials are often used, which must be separated in another process step. Hence, for all these reasons a plant may contain a large number of separation steps. With any separation step at least two flows result from one incoming flow. Unconverted feed components and auxiliary chemicals will generally be recycled, if possible.

Intermediates and by-products are, where possible, converted to useful products in separate reactors. All remaining flows are waste streams which must be brought into an acceptable condition before leaving the plant. Ideally only air and water are emitted. When all process steps and flows are put together a complicated network results.

The process designer bases the plant on the production capacity of finished products, composition (specification) of product and raw materials, and the permitted emission level of waste stream components that may contaminate the environment. These data determine the inputs and outputs of the installation. The process must further be based on sound economical considerations, safety requirement and labor conditions.

The consequence of this way of thinking is that a process designer regards a plant first and foremost as a collection of operations connected by a network of pipes. It requires a certain amount of 'thinking in systems' to design a process optimally. Therefore standardized symbols are used in process presentation which represent unit operations. (Figure II.2). The resulting network is called a 'flow chart'. Such a flow chart generally contains the mass and heat balances for all the operations. Clearly a flow chart can be quite complicated, the more so as every unit process can be regarded as a subsystem. For example, a distillation (Figure II.3.) is a combination of physical operations involving evaporation, condensation, heating and cooling.
One of the tasks of a process engineer is to choose from many possibilities, taking into account the many conditions set by product markets, geographical location of the plant, the social situation, legal regulations etc. This is not only important in choosing among existing processes but also in developing new processes. A prerequisite is that the final result must be economically attractive.

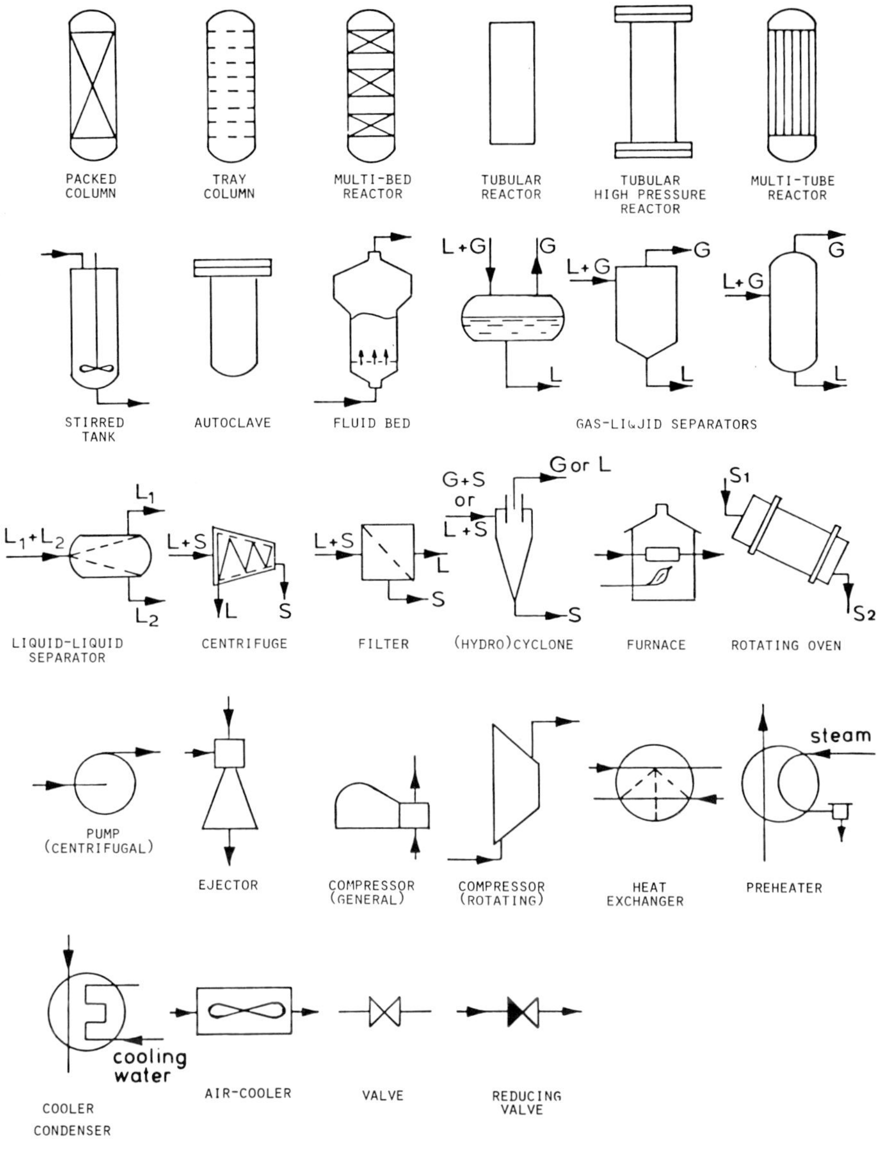

Figure II.2. Symbols used in flow charts

These activities are not merely a straightforward application of the scientific disciplines on which chemical technology is based (chemistry, physical transport processes, unit operations and reactor design). It is necessary to select relevant knowledge from these fields, combine different aspects and interpret these quantitatively. This means integration of knowledge from various fields of science. With such complicated systems as chemical processes some

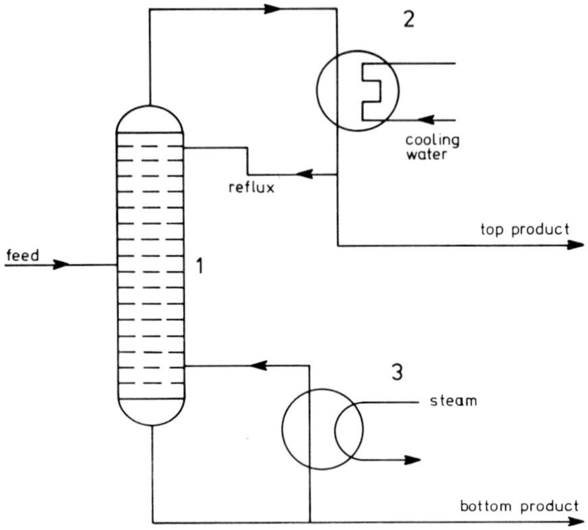

Figure II.3. Flowchart of distillation

1. distillation column

2. condenser

3. reboiler

of the questions will inevitably be answered in a semi-quantitative or even qualitative way. In those cases experience is a necessary asset to assist sound intuition and judgement.

A special complication is the fact that the chemical industry together with suppliers of raw materials and customers for products constitutes an extensive dynamic system of a higher order. Changes in the raw material situation, market relations, availability and prices of energy and, of course, politico-economical factors alter the picture constantly. A striking example is the expected dearth of oil in the future as a feedstock and a source of energy.

A checklist of basic questions

Because so many different chemical processes exist it is necessary to structure the analysis of processes. Table II.1. shows an approach which lists certain basic questions and describes the need for essential data.

Table II.1. Analysis of processes

Basic questions

1. Is continuous or discontinuous processing to be preferred ?
2. Which regions of process conditions are optimal ?
3. Which process conditions are dangerous ?
4. If the process is continuous: is spread in residence time acceptable ?
5. How is the reaction temperature reached ?
6. Which type of reactor (based on the answers to questions 1-4) is to be preferred ?
7. Is pretreatment of the reactor feed necessary ?
8. How is the reaction mixture processed ?
9. Have special measures to be taken in relation to coproducts and waste ?

To answer these questions information is necessary about:

the total process:
- objective(s)
- capacity
- frequency of application
- availability of raw materials
- destination of products
- alternative routes
- safety aspects

the reaction system:
- stoichiometry of the main reaction
- stoichiometry of parallel or series reactions
- composition at chemical equilibrium
- chemical kinetics and catalysis
- influence of impurities
- phases in reaction system
- phase in which the chemical reaction occurs
- relative movement of the phases
- transport limitations of mass and heat
- corrosivity

the other process units:
- feed requirements
- product specification
- materials specifications

- safety aspects
- nature and destination of by-products

Chemical process development

For the design of chemical plants calculations are made based on experimental data. Different classes of data may be distinguished:

- Properties of matter found in the literature. These have been measured experimentally by others and it is always necessary to check with different sources. Where in doubt they should be measured again.

- Properties of products and materials specific to the process and its conditions. These have to be measured under the anticipated plant conditions. Often tests of long duration are necessary.

- Kinetic data pertaining to the process. It should be realized that the chemical kinetics may be 'falsified' by the rate of physical transport processes, even to the extent that the latter determine the production rate. Chemical kinetics are determined by molecular properties alone and can be measured in laboratory equipment. It is often necessary to determine the kinetics under steady state conditions in a continuous reactor. If physical rate processes influence the kinetics the size of the equipment is a relevant parameter and, in general, additional research is needed to study these physical processes in model experiments. It is wise to carry out the final tests of a complex process in equipment on a larger scale. Quite often kinetic data from existing full-scale installations may be obtained.

- Data on the complex set of phenomena that play a part in 'handling' two-phase or multi-phase mixtures, highly viscous liquids etc. It is often necessary to study these aspects in large-scale equipment and it is not uncommon to arrange tests with large batches of experimental products or feedstocks with suppliers of equipment before deciding on a process design.

Process development in a laboratory is naturally always on a small scale but is not always limited to batch experiments in glassware. Continous experiments, not only in glassware but also in metal equipment can give a wealth of data under conditions expected for the plant. The set-up is often called a 'mini-plant' and is generally a bench-scale unit. It is essential that, even in this laboratory phase, the research work must be directed towards well-defined objectives, including different possibilities of process design and potential problems of a commercial installation. Gaps in the available knowledge and special problems are then discerned at an early moment and can receive extra attention.

This phase is generally followed by development in a pilot plant in which a number of operations are tried out on a scale larger than that of the mini-plant. This is especially advantageous when as yet incompletely understood process steps have to be applied. In this phase scaling-up problems can be looked into more closely.

In all phases of development economic evaluations are carried out. In the first phase they may be tentative but it is necessary to estimate economic prospects before a decision to scale up is made. Table II.2. summarizes the whole procedure.

Although development is a costly affair, in particular the building of a pilot plant, it must be understood that scaling-up problems can hardly be avoided. They occur because the characteristic parameters for apparatus depend in different ways on its dimensions. Capacity, stirring intensity, heat transfer, pressure drop etc. vary differently with dimensions. Design is always an extrapolation.

Other functions of a pilot plant may be:
- study of lifetime of apparatus and catalysts
- recirculation problems of unconverted feed and for intermediates
- fouling and corrosion (highly corrosive systems may require special construction materials)
- sensivity for change in process conditions
- improving economic evaluations
- production of larger amounts of product to explore markets
- comparing raw and auxiliary materials from different sources (or suppliers)
- demonstrating the process to potential buyers.

Table II.2. Process development

PLACE / TIME	PROCESS OR PRODUCT RESEARCH	PILOT PLANT	ENGINEERING BUREAU	PLANT	SALES ORG.	PATENTS AND LICENSING
SCOUTING	Idea Literature Laboratory research	Estimation of costs		Exploration of markets		Research
DEVELOPMENT	Systematic research Fundamental research Materials	Semi-technical research Design of pilot plants	Appraisal of costs and profits	Market analysis		Filing of patents
REALISATION	Supplementary research Measuring techniques	Trouble shooting Process dynamics Economy	Plant design	Construction	Exp. products	Filing of patents
PRODUCTION	Trouble shooting	Start-up of production	Design for other parties	Production Debottlenecking	Sales	Exploitation of know-how

REFERENCES

General

Baasel, W.D., 'Preliminary chemical plant design', Elsevier Scient. Publ. Co., Amsterdam 1976.

III. RAW MATERIALS AND PRODUCTS

General aspects

The most important raw material for the production of organic intermediates is petroleum (crude oil and natural gas). This has not always been the case; before the second world war coal was the main source in industrial organic chemicals, especially in Europe. Benzene is now the only base chemical produced in significant volumes from coal. The vast development of oil is demonstrated in Table

Table III.1. Petroleum as a raw material for the chemical industry[1]

	1950	1960	1970	1985 (est.)
% of organic chemicals based on petroleum	44	64	90	98
Production of organic base chemicals[2], 10^6 tons	2.3	9.7	46.7	125
Crude oil production, 10^6 tons[3]	476	855	1879	3000
Nat. gas production, 10^6 t.o.e.[4]	159	355	728	1200
Nat. gas liquids production, 10^6 tons	18.1	34.5	73.5	120
Non-energetic naphtha production, 10^6 tons	4.2	8.8	64.1	130

1 world production, excluding communist countries

2 ethylene, propylene, butadiene, benzene, toluene, xylenes

3 here and throughout the book, the word 'ton' signifies the metric ton of 1000 kg

4 tons of oil equivalent

Source: World Energy Supplies 1950-1974, United Nations Statistics

III.1. The tonnage of inorganic chemicals produced from petroleum is also very large. Main products are ammonia from natural gas, sulfur from sour natural gas and the oil refineries and carbon black from aromatic oil fractions.

The reasons for this development are as follows:

1. Up to 1973 oil was cheaper than coal. In particular in Western Europe, where coal generally occurs in zones which are difficult to exploit, the steep rise in the costs of wages made coal expensive. But even in the USA and South-Africa, where coal can be open-cast mined, oil was cheaper.
2. All organic intermediates contain much more hydrogen than coal. Because oil also has a higher H/C ratio than coal it is economically more attractive. To produce such intermediates from coal would mean increasing the amount of hydrogen by gasification, but this means use of extra raw material and energy.
3. Oil has the advantage of being a liquid. Transportation and processing of fluids is always easier and cheaper. Solids must be transported chargewise of pneumatically. Conveyor-belts may also be used, but compared with these pumping of oil through pipelines is simple. Another drawback of the use of solids is that a pretreatment is often required to achieve a proper particle size distribution. Hence continuous automatic processing of oil is easier than coal.
4. The composition of oil is less variable than the composition of coal. Moreover, the different main components of oil are easier to separate, usually by distillation. Coal contains highly condensed structures which are difficult to separate and are insoluble in any liquid.

In the future it can be expected that the relative scarcity of oil will lead to a larger role for other raw materials. Also, in several countries, there is a trend towards lessening their dependency on oil-producing countries.

Crude oil is a complex mixture of mainly hydrocarbons. Hetero compounds are also found, e.g. hydrocarbons containing one or more S, O or N atoms which may be incorporated in the molecular structures in different ways.

Low-boiling fractions contain mostly alkanes. With rising boiling point the proportion of cyclanes (naphthenes) and aromatics increase. Olefins rarely occur in mineral oil. The content of sulfur and nitrogen compounds is highest in the heavy fractions. Because sulfur is mainly present as homologs of benzothiophene and more complex aromatic structures only small amounts of sulfur are found in fractions boiling below 200 ^{o}C (at 0.1 MPa). Residual fractions, i.e. parts which cannot be distilled without thermal decomposition,

Table III.2. Composition of some crude oils

Source	Kuwait	Libya	Venezuela I	Venezuela II
Composition (wt%)				
S	2.54	0.04	2.72	1.40
V	0.0024	0.00001	0.10	0.018
C_1 - C_3 incl.	0.54	0.19	0.00	0.21
C_4 + C_5	3.45	2.19	0.22	2.30
Tops, < 100 °C*	7.9	7.6	0.8	6.3
Naphtha, 100-150 °C	6.2	9.3	1.0	6.4
Kerosene, 150-250 °C	14.5	21.6	5.9	14.7
Gas oil, 250-370 °C	19.8	27.0	18.7	20.9
Residue, > 370 °C	49.6	33.6	73.6	50.7

*Temperatures indicate boiling ranges at 0.1 MPa

are often rich in porphyrin complexes of V and Ni. The presence of these metals and heterocyclic compounds affect the processing and application of petroleum. Table III.2. gives an impression of the composition of 4 oil samples.

Distillation is by far the most important method of obtaining primary products from crude oil. Up to a boiling point of about 370 °C (0.1 MPa) distillation can be carried out without thermal decomposition if short residence times are used. To separate heavier fractions vacuum distillation is applied to prevent cracking and formation of coke-like material. Figures III.1. and III.2. show the principal schemes for these processes which are designed for capacities of up to 40000 ton crude per day (single train). Such large scale operation makes it possible to save energy by intensive heat exchange and saves capital by low investment per ton of oil.

For the chemical industry cracking of naphtha and gas oil is an important operation to produce raw materials. Although in Chapter X this is treated in more detail, a simplified scheme is presented in Figure III.3. which shows the relation between an oil refinery and the chemical industry.

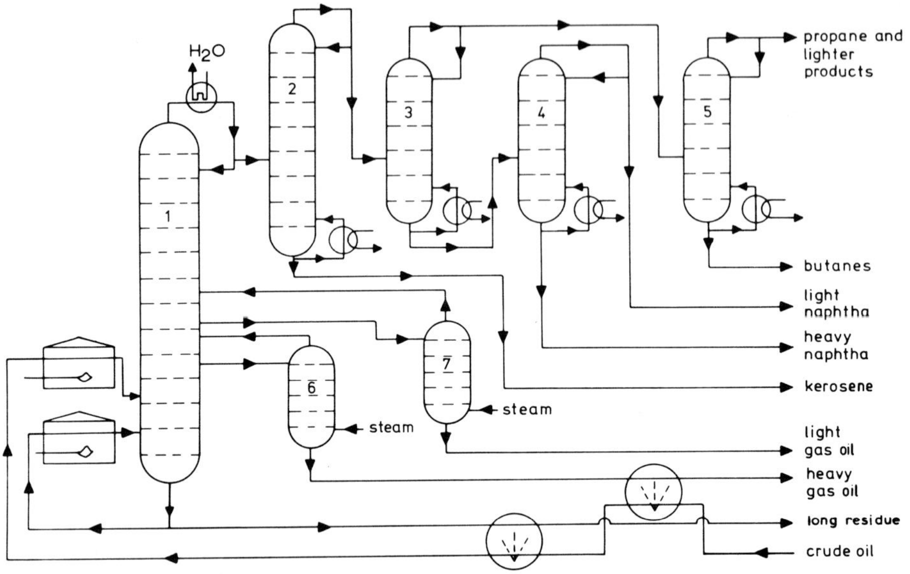

Figure III.1. Atmospheric distillation

1. *crude oil column*
2. *kerosene column*
3. *debutanizer*
4. *gasoline splitter*
5. *depropanizer*
6. *stripper*
7. *stripper*

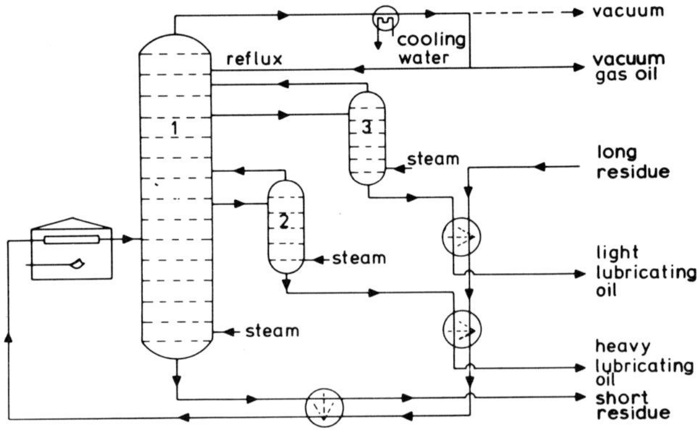

Figure III.2. Vacuum distillation

1. *vacuum column*
2. *stripper*
3. *stripper*

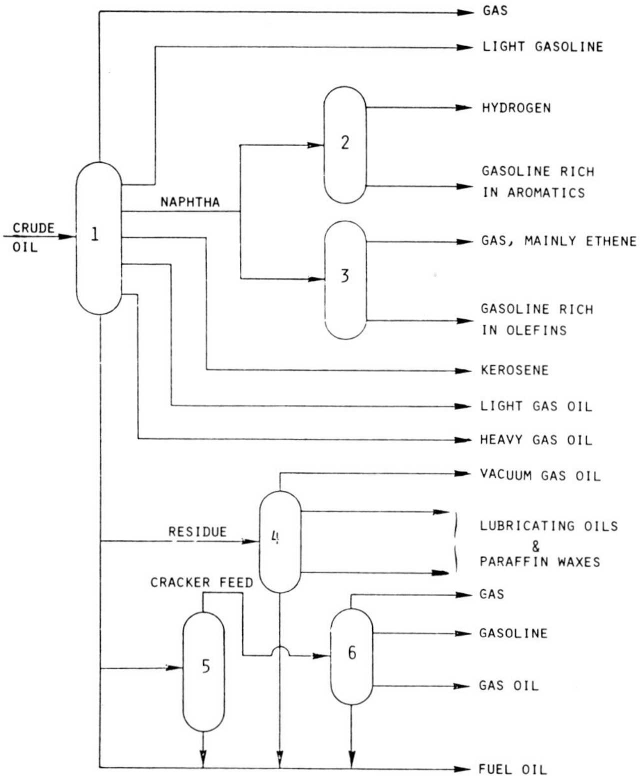

Figure III.3. A refinery as a producer of raw materials for the chemical industry

1. primary distiller

2. platformer

3. naphtha cracker

4. vacuum distiller

5. vacuum flasher

6. cat cracker

Natural gas

Natural gas can be found together with petroleum or in separate fields. In the former case the gas not only contains methane but also higher alkanes and if the oil contains sulfur the gas contains hydrogen sulfide. An impression of the composition of some natural gases is given by Table III.3.

Table III.3. Composition of some natural gases (vol%)

Source	CH_4	C_2H_6	C_3^+	H_2S	$CO_2 + N_2$
USA, Mid-continent	96.0	-	-	-	4.0
USA, Pennsylvania	67.6	31.3	-	-	1.1
USA, West-Texas	58.7	16.5	18.4	6.4	-
Italy, Cortemaggiore	91.0	4.4	2.9	-	1.7
Netherlands, Groningen	81.0	2.5	1.1	0.4 ppm	15.4

To use natural gas it is generally necessary to desulfurize and dry it. Higher alkanes are condensed. Apart from its use as an energy source natural gas is a raw material for the production of synthesis gas and chlorinated hydrocarbons. Synthesis gas is an intermediate for two important bulk chemicals: ammonia and methanol. On a smaller scale it is used for hydroformylation (e.g. by the Oxo-process).

Coal

Coal has been formed from vegetable material under the influence of temperature and pressure by which carbohydrates and lignin were converted. Coal consists mainly of aromatic ring structures which are peri-condensed and connected by C, O, S and N bridges. Heterocyclic structures also generally occur. Figure III.4. shows a schematic representation of a typical coal structure:

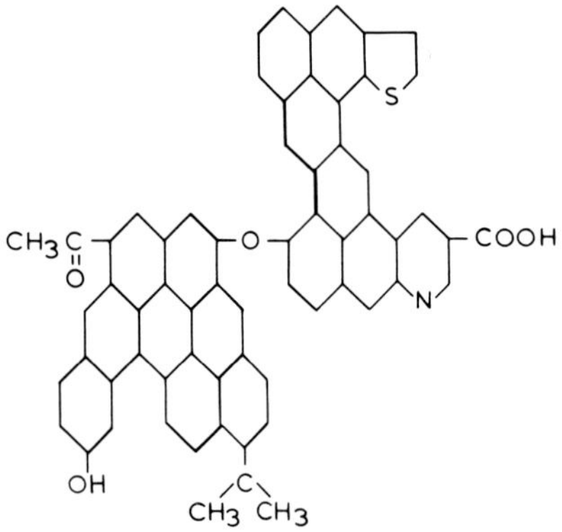

Figure III.4. An impression of the composition of coal

With the progression of the coal-formation process CH_4, CO_2 and H_2O are formed; at the same time the aromatic structure content rises and the number of functional groups and side chains decreases. The molecular weight increases.

Coal is converted on a large scale to produce coke for the steel industry. This means pyrolysis (with exclusion of air) at high temperatures (e.g. 1000 ^oC). In this way the coal is cracked. The by-product gases contain aromatics (benzene, toluene, xylenes etc.), hydrogen (+CH_4, CO_2 and H_2O) and tar. The nitrogen in the coal forms NH_3 and HCN, and sulfur is released as H_2S. The pyrolysis takes about 24 hours in chamber-ovens.

Coal can serve as a base material for the chemical industry in several ways:

1. products made by coal pyrolysis can be used to produce a large number of predominantly aromatic chemicals; in fact, coal pyrolysis products have formed the basis of the industry since it first began to develop in the second half of the 19th century;
2. synthesis gas can be made from coal by gasification with steam and air or oxygen; further conversion into liquid or gaseous hydrocarbons is possible by catalytic conversion in the Fischer-Tropsch process;
3. similarly, synthesis gas can be converted into methanol, which is further processed into liquid hydrocarbon mixtures high in aromatic content;
4. hydrogenation of coal to liquid products is possible; these products can then serve as feedstock for chemical processes in a similar manner as petroleum oils.

The above processes are discussed in more detail in Chapter XIV.

Other raw materials for organics

Apart from the raw materials mentioned above others are used, albeit on a smaller scale, such as vegetable materials like fatty oils, starch, sugar and molasses, wood and straw. With these materials less drastic conversions are applied in comparison with coal and oil, to make as much use of the chemical structures in the feed as possible. On the other hand, the product spectrum then is much narrower and production volumes are generally smaller, in other words the products are much less important as 'building blocks' of the chemical industry as compared with ethene, propene, benzene and the like. Apart from coal and oil other fossil deposits could play an increasing role in the energy conversion industry as well as for chemical purposes. This applies particularly to tar sands and oil shale of which vast deposits are located in Canada and the USA. Utilization is, however, likely to be costly from the point of view of energy consumption for processing and also for ecological reasons. Pyrolysis, extraction and underground gasification are being considered for development.

Inorganic raw materials

The inorganic chemical industry is based on minerals as well as on air and water. Many of the minerals are processed for making the corresponding metal or metal compounds; this field is not treated in this book and the reader is referred to texts on metallurgy. Other minerals are converted into products like building materials and pigments. There is, however, a group of raw materials from which a limited number of rather important inorganic intermediates is made that find widespread uses. These are:

- air, the source of oxygen and nitrogen, can be separated via liquefaction and distillation. It is also possible to let the oxygen react, e.g. to CO_2 and water, followed by removal of the oxygen-containing reaction product. This is practiced in making ammonia and its derivatives;
- sulfur is produced from sulfates or sulfur deposits and is also recovered as by-product from oil, coal or natural gas processing. The main product made from it is sulfuric acid, the most important inorganic intermediate in terms of production volume;
- sodium chloride is the starting material for making soda ash, caustic soda and chlorine and its derivatives;
- calcium carbonate is converted into CaO and CO_2;
- silica is applied for making adsorbents as well as synthetic compounds containing Si, such as silicones;
- bauxite, mainly Al_2O_3, forms the basis for alumina and aluminum and for a limited number of adsorbents and aluminum compounds, e.g. the sulfate which is used in papermaking and water treatment;
- water is used as feed, solvent or auxiliary material in very many chemical processes.

Products

Products can be classified as intermediates, i.e. building blocks for more complex semi-manufactured products, or finished products resulting from these. The last processing step, however, is not often carried out in the chemical industry but rather in what could be called the supply industry. A further rough distinction is between specification products, performance products and materials.

Specification products are characterized by a quantitative description of the composition, mostly on a chemical basis. They consist of a single homogeneous phase (or pseudo-homogeneous, e.g. a suspension): average and local concentrations of the product's constituents are equal. Such products are sold for many different purposes. Examples include essentially all gaseous and liquid, and a few solid chemicals.

Performance products are not only specified in terms of composition but also in terms of performance, very often measured physically, mechanically or in performance tests in which the conditions of use are simulated as closely as practicable. Examples are gasoline, lubricating oil, detergents and catalysts.

Materials are performance products with the special characteristic that they are used in construction. They are often solids, of else they solidify on application: metals, plastics, fibers, glues, etc.

The relations between raw materials, intermediates, semi-manufactured products and finished products are complex because

- a raw material can give a broad spectrum of intermediates;
- the ratios in which the intermediates are produced depend on process conditions;
- certain products may be made from different raw materials;
- the specifications of the intermediates depend on the application and should often be adjusted to the end product specification (e.g., polyvinylchloride for the production of sheets or packaging material differs from PVC for construction purposes).

One example of the problem of choosing from different process routes is the production of acetaldehyde. Table III.4. illustrates possibilities. Several of the possible routes are not used nowadays. Ethylene oxidation with the Wacker process and ethanol oxidation are generally used, acetylene being too expensive. The yield of acetaldehyde from butane oxidation is usually low and several other products are also produced in this case.

Examples of intermediates from oil and natural gas are given in Tables III.5. and III.6. For inorganic products only a few are mentioned (Table III.7.).

Table III.4. Possible routes to acetaldehyde (AA)

carbohydrates	fermentation	ethanol	dehydrogenation or oxidation	AA
naphtha	cracking	ethylene	hydration	AA
natural gas	separation	n-butane	oxidation	AA
coal	calcium carbide	acetylene	hydration	AA
		acetylene → methyl-vinyl ether		AA

The 'product-trees' presented here are specially useful in the development of new processes and products. See for instance 'Chemical Origins and Markets', Stanford Research Institute 1977.

In the organic intermediates section the polymers are the most important with respect to volume. Not only are these used as plastics for construction purposes, packaging and synthetic rubbers but also in the field of textile fibers. Other important products of the chemical industry are detergents, dyestuffs, paints and pharmaceuticals. The inorganic industry produces fertilizers, pigments, ceramic materials and catalysts. Another important section, the metallurgical industry, is not considered here. Although the basis for the production of steel, alloys etc. has much in common with the production of bulk chemicals, the processing is strongly dependent on product specifications. For the same reason no attention is paid to lubricants, paper, glass etc.

Table III.5. Intermediates from oil and natural gas

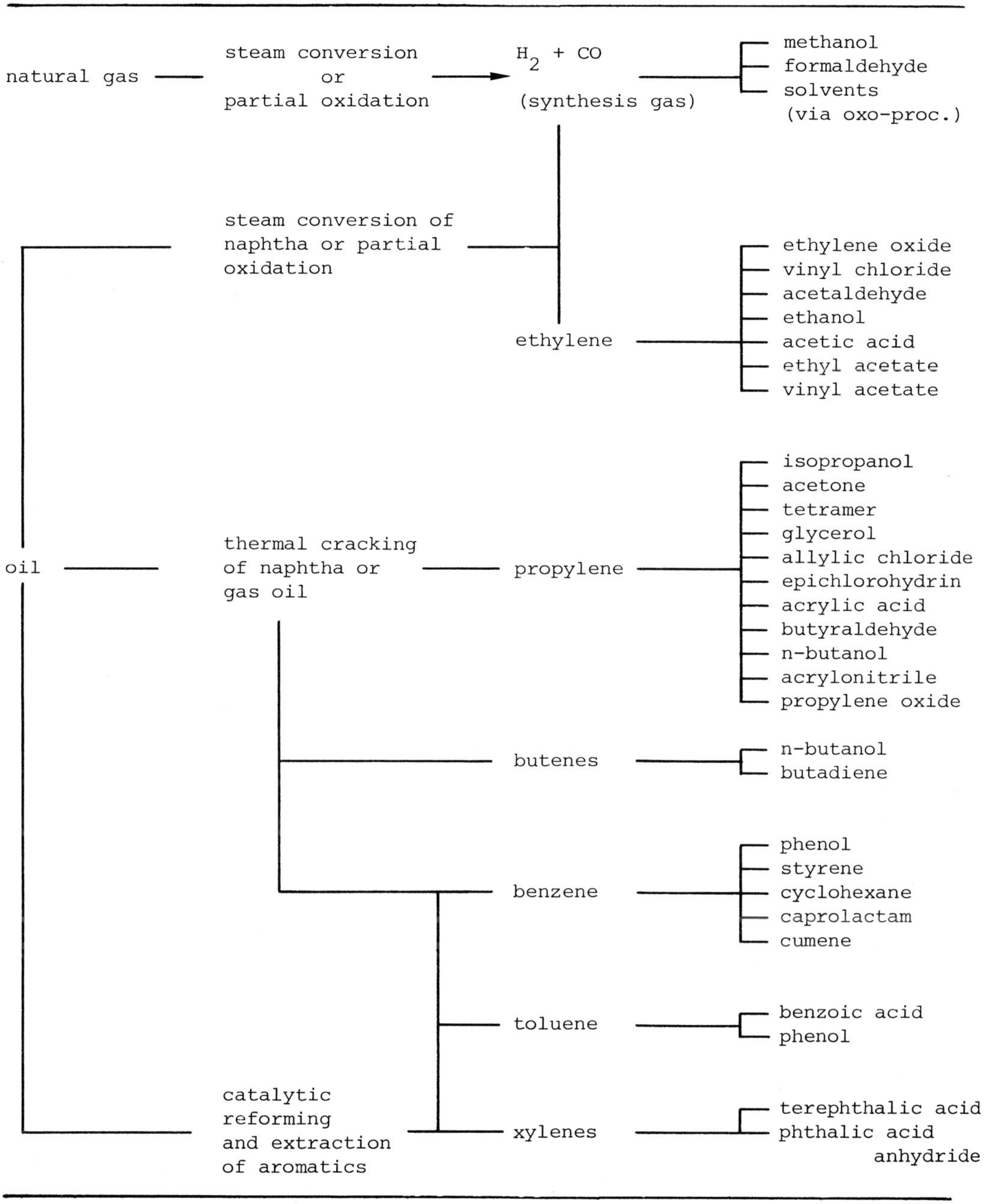

Table III.6. Intermediates from methane

methane	steam conversion or partial oxidation	synthesis gas (CO, H_2, CO_2, possibly N_2)	ammonia urea methanol formaldehyde liquid HC
	nitration		nitromethane
	sulfurization		carbon disulfide
	chlorination		methyl chloride dichloromethane chloroform carbon tetrachloride
	thermal conversion	acetylene	vinyl chloride vinyl acetate acrylonitrile acetaldehyde acetic acid ethanol butadiene

Table III.7. Inorganic intermediates

air	nitrogen oxygen	NH_3	urea nitric acid nitrates hydrocyanic acid
sulfur	SO_2		sulfates sulfites sulfides
	H_2SO_4 (from SO_2)		
phosphate rock	(H_2SO_4)		phosphoric acid phosphates.
sodium chloride			chlorine sodium hydroxide chlorate hypochlorite

REFERENCES

General

1. Krevelen, D.W. van, 'Coal' (Typology, chemistry, physics, constitution), Elsevier Scient. Publ. Co., Amsterdam 1961.

2. Given, Editor P.H., 'Coal Science', Adv. Chem. Ser. Nr. 55, Am.Chem.Soc., Washington D.C. 1966.

3. 'Kirk-Othmer Encyclopedia of Chemical Technology', 3rd Ed., J. Wiley and Sons, New-York 1978-

4. 'Ullmanns Encyklopädie der technischen Chemie', 4th Ed., Verlag Chemie, Weinheim/Bergstr., W.-Germany 1972-

5. Shreve, N.R., 'Chemical Process Industries', 3rd Ed., McGraw-Hill, New-York 1967

6. Weissermel, K., H.J. Arpe, 'Industrial organic chemicals', Verlag Chemie, Weinheim/Bergstr., W.-Germany 1976.

Periodicals

1. 'Chemical Engineering', McGraw-Hill, New-York.

2. 'Hydrocarbon Processing', Gulf Publishing Co., Houston.

3. 'Chemical and Engineering News', Am.Chem.Soc., Washington D.C.

4. 'Petroleum and Petrochemical International', Petroleum Publ. Co., Tulsa (Okla.)

5. 'Chemical Engineering Progress', Am.Inst.Chem.Eng., New-York

6. 'Industrial and Engineering Chemistry', Am.Chem.Soc., Washington D.C.

7. 'Modern Plastics', McGraw-Hill, New-York

8. 'Environmental Science and Technology', Am.Chem.Soc., Washington D.C.

9. 'Journal of Engineering for Power', Am.Soc.Mech.Eng., New-York.

10. 'Combustion Science and Technology', Gordon and Breach, New-York.

IV. PRODUCTION OF AMMONIA FROM NATURAL GAS

Introduction

Ammonia is an important intermediate for the manufacture of nitrogen-containing fertilizers. It is indispensible for agriculture in connection with the production of vegetable proteins, which, in turn, are necessary to produce animal proproteins because most organisms are unable to convert molecular nitrogen from the atmosphere directly. Among the exceptions are certain free-living microorganisms and some micro-organisms living in symbiosis with plants, examples of the first type are blue-green algae and *azotobacter*; of the second type *rhizobium*, which is often associated with legumes such as clover, peas and soybean. Although nitrogen fixation by these organisms is utilized to a large extent for protein production as well as fertilization, ammonium and nitrate fertilizers made industrially are still required in large quantities to increase productivity, the more so because the amount of chile salpeter, a natural material that can be digested by most plants, is very limited.
One of the process routes to ammonia is described in some detail below. The subsequent chapter deals with other processes for ammonia; all of these consist of a system for the production of ammonia synthesis gas followed by an ammonia synthesis section. Differences among the various routes concern the raw material, the way in which the synthesis gas is produced, and the manner in which ammonia is recovered from the product gases.
This chapter concludes with a short description of methanol manufacture, which is done by processes resembling ammonia manufacture in many respects.

Outline of methane based ammonia process

A block diagram of the process for making ammonia from methane or methane-rich natural gas is shown in Figure IV.1. The reaction

$$N_2 + 3\ H_2 \rightleftarrows 2\ NH_3 \quad (\Delta H_s^o = -45{,}3\ \text{kJ/mole}\ NH_3)$$

requires a synthesis gas containing hydrogen and nitrogen in a molar ratio between 2½ and 3, (see page 42) with minimum concentrations of inert or harmful contaminants like methane, oxygen compounds and noble gases. The following operations are necessary to make ammonia from methane:

- *conversion of methane with steam* over a nickel catalyst under pressure and at elevated temperature in the *first reformer reactor*, from which a product gas containing CO, CO_2, H_2 and residual methane and water vapor is obtained;
- *partial oxidation* of the gas from the first reformer in a *second reformer* reactor with the amount of air needed to give a H_2/N_2 molar ratio of 2½-3, using another nickel catalyst and still higher temperatures;
- after cooling in a waste heat boiler and by heat exchange, CO is converted with steam to CO_2 and H_2 in the *CO shift* reactor;
- the shift convertor is followed by further heat exchange and cooling to temperatures between 30 and 100 ^{o}C necessary for water condensation and for *removal of carbon dioxide*, usually by washing the gas countercurrently with an absorption solution. The solution containing the CO_2 is regenerated and recycled;
- since the crude synthesis gas obtained after removal of carbon oxides still contains traces of oxygen compounds which will deactivate the iron catalyst used for ammonia synthesis, the next step is virtually complete conversion of CO and CO_2 to methane (*methanation*) with the hydrogen in the gas, using a nickel catalyst;
- water is then removed by *adsorption*, usually by means of *molecular sieves*, i.e. crystalline synthetic aluminosilicates having a very high capacity for water at ambient temperatures and very low partial pressure of water vapor;
- the gas obtained from the previous operations is now *compressed* to the high pressure needed for ammonia synthesis. It consists of hydrogen and nitrogen in the required ratio and contains only traces of CO and CO_2 (about 10 ppm) and even less water (1-5 ppm). The methane concentration depends on the process conditions applied but it is seldom higher than ½ vol %.
  The concentration of noble gases, about 0.2 vol %, is directly related to the N_2 content because these gases also originate from the air used for partial oxidation in the second reformer.
- the final catalytic conversion to ammonia cannot be achieved with high conversions because of the unfavorable equilibrium at the high temperatures required to give adequate rates over the iron catalyst used. Ammonia is, therefore, condensed out of the partially converted gases, most of which are recycled to the convertor. A smaller part must be *vented* to prevent build-up of methane, noble gases and excess nitrogen.

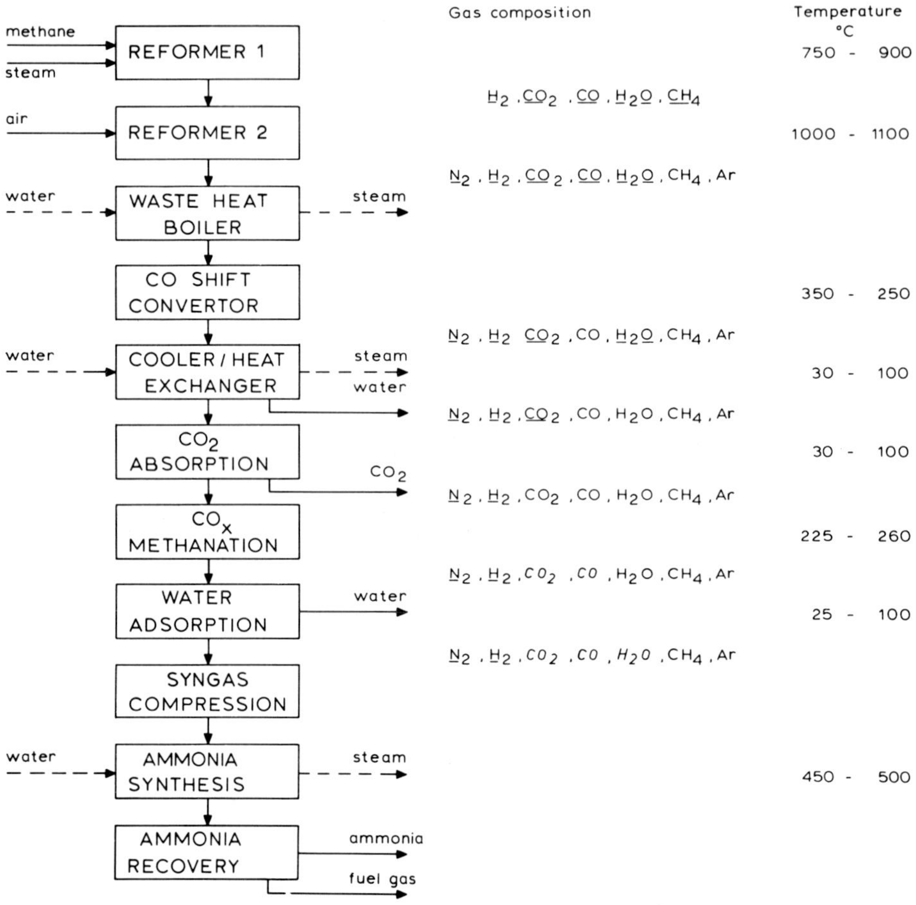

Figure IV.1. Block diagram of ammonia synthesis from methane
Major components (> 1 vol%):underlined
Trace components (1-20 ppmv):in italics
Ar:includes other noble gases.

Figure IV.1. indicates the changes in gas composition throughout the process as well as the approximate temperatures applied in some of the operations.

Some details about the process steps

Reformer 1

The following reactions proceed in the first reformer:

$$CH_4 + H_2O \rightleftarrows CO + 3H_2 \quad (\Delta H_s^o = +206.6 \text{ kJ/mol } CH_4)$$

$$CH_4 + 2H_2O \rightleftarrows CO_2 + 4H_2 \quad (\Delta H_s^o = +165.5 \text{ kJ/mol } CH_4)$$

Equilibrium is reached by applying an active Ni/Al_2O_3 catalyst at temperatures of about 750-900 ^oC and pressures ranging from 1 to 4.5 MPa. Since the reactions are endothermic, much heat must be supplied to keep the gases at the high temperatures required for fast conversion. Care must be taken to avoid the formation of carbon, which would interfere both with the heat transfer to the reacting gases and with the catalyst activity; the carbon can be formed by the reactions:

$$CH_4 \rightleftarrows C + 2\,H_2 \quad \text{(endothermic methane decomposition)}$$

$$2\,CO \rightleftarrows C + CO_2 \quad \text{(exothermic; Boudouard's reaction)}$$

If carbon is not formed, it is possible to calculate the composition of the product gases from thermodynamics if it is assumed that equilibrium is established. The following equations are available for this purpose:

- the equilibrium relations for CO and CO_2 formation:

$$K_{CO} = \frac{[CO]\,[H_2]^3}{[CH_4]\,[H_2O]} \quad \text{and} \quad K_{CO_2} = \frac{[CO_2]\,[H_2]^4}{[CH_4]\,[H_2O]^2}$$

To a first approximation, the equilibrium constants depend on the temperature according to relations such as:

$$\ln K = A + \frac{B}{T}$$

where T is the absolute temperature and A and B are constants. Three more relations can be used in the calculation, viz. mass balances over the reactor:

- the *carbonbalance* : $[CH_{4,\,in}] = ([CO] + [CO_2] + [CH_4])_{out}$

- the *hydrogen balance* : $(2[H_2O]_{in} + 4[CH_4]_{in}) = (2[H_2O] + 2[H_2] + 4[CH_4])_{out}$

- the *oxygen balance* : $[H_2O]_{in} = ([CO] + 2[CO_2] + [H_2O])_{out}$

The five equations contain four independent variables (temperature and pressure at the reactor outlet, moles CH_4 and moles steam) and the five concentrations in the product gas, expressed in moles per unit volume, as the dependent variables. The number of independent variables can be reduced to three by expressing all concentrations as ratios to the molar quantity of steam (or methane) in the feed. Solution is possible if the process conditions, i.e. independent variables, are known; some numerically obtained results are listed in Table IV.1., for a case in which the feed consisted of equimolar quantities of the two reactants. The volumetric change due to reactions was neglected.

The data in the table show that the equilibrium conversion of methane increases when the temperature is raised at otherwise unchanged conditions. A similar effect is noted for a decrease in pressure. Both conclusions are a direct consequence of the principle of Van 't Hoff - le Châtelier, which states that a change in a variable determining the state of a system in equilibrium causes a shift in equilibrium position that tends to counteract the change. Another example of this rule is that addition of steam- in other words the use of a higher steam/methane ratio in the feed- results in a higher methane conversion at equilibrium.

Table IV.1. Calculated methane contents of equilibrium gases as a function of reaction conditions *

Pressure	Temperature ** required to give indicated CH_4-content in dry gas:				
(MPa)	5,0 vol %	2,0 vol %	1,0 vol %	0,5 vol %	0,2 vol %
0.1		700			800
1.0	800	870	910	950	1000
2.0	870	950	1000	1030	1100
4.0	940	1020	1080	1080	1200

* feed: ratio of CH_4 to steam = 0.5

** in ^oC

Because methane/steam reforming is an endothermic process performed at very high temperatures, it is necessary to use a number of tubular reactors placed in parallel to provide sufficient surface area for heat transfer. The tubes are mounted in a furnace in such a way that they can expand more or less freely when heated to the process temperature. The location of the burners used for this purpose must be selected with care to ensure that even heating by direct radiation is ensured. These precautions serve to minimize build-up of tensions in the material. Different modes of firing are possible: top-fired, bottom-fired, etc. The tubes are made of centrifugally cast Cr/Ni-alloy free of blow-holes; they are very costly and should be operated at conditions where a satisfactorily long life is achieved. This implies that the temperature and the process pressure should not be unduly high, as may be seen from the data in Figure IV.2. Another factor is the use of a clean fuel free of ash, sulfur or polyaromatic hydrocarbons, in order to avoid deposit formation or corrosion, both of which would reduce heat transfer and thus shorten tube life. Usually, natural gas or naphtha are employed.

If a certain process condition gives too short a tube life the process variables must be adjusted. If this is done by lowering the temperature, it is also necessary to lower the pressure or to increase the steam/methane ratio to obtain the same methane conversion.

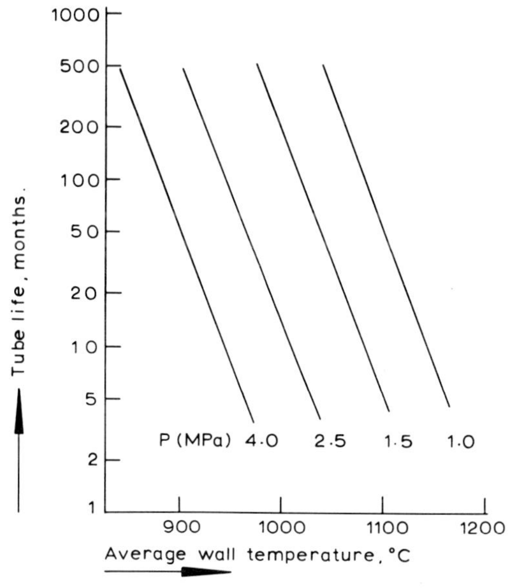

Figure IV.2. Expected reformer tube life as a function of process conditions.

Both these corrections result in higher costs since more compression energy or steam must be provided. The costs must be weighed against the savings resulting from longer tube life.
Steam has still other functions besides that of being reagent for methane conversion. A certain minimum steam dosage is required to prevent carbon deposition on the catalyst; with methane feeds the minimum is 1,0 mole steam/ mole methane. The use of some excess steam is favorable because it can remove any carbon deposited under temporarily less favorable conditions, by gasification to carbon oxides and hydrogen:

$$C + H_2O \rightarrow CO + H_2$$

Addition of promotors to the catalyst reduces carbon deposition and/or increases the rate of gasification of carbon deposits by steam. Preferred feeds to the reformer section contain very low concentrations of sulfur; in general, < 1.0 ppmv H_2S and no other sulfur compounds are specified. If this specification is not met it is necessary to desulfurize the feed; it is customary to place a guard bed containing ZnO upstream of the first reformer, to remove any traces of H_2S that could poison the reforming catalyst by irreversible sulfide formation. The conditions of the first reformer are selected in such a way that the product gas contains 8-8½ vol% methane, if ammonia synthesis gas must be produced. This amount of methane is then converted almost completely to carbon oxides and hydrogen in the second reformer. Since the outlet temperature of reformer 2 determines the residual methane content as is discussed on page 32 , the conditions to be used in the first reformer can only be determined after the outlet temperature and pressure of reformer 2 have been established.

The second reformer

In the second reformer the product gas from reformer 1 is oxidized with the quantity of air needed to give a H_2/N_2 ratio of 2½-3 in the final synthesis gas. This is carried out over a suitable Ni/Al_2O_3 catalyst to ensure that the conversion of oxygen is complete and to establish the same equilibria between CH_4, H_2O, H_2 and carbon oxides as in the first reformer. The oxidation is exothermic; the reactor can in this case be constructed as an adiabatic fixed bed of catalyst because external heating is not required. The reactor, as well as the connecting line between the two reformers, should be well insulated to ensure that all the heat evolved is absorbed by the reacting gases. This is desirable because of the more favorable equilibrium position at higher

temperature. The outlet temperature can be well over 1000 °C.
The conditions at which the two reformers must operate can be approximated in the following manner. If a product gas is required containing 0,2 vol% methane after reformer 2, at a process pressure of 2.0 MPa, it follows from Table IV.1. that the outlet temperature of the reformer section should be about 1100 °C. A heat balance calculation over reformer 2, assuming adiabatic behavior, enables the temperature rise over this reformer to be calculated; it amounts to about 150 °C. This implies that the process gas temperature at the outlet of reformer 1 should be 950 °C; since the wall temperature is up to 100 °C higher the tube wall temperature should approach 1050 °C. At these conditions, it is found that reformer tube life is about one year (Figure IV.2.), which is too short to be acceptable. To correct this, the calculation is repeated after introduction such of changes as

- lowering of pressure
- lowering of temperature
- relaxation of the methane specification
- increasing the steam dosage to reformer 1

Part of the adjustment can also be effected by raising the temperature to which the air fed to reformer 2 is preheated.
However, it is not possible to optimize the reformer section without taking into account the effects of changes in conditions on other units of the entire ammonia plant. This point is discussed further at the end of this chapter.

CO shift reaction

In the CO shift converter, CO is converted with steam to CO_2 and H_2:

$$CO + H_2O \rightleftarrows CO_2 + H_2 \quad (\Delta H^o_s = -41 \text{ kJ/mole})$$

The reaction, the water gas shift reaction, produces valuable hydrogen and also reduces the concentration of CO, a potent poison for the ammonia synthesis catalyst. The equilibrium CO conversion is, however, not very favorable: equimolar mixtures of H_2O and CO are converted at equilibrium for 85% at 300 °C, 75% at 400 °C and 65% at 500 °C. As expected for this exothermal reaction, the equilibrium conversion declines when the temperature is raised.
Since the crude gas from reformer 2 contains appreciable amounts of hydrogen and carbon dioxide, it is necessary to use excess steam and as low a CO shift convertor outlet temperature as possible. Sufficiently high rates of conversion at low temperatures can only be obtained by catalyzing the reaction.
The shift catalyst developed for the original Haber-Bosch ammonia process, Fe_2O_3, is still widely used for the CO shift, but modern catalysts also contain

Cr_2O_3 as a stabilizer, i.e. to produce high activity over a considerable period of time. The reactor preferred here is an adiabatic fixed-bed reactor because of its reliability and relatively low cost. The process gas temperature goes up as a result of the heat produced by the shift reaction, which would result in rather low conversions at the reactor outlet if nothing is done to improve this. A possible solution is to apply a convertor consisting of two or more catalyst beds in series, with cooling between the beds. Most of the CO shift proceeds between 350 and 450 °C in the first and often smallest bed; subsequent beds are longer to ensure that the conversion goes to completion even at low inlet temperatures and reactant concentrations. Cooling can either be by water circulating through pipes to produce steam, or by injecting water into the gas stream. In the latter case the cooling effect of the evaporating water is supplemented by the favorable influence of water on the equilibrium conversion of CO.

However, even with these improvements it is not possible to convert CO to a sufficient extent. Two methods can be applied to improve the conversion, viz.:

(1) Removal of CO_2 after the first shift catalyst bed, by cooling the gas and then washing it with an alkaline liquid. The gas is reheated and subjected to the second stage shift conversion, which is followed by another CO_2 removal step. This implies that the gas must be cooled twice from about 450 °C to the temperature of the washing operations, 30 - 100 °C. This solution has been abandoned since heat exchange of large gas flows is an expensive operation requiring large heat exchangers and because a second absorber for removing CO_2 is required.

(2) Application of the low temperature shift *(LTS)* catalyst, Cu/ZnO. This is a much better solution because equilibrium can then be reached at temperatures as low as 250 °C. Drawbacks of LTS catalyst are that it is much more sensitive towards catalyst poisons like H_2S and to overheating resulting in activity loss by sintering. Cu/ZnO is also much more expensive than the Fe_2O_3/Cr_2O_3 high temperature shift (*HTS*) catalyst. Thus, LTS catalyst is not applied in the first bed of a shift convertor, the more so because the temperature rise over this bed can be as much as 100 °C.

Modern shift convertors like the example of Figure IV.3. consist of a bed of HTS iron oxide catalyst, a guard bed of ZnO to retain any H_2S released from the HTS catalyst by reduction of sulfate present in this catalyst when fresh, a cooling section and a third bed of LTS. If the feed is much cooler than the product gas a heat exchange section may be added. Such convertors can reduce the CO content of crude synthesis gas to 0.1-0.2 vol% if the inlet temperature of the LTS bed is about 225 °C. The operating temperature of the LTS catalyst

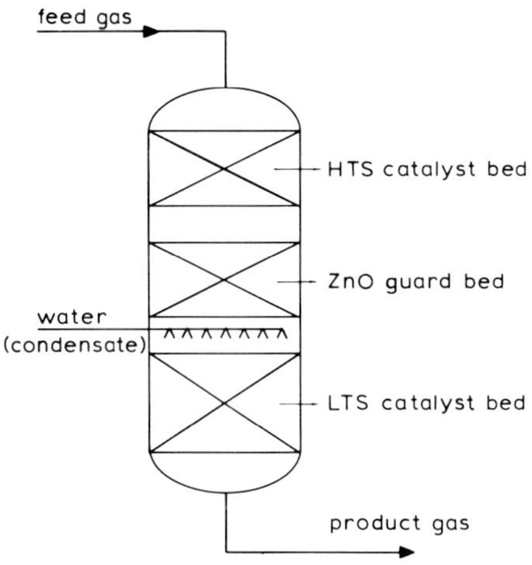

Figure IV.3. CO shift converter

should not be much lower because of the risk of condensation of water vapour in the catalyst pores, which would cause loss of activity. Since the pressure in the synthesis gas preparation units of methane-based ammonia plants is seldom higher than 4.0 MPa, this difficulty is rarely encountered.

Carbon dioxide removal

CO_2 can be removed from the raw synthesis gas by washing with an alkaline liquid such as an aqueous solution of an alkanolamine. This type of process is discussed more fully in chapter VI.

In the simplest unit for CO_2 removal (Figure IV.4.) the crude gas is contacted countercurrently with the washing liquid in an absorber. With alkanolamine solutions the inlet temperatures of gas and liquid are mostly between 25 and 40 ^{o}C. The crude gas from the CO shift convertor is first cooled to remove excess water in such a way that no water is lost from or taken up by the absorption system. The fat solution containing the CO_2 is sent through a throttle valve to a flash tank where physically dissolved gases desorb by pressure reduction to about 0.1 MPa. The amount of gas set free is too small to return it to the process after recompression. The solution then goes to the regenerator, where it is freed of CO_2 by heating to about 100 ^{o}C. The top product gas consisting of CO_2 and water is cooled to condense the water; the condensate is recycled as reflux. The CO_2 can be used for other purposes, such as the manufacture of urea or dry ice. With methane-based ammonia plants it is also possible to

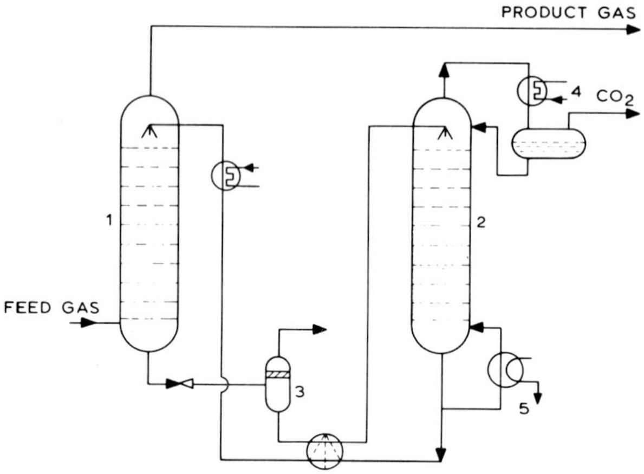

Figure IV.4. Principle of CO_2 removal by absorption in an aqueous alkanolamine solution.

1. *absorber*
2. *regenerator*
3. *flash vessel*
4. *cooler/condenser*
5. *reboiler*

vent the CO_2 to the atmosphere because it does not contain H_2S, a contaminant which, if present, would also be washed out by the alkanolamine solution. Absorbers and regenerators of large ammonia plants are usually plate columns. Packed columns are less suitable in this application because they would require costly liquid distributors to ensure adequate gas/liquid contact at the relatively low liquid loads applied here. Another advantage of plate columns is that maintenance and cleaning are somewhat easier. However, in operations where the liquid load is higher, packed columns can perform quite satisfactorily and are preferred for smaller throughputs because of their lower cost.

Methanation

The product gas from the CO_2 absorber still contains enough CO and CO_2 (0.1-0.2 vol% of each) to cause rapid poisoning of ammonia synthesis catalysts. The carbon oxides are removed down to concentrations of 10-20 ppmv by conversion over an adiabatic bed of Ni/Al_2O_3 methanation catalyst at inlet temperatures between 225 and 250 ^{o}C:

$$CO_x + (2+x)H_2 \rightleftarrows CH_4 + x\ H_2O$$

Since the reaction, the reverse of the methane/steam conversion, is rather exothermic the temperature rise over the methanator amounts to 50-70 ^oC per percent CO_x in the gas. A sudden increase in CO_2 content of the feed gas such as is caused by an upset in the CO_2 absorption unit would lead to a very rapid rise in methanator temperature. Undesirable consequences are loss of catalyst activity by sintering due to overheating and too high a CO_2 content of the product gas. Fast-acting control equipment is used to ensure that the off-specification gas is vented within a few seconds to protect the ammonia synthesis catalyst. If the methanation temperature is below 200 ^oC the carbon oxides may well not be efficiently removed. At still lower temperatures, CO tends to react with the Ni component of the catalyst to form Ni-carbonyl. Operation below about 150 ^oC is not possible since at that temperature increased surface mobility of Ni caused by corrosive CO chemisorption leads to rapid activity loss.

Drying

After methanation the only ammonia synthesis catalyst poison left in the gas is water. The gas is usually dried by adsorption of the water on silica gel, activated alumina or molecular sieves. In spite of their high cost, molecular sieves have displaced the older types of adsorbents because of their high capacity for water even at relatively high adsorption temperatures. If molecular sieves are used it is not necessary first to remove the bulk of the water by drying with a hygroscopic liquid such as triethylene glycol in a system resembling that of Figure IV.4. The superior properties of molecular sieves for drying operations follow from the adsorption isotherm of Figure IV.5., which indicates that molecular sieves can adsorb much water even if very low partial water vapor pressures are required. Alumina and silica have a much lower capacity for water if deep drying is necessary; these cheaper adsorbents are more suitable if some small amount of water remains in the gas, the more so because they can be regenerated at lower temperaures than molecular sieves (Figure IV.5b.).

Drying is accomplished by passing the gas over fixed beds of the adsorbent. Since regeneration is necessary from time to time when the adsorbent is saturated with water, more than one bed must be installed. Regeneration is done by pressure reduction combined with heating to 150-200 ^oC to desorb the water. Further information on cyclic adsorption operations is given in chapter VI.

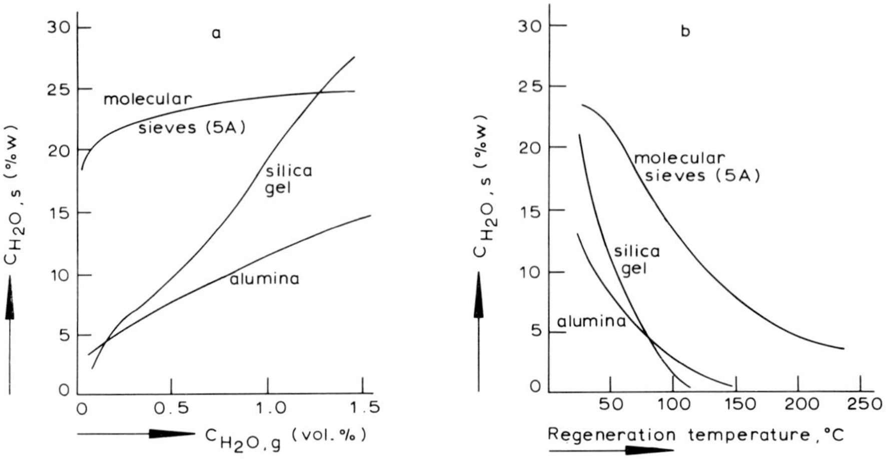

Figure IV.5. Properties of adsorbents

a. adsorption isothermes (25 °C)

b. adsorption isobars

(P_{H_2O} = 10 mmHg)

Compression

The first generation of ammonia plants was equipped with reciprocating compressors. These machines compress the synthesis gas to the requisite high pressure in a number of stages, usually three, with cooling between the stages. Advantages of reciprocating compresssors are that very high pressures can be reached and that the gas throughput can be controlled over a very wide range. Disadvantages are that the reciprocating movement of the piston and other discontinuously moving parts produce noise and wear and cause high maintenance cost.

With turbocompressors, the movement is continuous and rotary; the gas is compressed by a number of blades mounted on a common shaft. This rotor is also divided into stages, with interstage cooling. Turbocompressors are preferred to piston compressors because of their lower cost for high capacities, virtual absence of noise and vibration, oil-free operation, greater simplicity and reliability. The latter is particularly important for large single-train ammonia units: shutdowns caused by compressor failures lead to very high loss of production and are economically unacceptable.

However, the turbocompressor also has disadvantages related to energy losses between the blades of the rotor. Since these losses are particularly noticeable at high pressure gradients, turbocompressors become uneconomical above about 32.5 MPa. Losses also increase when the speed is reduced to diminish the throughput; therefore, this type of compressor cannot be applied in as wide a range of throughputs as reciprocating compressors. In spite of these drawbacks and the somewhat lower overall efficiency of turbocompressors these machines are now exclusively used in modern ammonia plants.

Ammonia synthesis

Ammonia is formed by the equilibrium reaction

$$N_2 + 3\ H_2 \rightleftarrows 2\ NH_3 \quad (\Delta H^o_{r,s} = -45.3\ \text{kJ/mole}\ NH_3)$$

In view of the exothermic nature of the reaction and the volume contraction by conversion to NH_3 it is necessary to operate at relatively low temperatures and high pressures; this is confirmed by Figure IV.6., where the NH_3 content of the equilibrium mixture is plotted as a function of pressure, with temperature as the running variable. Economically favorable rates are obtained by catalyzing the reaction: iron catalysts give good conversions at about 30 MPa and 500 ^oC in reactors of acceptable dimensions. Another means of increasing the rate is to use some small excess of nitrogen because nitrogen has more influence on the rate than hydrogen in the range of space velocities* used in practice. A H_2/N_2 molar ratio between 2.6 and 2.9 proves to be the optimum. It is, however, always necessary to check whether the advantage of using excess

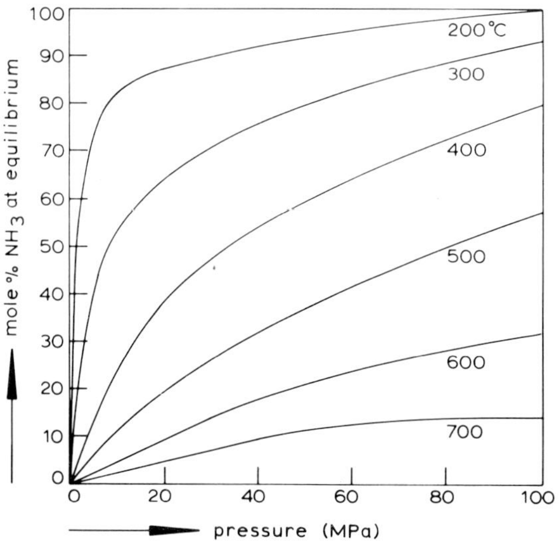

Figure IV.6. Equilibrium yield of NH_3
feed : H_2 / N_2, 3 :1 mole/mole

*Space velocity $\frac{1}{\tau}$ = volumes of gas (standard temperature and pressure) fed to the reactor per unit volume of reactor and unit of time; τ= space time.

nitrogen is economically attractive. Figure IV.7. illustrates the effect of working with non-stoichiometric feeds on the conversion to ammonia.
The function of the catalyst is to activate the particularly stable bond in the nitrogen molecule. This high-energy bond (950 kJ/mole) is broken or substantially weakened by chemisorption on the catalyst surface, i.e. chemical interaction with surface iron atoms. Further reaction with hydrogen atoms formed from hydrogen molecules by dissociative chemisorption is then possible. The experimental finding that maximum conversions to NH_3 are obtained at H_2/N_2 molar ratios slightly below 3 at industrial conditions is in line with the postulate that nitrogen chemisorption is the rate-determining step in ammonia synthesis. However, opinions among catalytic experts as to the exact mechanism of of ammonia synthesis over Fe catalyst still differ although the process has been in industrial operation for more than 60 years. In general, exact knowledge on reaction mechanisms is rarely needed for succesful process development, but it can be very helpful in process optimization.

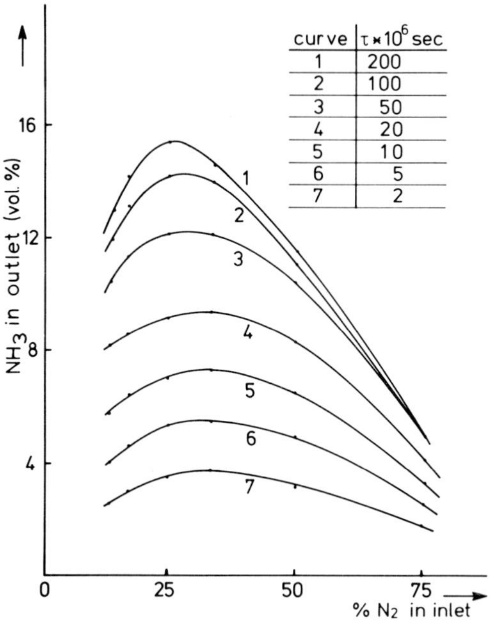

Figure IV.7. Effect of non-stoichiometric feeds on conversion to ammonia

Catalysts for chemical processes should be active, stable, selective and mechanically strong enough to withstand the static pressure of the catalyst bed as well as sudden pressure changes in the reactor. For the ammonia catalyst, high selectivity is not a problem because parallel or subsequent reactions do not occur. Activity and stability are improved by the addition of small amounts of stabilizers and promotors such as CaO, K_2O, Al_2O_3 and, less commonly SiO_2. It is necessary to prepare the catalyst very carefully to obtain the right catalyst texture, i.e. adequate pore volume, pore size and size distribution, and size of primary iron crystallites. This is because the rate of ammonia formation is not only a function of the chemical reaction steps on the catalyst surface, but also of the rate at which the reactants are transported through the pores to the inner surface, in particular with catalysts prepared from a melt. This type of transport limitation by pore diffusion has much more effect on the overall rate of formation of ammonia than film diffusion, i.e. mass transport to and away from the outer surface of the catalyst particle. In summary, the process rate is a complex function of a number of chemical and physical phenomena.

Water poisons ammonia synthesis catalysts by being very strongly chemisorbed on the iron surface. Its effect is essentially reversible at relatively low temperatures but it becomes more and more irreversible at higher temperatures. Poisoning by water can be illustrated by the fact that the rate of ammonia formation at 400 ^{o}C drops by a factor of about 3 when 25 ppm water is added to the feed gas. Other poisons like CO, CO_2 and oxygen are supposed to be hydrogenated to water . These types of poisoning effect were initially not recognized by Haber and Bosch when developing the process in the decade 1900-1910. They recommended Os and U for the first ammonia plant (1913). However, further development work soon showed the importance of exhaustive gas purification; the much cheaper but more sensitive Fe catalyst could be used in subsequent plants with improved synthesis gas preparation units. When designing ammonia reactors the following factors should be taken into account:

- construction costs — amount of catalyst needed
- pressure drop — control properties
- effective use of the heat of reaction

At the high temperatures and pressures needed here, the iron catalyst must be applied in the form of fixed catalyst beds. Good temperature control is essential because of the less favorable equilibrium conversion at higher temperatures. The older types of ammonia convertor contained many parallel tubular reactors containing the catalyst (Figure IV.8a.).

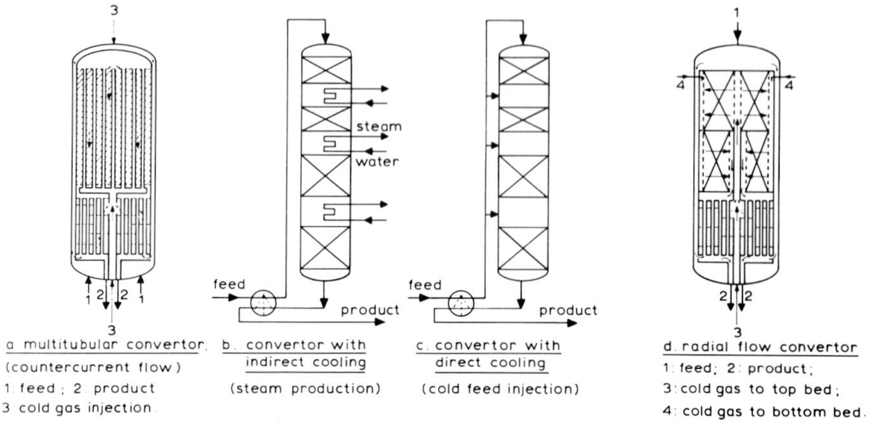

Figure IV.8. Four types of ammonia synthesis reactor

Feed gas preheated by heat exchange with product gas was brought to the reaction temperature by first passing it cocurrently or countercurrently along the catalyst tubes for further heating. A small quantity of feed was injected without heat exchange with the reactor product to ensure that adequate temperature control was possible. This type of reactor is no longer used for ammonia synthesis because of its high cost.

More modern reactors are built as multi-bed convertors (Figures IV.8b. and c). These contain a number of adiabatic fixed beds in series, with cooling between the beds. Cooling may either be indirect by water, producing steam or direct by injecting cold feed gas between the beds. The latter method, which is possible only when subsequent selectivity-decreasing reactions do not occur, has the disadvantage that the conversion per pass decreases somewhat, thus requiring more catalyst per unit weight of ammonia produced. The cost of construction is, however, lower than for indirectly cooled multibed reactors. Both types of multi-bed reactor are very easy to control.

Multitubular and multi-bed reactors are filled with rather large catalyst particles (6-10 mm) because otherwise the pressure drop across the reactor would be too high. However, large particles have the disadvantage that pore diffusion limitation of the process rate is more serious than with smaller particles, the diffusion pathway being longer. This has led to the design of a radial flow reactor with relatively thin catalyst layers in which 2-3 mm particles can be applied. This is shown in Figure IV.8d. in which the gas flows from the centre of the reactor to the wall in the top section and in the opposite direction in the bottom section. The higher process rate obtained with smaller particles at satisfactorily low pressure drop imply that smaller reactors and less catalyst can be used for a given capacity. A heat exchange section below the beds is used to transfer the heat from the product gases to the fresh feed. Good temperature control is again possible by varying the proportion of cold feed gas which bypasses the heat exchanger and is led directly to the top bed. A drawback is that the effects of catalyst poisoning or sintering are more readily noticed than with large particles because the reactor contains less catalyst. The reactor effluent is sent to the recovery section, where ammonia is recovered by condensation. Water can be used as the cooling agent if the pressure is above about 35 MPa. This is not possible with most modern units in which turbocompressors are installed because their pressure does not exceed 32,5 MPa. In such plants recovery is effected by cooling the product stream in evaporation condensers cooled with evaporating liquid ammonia, i.e. part of the product stream. If this is done, the NH_3 concentration can be lowered to about 1 vol% at about - 30 ^{o}C. The ammonia evaporated in the cooling section can sometimes be sold or used in the gaseous state. If this is not possible, it should be recompressed and liquefied.

Since the conversion to ammonia is mostly between 10 and 20%, the unconverted synthesis gas components should be recycled to the reactor after recompression to compensate for the pressure drop across reactor and condensor. Some of the gas should be vented to prevent build-up of inerts like CH_4 and noble gases; the purged gas should be washed free of ammonia to prevent environmental pollution. In some of the newer ammonia plants the purged gas is separated by pressure swing adsorption into a hydrogen-rich fraction for recycle and a fraction rich in inerts (see Chapter VI.). It is also possible to separate pure noble gases from the purge gas.

Integral process

The various elements discussed above should be integrated to obtain an organized system (Figure IV.9.) in which the operating conditions of the various elements are such that optimum plant performance is attained.

The previous discussion of the elements indicates that a change of the conditions in one of them may have effects that are felt throughout the entire process. If the economic consequences of such changes are to be quantified, it is necessary to construct a mathematical model of the entire plant, i.e. a set of mathematical equations containing mass and energy balances of the various units as well as an objective function by which the total cost of ammonia per unit weight can be calculated. Numerical methods are used to find the economic

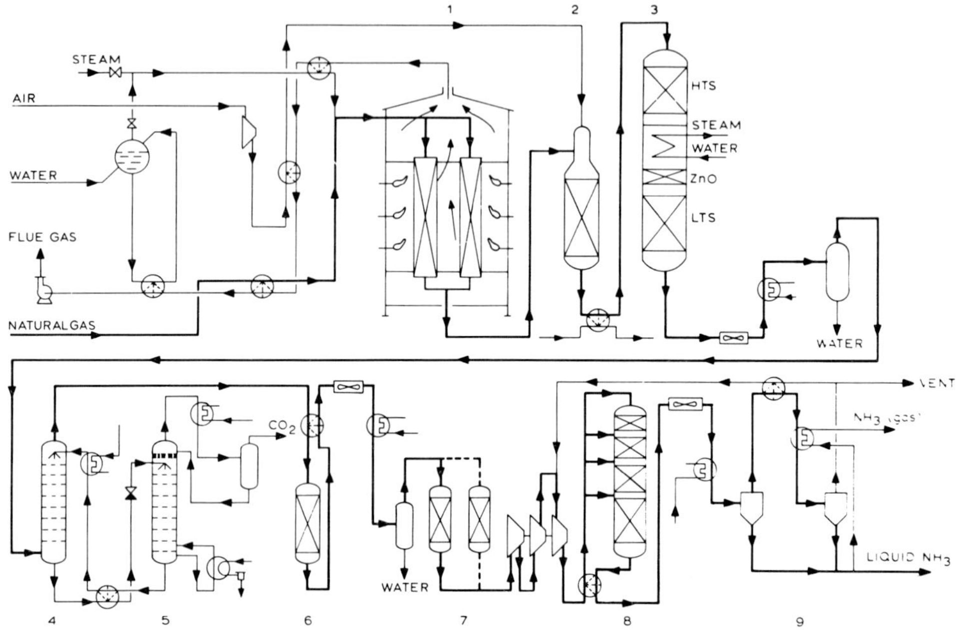

Figure IV.9. Ammonia plant based on natural gas

1. first reformer

2. second reformer

3. CO shift convertor

4. CO_2 absorber

5. regenerator

6. methanator

7. dryers

8. NH_3 reactor

9. condensation unit

optimum for a given set of independent process variables and cost factors. An example of model calculation results is given in Figure IV.10., in which the relation between ammonia production cost and synthesis pressure is shown for an assumed set of process and cost conditions. A shallow cost minimum is found between 20 and 30 MPa. At pressures above the optimum, the energy saving gained from the higher pressure in the recovery section and the smaller volume of unconverted feed to be recycled are more than offset by the higher cost of equipment and fresh feed compression. Below the optimal pressure the reverse holds. Results at pressures above 32.5 MPa are not significant because it was assumed that a turbocompressor is used. The exact position of

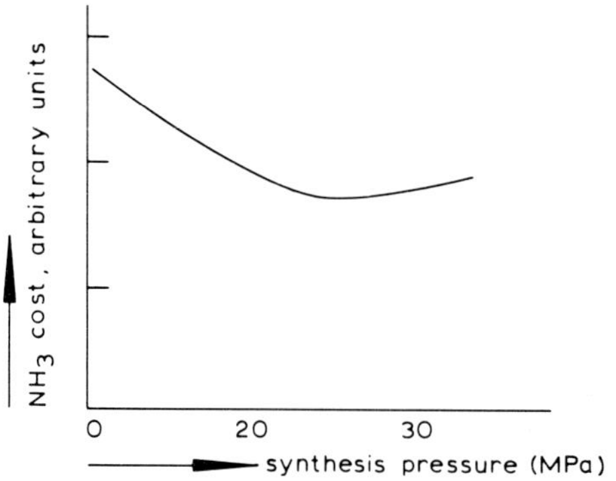

Figure IV.10. Qualitative example of result of pressure optimation calculations

the minimum depends on many of the assumptions made, e.g. the price of feed-stock and utilities, the percentage of the product to be delivered in liquid form, capital costs, plant capacity, reformer pressure and the like. Such process models are often used to assess the sensitivity for changes in technical and economic parameters. In the example of Figure IV.11., curve a, the cost of producing ammonia decreases when plant size increases. Most of the benefits are obtained by increasing the capacity of single-train units from about 200 to about 700 tons per day. Further increases not only have much less economic

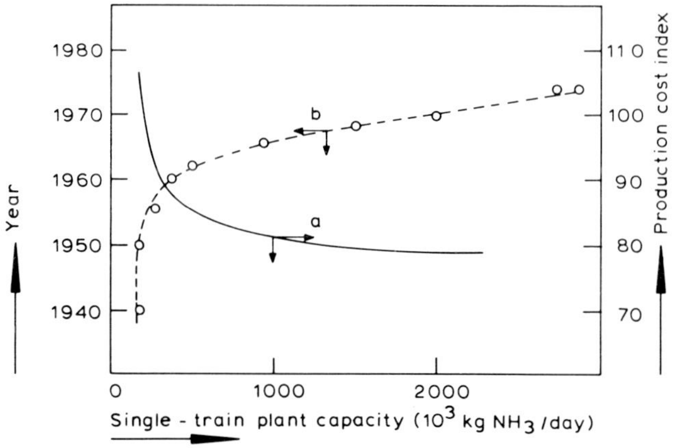

Figure IV.11. Plant size and production cost index ammonia manufacture
b= size :daily capacity of the largest plant built in a certain year.
a= index :calculated production cost (1972) :cost of 200 t/d NH_3 plant.

effect but are also disadvantageous in that the plant would have to be made extremely reliable, the economic damage of even a few days' production loss being considerable. Reliability can be increased by providing spares for critical equipment for use in emergency situations and extra storage facilities for intermediates, but at the expense of higher investment costs. Thus, there is an economic limit to the growth of ammonia plant size, which is probably between 1000 and 1500 tons/day production capacity. This conclusion is confirmed by the fact that only a few larger plants have so far been built. Figure IV.11. (curve b) also shows how the size of single-train ammonia units has increased over the past 40 years.

Related industrial processes

Several process steps used in making ammonia from methane are applied for many other purposes. Drying and CO_2 removal by washing with an alkaline liquid are among the most common operations in the process industry. Hydrogen and methanol synthesis gas are often produced via steam reforming. Methanation of traces of carbon oxides is a necessary element of process systems for making hydrogen via steam reforming of hydrocarbons or coal gasification. The same methanation reaction is also involved in the manufacture of substitute gas (SNG) from coal or heavy fuel oil (cf Chapter XIV), but in that case methane synthesis and not purification is the objective. This has important consequences for the design of SNG methanators: adiabatic reactors cannot be used without dilution of the feed gases because of the large amounts of heat evolved.
The technology of methanol synthesis resembles that of ammonia manufacture in many respects. Here, too, the main reaction is exothermic and the conversion is limited by equilibrium:

$$CO + 2\ H_2 \rightleftarrows CH_3OH\ (\Delta H^o_s = -\ 91\ kJ/mole\ CO)$$

Favorable conversions in reactors of acceptable dimensions are obtained when catalysts are used. Until about 1960, ZnO/Cr_2O_3 was in general use; at 30 - 35 MPa and 350 - 400 ^oC about 20% conversion per pass could be obtained in reactors of the type shown in Figure IV.8a.. As is done in ammonia plants, most of the unreacted feed is recycled and a small amount vented.
There are also important differences between the two processes. Methanol synthesis gas is produced in a single steam reforming step and the subsequent gas purification is limited to the removal by condensation of excess water.

ZnO/Cr_2O_3 catalysts are somewhat less sensitive to sulfur than ammonia catalysts. Another difference is that side reactions occur:

$$CO + 3\ H_2 \rightleftarrows CH_4 + H_2O \qquad (\Delta H_s^o = -\ 205\ kJ/mole\ CO)$$

$$2CO + 4\ H_2 \rightleftarrows C_2H_5OH + H_2O \qquad (\Delta H_s^o = -\ 128\ kJ/mole\ CO)$$

The above side reactions must be suppressed by kinetic means, i.e. by careful catalyst formulation. Moreover, their very exothermic nature requires careful temperature control. This is the more necessary because the activation energy for methane formation is higher than that of the methanol synthesis reaction: if the temperature rises, more methane is formed which causes the heat of reaction to rise etc.. Another complication is ether formation:

$$2\ CH_3OH \rightleftarrows CH_3OCH_3 + H_2O$$

Since the equilibrium amount of dimethyl ether is usually formed it is possible to recycle the ether after separation from the liquid reaction products. If, however, the methanol is used to obtain formaldehyde by catalytic oxidation (Chapter XI) the ether can remain in the crude methanol because it also gives formaldehyde in good yields.

In the early 1960s, ICI introduced a better catalyst, Cu/ZnO, which is much more active at lower temperatures and also more selective. The first plants in which this new catalyst was applied operated at about 250 ^oC inlet temperature and 5.0 MPa. Subsequent work led to improvements in the thermal and chemical stability of Cu/ZnO, with the result that the process pressure could be raised to about 15 MPa. This causes increased conversion as well as a higher heat production, which necessitated re-design of the reactor. This is more important because Cu/ZnO sinters relatively rapidly near 300 ^oC. This catalyst is also very sensitive towards traces of sulfur compounds and its use became possible only after improved gas desulfurization techniques had been developed.

As compared with the older ZnO/Cr_2O_3 catalyst, the following advantages of applying Cu/ZnO are claimed:

a. reactor and separation section are cheaper because of the lower pressure;
b. less energy is consumed, for the same reason;
c. the liquid product is easier to purify because it contains less dimethyl ether, heavy ends and ketones.

The synthesis gas for methanol manufacture is produced by steam reforming over Ni/Al_2O_3 at lower pressure and a higher temperature than is needed for NH_3 synthesis gas. Otherwise, the methane conversion would be too low. The CO/H_2 ratio of gas produced by steam reforming of methane is lower than required; if nothing is done to lower this ratio, the excess hydrogen must be vented from the synthesis loop, which is economically unattractive. A better solution is to add CO_2 from an external source to bring the CO/H_2 ratio to the desired value of about 2. This is possible because the catalysts for methanol synthesis also catalyze the (reverse) CO shift reaction.
Apart from correcting the CO/H_2 ratio, this reaction also diminishes the heat evolved because it is endothermic. The CO_2 obtained as a by-product in ammonia synthesis gas purification can be used for this purpose. This example shows that it is often profitable to locate ammonia and methanol plants in the same area. Another source of CO_2 is the reformer flue gas. A simplified flow sheet of a methanol plant is shown in Figure IV.12. The principal application of the product is for the manufacture of formaldehyde by catalytic gas-phase oxidation. The use of methanol as a component of motor fuel has also been contemplated, because of its attractive performance properties in gasoline engines and relative ease of transportation.

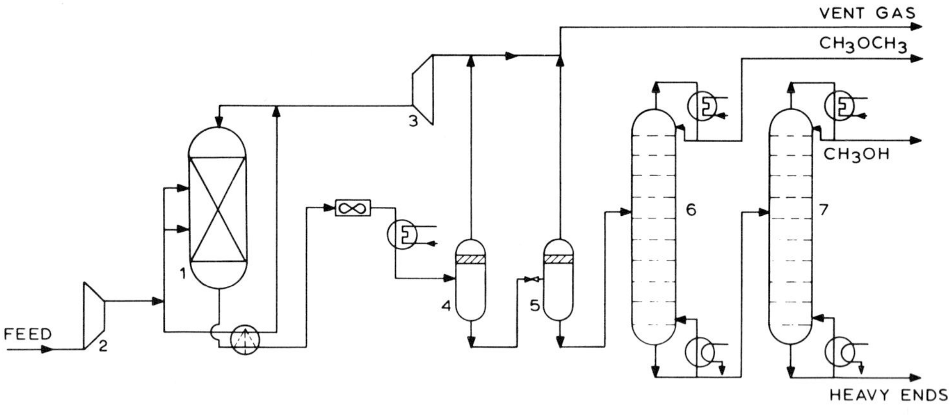

Figure IV.12. Methanol synthesis

1. *reactor*
2. *fresh feed compressor*
3. *recycle compressor*
4. *high pressure separator*
5. *flash vessel*
6. *dme column*
7. *methanol column*

However, improvements in e.g. the transportation of natural gas, gasoline manufacturing processes, as well as shifts in petroleum refining pattern have led to lower prices of hydrocarbon fuels as compared with methanol. Further improvement of methanol processes and shifts in the price structure of petroleum stocks may cause this picture to change in the more distant future.

REFERENCES

General

1. Vancini, C.A., 'Synthesis of Ammonia', Transl. by L. Pirt, Ed. D.J. Borgars Mcmillan, London 1971.

2. Slack, Editor A.V., 'Ammonia', Vol. 1-4, Marcel Dekker Inc., New-York 1973

Special

1. Andrews. S.P.S., Chem.Ind. (London), 826-830 (1965).
   Figure IV.2. is reproduced from this article with permission.

2. Uchida, H., M. Kuraishi, Bull.Chem.Soc. Japan, 28, 106-112 (1955).
   Figure IV.7. is adapted from Figure 7. in this article with permission.

3. Strelzoff, S., L.C. Pan, Adv.Petr.Chem.Refin., 9, 283-350 (1964).
   Table IV.1. is reproduced from this article with permission.

V. OTHER METHODS OF AMMONIA MANUFACTURE

Most of the world production of ammonia takes place in natural gas-based plants. Some of these have a structure differing from that described in the previous chapter: it is possible to replace one or more operations of the process sequence by others. Moreover, about one third of the production is based on feedstocks other than methane, such as refinery gas, propane, naphtha, fuel oil and coal. Some of the alternatives for ammonia manufacture are discussed below. All of these apply the iron-catalyzed high-pressure synthesis. Quite different possibilities of obtaining ammonia are briefly mentioned at the end of this chapter. The discussion below is intended as an illustration of the rule that several solutions often exist for any given problem. Which one of these is preferred in a specific case depends on many technical, economic and even social factors, many of which are time-dependent.

Cryogenic purification of synthesis gas

The first generation of ammonia plants was based on coke oven gas as the feedstock. The gas was purified by removing CO by the shift reaction and CO_2 by a washing operation; after drying it was cooled to about -190 ^{o}C for partial liquefaction. The hydrogen-rich gas phase was then washed countercurrently with refluxing nitrogen to remove impurities like methane, argon and the rest of the CO. Water and CO_2 were removed prior to the washing operation by freezing. Traces of catalyst poison were usually absorbed on a fixed bed of used ammonia catalyst at essentially synthesis conditions. For several decades cryogenic separations at very low temperatures were too costly to apply them in making a bulk product like ammonia. Recently, however, this has changed to some extent because better designs of cryogenic separation units have become available through the use of much improved insulating materials. Moreover, it is economic to apply extensive heat exchange in the very large plants that are now standard throughout the world. As a result, there are several ammonia plants in operation in which a low-temperature gas purification step is incorporated. A block-diagram of such a plant is shown in Figure V.1.

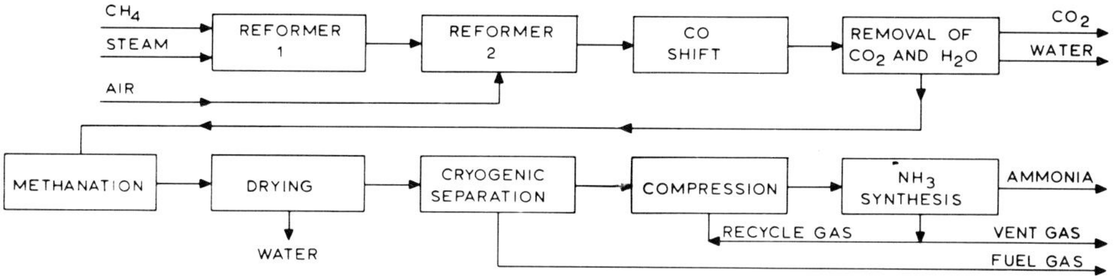

Figure V.1. Block diagram of ammonia plant containing a cryogenic separation step

Table V.1. Boiling points of components of crude synthesis gas at 0.1 and 2.0 MPa

Component	N_2	Ar	CO	CH_4	CO_2
boiling point at 0.1 MPa (K)	77.3	87.5	81.8	111.6	195
boiling point at 2.0 MPa (K)	115.5	131.8	123.4	164.6	--

The reformer section is similar to that of the previous chapter, but the crude synthesis gas now contains somewhat more nitrogen. The gas is freed of CO, CO_2 and water in the usual manner and then sent to a cryogenic separation unit (Figure V.2.). It is first cooled to about -140 ^oC by heat exchange with cold gases and then passed through an expansion turbine. This brings about partial liquefaction and further cooling to about -175 ^oC at a pressure of but a fraction of a MPa, along with power recovery. The liquid phase is rich in methane and nitrogen and also contains about half of the Ar and CO present in the feed (for boiling points see Table V.1.). After further cooling by heat exchange, the gas/liquid mixture enters a packed separation column in which the gas phase is washed countercurrently with refluxing nitrogen. The result is that virtually all the methane and the bulk of the CO and Ar are removed.
The temperature in the top condenser should be lower than that of the bottom liquid to condense part of the nitrogen as reflux. This is done by expansion of the liquid phase through a throttle valve (see Figure V.2.). The gas emerging from the coolant side of the condenser is used as fuel, after having served for precooling the feed gas to the separator.

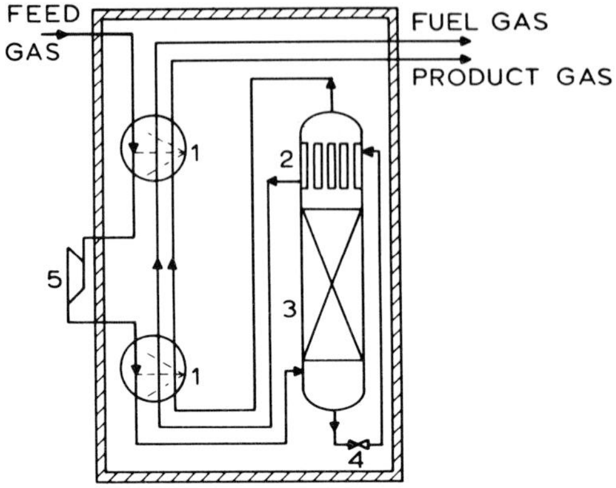

Figure V.2. Cryogenic separation unit

1. heat exchangers

2. condenser

3. separation column

4. throttle valve

5. expansion turbine

The ammonia process containing a low-temperature separation stage has the following advantages over the conventional processing scheme described in the previous chapter:

- the outlet temperature of reformer II can be somewhat lower because the methane concentration in the crude synthesis gas is less critical, the methane being removed almost completely in the cryogenic separation. The adiabatic temperature rise over this reformer is, if anything, a little larger because rather more air is fed to reformer II to obtain the excess nitrogen required for the low-temperature washing operation. Thus, the temperature of the *first* reformer is rather lower than with conventional processing, i.e. about 100 ^{o}C. This means that longer tube life is obtained and that the construction of reformer I is simpler. An alternative is to apply a higher pressure in reformer I without risking inadequate tube life; this saves on compression costs.
- fuel costs are lower because the conversion in reformer I can be lower. The indirect heat supply to reformer I is replaced partly by direct heat supply by partial oxidation in reformer I.
- CO shift and methanation are somewhat less critical.
- since the synthesis gas contains fewer inerts it is possible to reduce the inerts content of the feed to the ammonia reactor at the same purge ratio;

alternatively, less gas needs to be purged at the same inerts content in the combined feed to the reactor. The result is a saving in compression costs in both cases.

- the installation is a little more flexible because small but significant disturbances in the operation of the units upsteam of the cryogenic unit can be corrected relatively easily.
- cryogenic processing is advantageous when the natural gas feed contains some nitrogen, which is often the case.

Cryogenic separations have also disadvantages. Since the feed gas should contain more nitrogen, more partial oxidation is needed resulting in larger volumes of feed gas. Moreover, the plant is more complex and investment costs for ammonia units with a cryogenic separation unit and allied equipment like controllers and a cold trap for water and carbon dioxide are higher than for ammonia units based on conventional methane/steam reforming. Another reason is that spares for vital components of the cryogenic unit must be provided to minimize the consequences of equipment failures.

Recovery of ammonia by a water wash

The importance of optimizing the ammonia synthesis loop pressure has been underlined in the previous chapter. Figure IV.10. shows that the savings in fresh feed compression energy, obtained by working at a lower pressure, are less than the extra expense of having to re-compress the ammonia evaporated in the recovery section. This suggests the use of alternative methods for recovering ammonia from the reactor effluent.

One of these is the use of a water wash for ammonia recovery. This method is by no means new: it was applied in the first Haber-Bosch plant, but abandoned in the later designs because of the adverse effects on catalyst activity of water introduced via wet recycle gas. The advent of efficient gas drying procedures which allow drying to but a few ppm by volume of water has caused renewed interest in ammonia recovery by a water wash. Recently, SNAM has developed the process shown in Figure V.3. in the form of a block diagram. It contains the following new operations:

- the reactor product is first cooled to 20-50 ^{o}C by heat exchange and water cooling;
- a concentrated aqueous ammonia solution is then produced by a countercurrent water wash. This operation can be carried out in a film absorber in which the water flows downwards as a thin film over the inner surface of pipes;

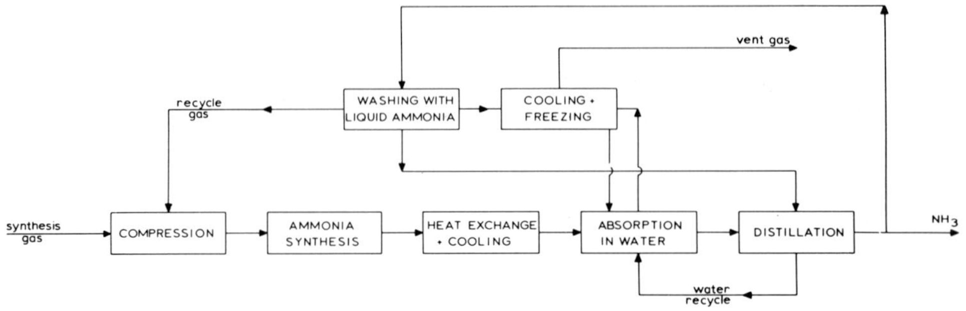

Figure V.3. Ammonia recovery by washing with water

the pipes are cooled externally to remove the large exothermic heat of absorption;

- almost anhydrous ammonia is recovered as the top product of the subsequent distillation. The bottom product, dilute aqueous ammonia, is recycled to the absorber;
- the top gas of the absorber is cooled to about -20 °C to condense the rest of the ammonia and nearly all the water;
- the small residual amounts of water are then removed from the gas by washing with liquid ammonia. The liquid is recycled to the distillation stage and most of the gas to the ammonia reactor.

The developers, SNAM, recommend this process in combination with the cryogenic synthesis gas purification described earlier in this chapter, and for fairly low reactor pressure (12-15 MPa). Advantages claimed are substantial savings in utility costs, greater simplicity of the plant which results in lower investment costs and the absence of a compressor for liquefaction of the ammonia used for cooling purposes in conventional recovery units. The new method is held to be especially attractive for a combined ammonia/urea plant (*cf* Chapter VIII). However, it should be noted that other equipment should be added.

Experience with this recovery alternative is limited. In general, acceptance by the industry of new variants of existing and well-established processes is slow, particularly with plants of the size of present-day ammonia units. A high degree of certainty that new types of plants will perform well and reliably must be provided because losses incurred by failures of very large plants are extremely damaging if they occur.

Other feedstocks for steam reforming

If desired, hydrocarbon feeds other than methane, such as naphtha and LPG (propane + butane) can be used for steam reforming over nickel catalysts. These feedstocks should comply with strict specifications. Aromatics and olefins must be absent and the atmospheric final boiling point should not exceed 200 ^{o}C. Otherwise, carbon deposition on the reforming catalysts will occur. The sulfur content should be very low to avoid the formation of nickel sulfide and consequent catalyst deactivation.

The main difference in operating conditions is that heavier feeds require a somewhat higher steam/hydrocarbon ratio, to minimize the risk of carbon formation. At suitable conditions the only hydrocarbon in the product gas is methane: heavier hydrocarbons can hardly be detected. The composition of the product gas can be calculated by the method shown in Chapter IV, provided that the H/C-ratio of the feedstock is substituted in the hydrogen balance. Most naphthas contain somewhat more sulfur than the very low value of about 1 ppm by weight that can be tolerated in the reformer feed. Therefore, a desulfurization step is added just before the first reformer. The feed is evaporated, mixed with hydrogen-containing gas and passed over Co-Mo-sulfides/ Al_2O_3 catalyst to convert the sulfur compounds quantitatively into H_2S. The hydrogen sulfide is removed in a subsequent step by passage over a guard bed containing ZnO or by stripping. Wet methanated synthesis gas can be used for this hydrogenation; the pressure at which this operation is carried out is the same as that of the first reformer. Naphtha is a more expensive feedstock than natural gas; its use is limited to areas where natural gas or LPG is unavailable.

Synthesis gas by partial oxidation of petroleum fractions

One of the disadvantages of steam reforming is that natural gas, LPG and particularly naphtha are rather expensive feedstocks. Much cheaper products like heavy fuel oil or even gas oil cannot be steam-reformed because catalyst life would be very short. This is the reason why Shell and Texaco have introduced similar processes for gasification of heavy oils by partial oxidation to obtain synthesis gases and hydrogen. Oxygen is used as the oxidant. Several process configurations are possible if ammonia synthesis gas is to be obtained. The two systems differ in the way in which the synthesis gas is purified.

In the process of Figure V.4.a. fuel oil is oxidized at 1200-1500 ^{o}C at

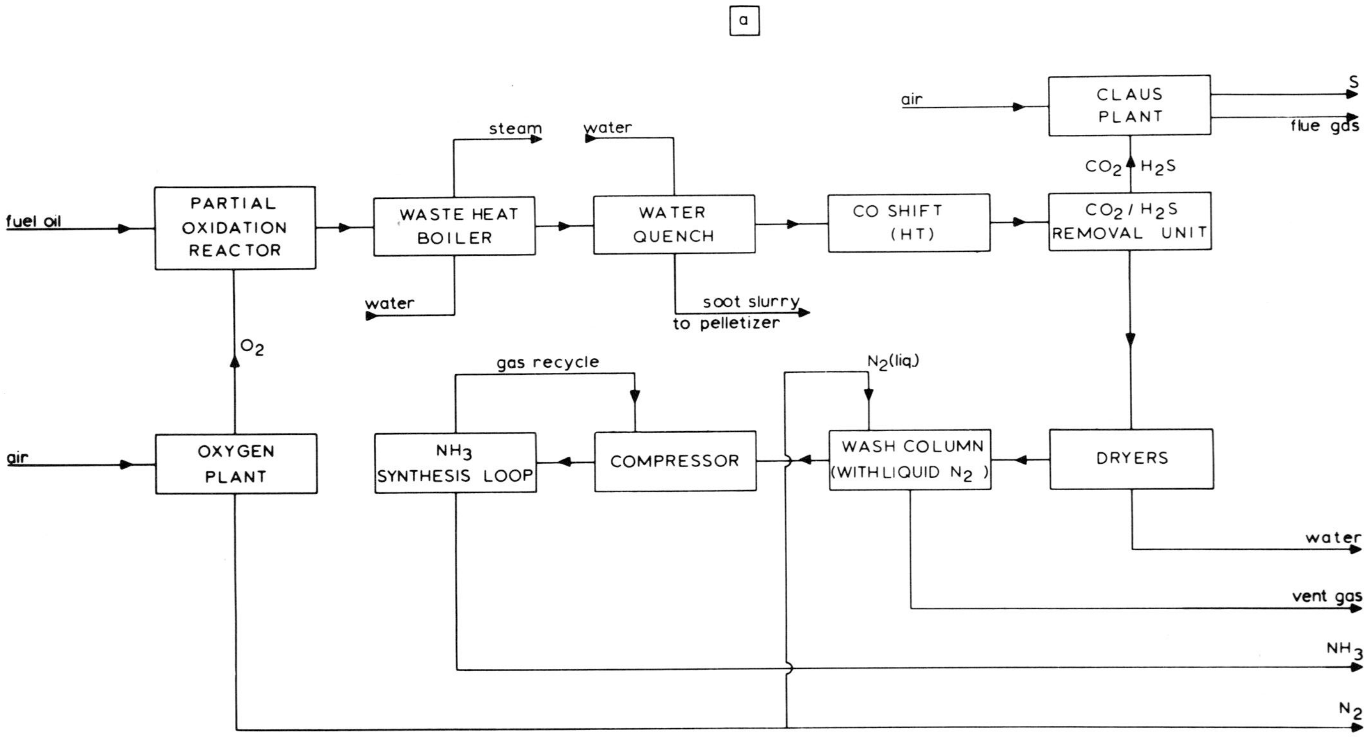

Figure V.4. Ammonia manufacture based on fuel oil - a) Gas purification containing a liquid nitrogen wash

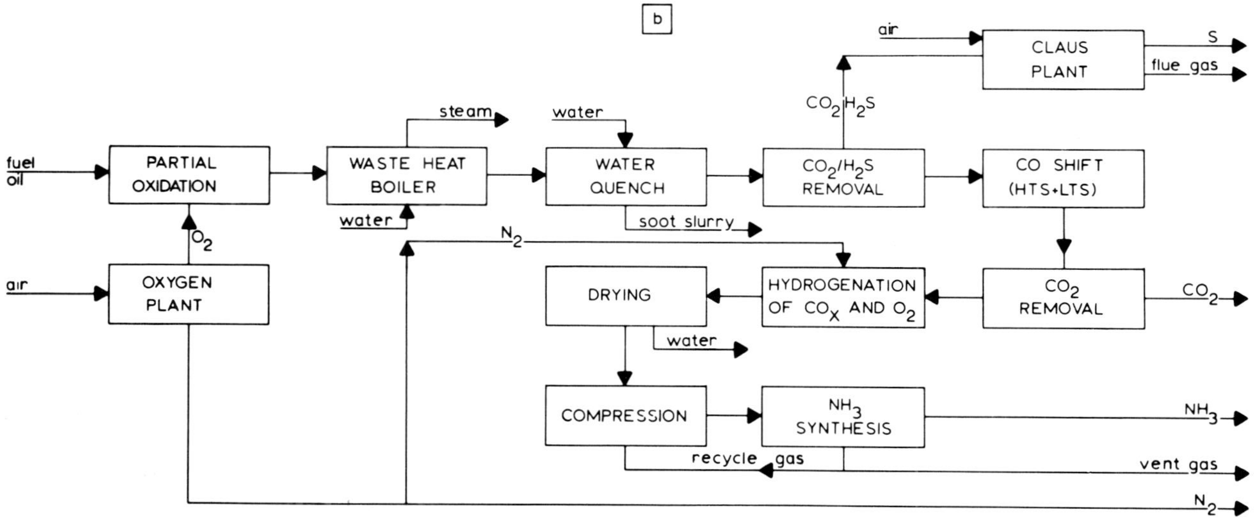

Figure V.4. Ammonia manufacture based on fuel oil - b) Gas purification containing a two-stage CO shift

pressures up to 10 MPa with oxygen produced from the liquefaction and distillation of air. The product contains much N_2, CO and CO_2, smaller amounts of H_2S and traces of COS and HCN. Although some steam is added, about 1% of the feed is converted into soot. From the reactor, which is lined with firebrick to withstand the high temperatures, the gases flow through a waste heat boiler where high-pressure steam is generated in quantities sufficient for the process as well as for some export to adjoining units. The next step is a water quench for soot removal; in the Shell Gasification Process the soot suspension obtained is fed to a pelletizer to convert it to easily recoverable carbon pellets with the aid of gas oil. The water quench also serves to convert the HCN quantitatively into formic acid. Most of the CO is removed in a shift reactor containing a sulfur-resistant catalyst operating at a rather high temperature. CO_2 and H_2S are washed out by an alkaline washing liquid (*cf* Chapter VI). After drying the gas is contacted countercurrently with liquid nitrogen (obtained in the air separation unit) to remove CO and methane. The former impurity is present in appreciable concentrations because of the rather high temperature of the preceding CO shift. The low-temperature separation is followed by compression and ammonia synthesis. H_2S is converted into sulfur in a Claus unit (Chapter VI); a plant of this type produces steam, nitrogen and sulfur as by-products.

The other form of this process (Figure V.4.b.) differs in the method used for gas purification. In this case the water quench is followed by removal of H_2S and CO_2. CO is converted in a HTS/LTS reactor. It is now necessary to remove the CO_2 formed in the shift reaction in a second washing operation. The amount of N_2 needed for NH_3 synthesis is then added; a methanator, which also removes traces of oxygen, and dryers are applied for final purification. The process of Figure V.4.a. is probably more attractive because the rather expensive low-temperature shift stage, the methanator and the second CO_2 removal step are not needed. Moreover, less gas/gas heat exchange is required. Another advantage of the first type of process train is that somewhat higher pressures (up to 8-10 MPa) are possible with it because it does not contain an LTS reactor. If such a reactor is used, the maximum operating pressure cannot be much above 4 MPa because otherwise water condensation in the pores of the LTS catalyst would result in severe deactivation. In general, the best process configuration depends on many factors, such as price of feedstock and utilities, possibility of using the excess steam in other applications, capital costs in relation to other expenses, etc. The above partial oxidations processes can also be applied for electricity generation in combined-cycle systems (see Chapter XIV).

Partial oxidation of coal

The first ammonia units of the Haber-Bosch type operated on synthesis gas made from coal or coke by gasification with steam. The composition of the gasification product is then determined by the equilibria:

$$C + H_2O \rightleftarrows CO + H_2 \qquad (\Delta H^o_s = +131.6 \text{ kJ/mole CO})$$

$$CO + H_2O \rightleftarrows CO_2 + H_2 \qquad (\Delta H^o_s = -41.1 \text{ kJ/mole } CO_2)$$

$$C + 2\,H_2 \rightleftarrows CH_4 \qquad (\Delta H^o_s = -75.0 \text{ kJ/mole } CH_4)$$

Gasification temperatures are above 1000 ^oC. Since the overall heat effect is endothermic, energy should be added, which is done by partial combustion of the coal. In practice, 40-45% of the coal feed must be burnt to supply this energy.

Further information on the gasification of coal is given in Chapter XIV. dealing with energy conversion processes. The only conclusion drawn here is that at the present time ammonia plants based on coal or coke cannot compete with those discussed earlier in this chapter at conditions prevailing in Western Europe or North America. If prices of oil and natural gas continue to rise more rapidly than those of coal or coke, this position may change. It does not seem likely that most of the ammonia produced will again be based on coal until well into the 21st century.

Economic aspects

The literature contains several examples of a comparison between the costs of making ammonia from the feedstocks natural gas, naphtha or heavy fuel oil. In general, the following conclusions are drawn:

- feedstock cost is the highest item for steam reforming of naphtha and natural gas; the second largest item is capital costs
- feedstock and capital costs contribute about equally to the cost of ammonia produced by the partial oxidation route
- capital requirements for partial oxidation are higher, mostly because of the investment for the oxygen plant
- the costs of making ammonia from natural gas are about the same as those of the partial oxidation route
- naphtha is somewhat less attractive as a raw material.

However, literature data of this kind should generally be approached with some caution. Firstly, such comparisons are dependent on the location assumed.

Furthermore, changes in price structure of feeds and products may appreciably affect the outcome of economic comparisons. Similar changes may come about when wages and salaries are raised. But perhaps the most uncertain point is that such cost comparisons are often made by licensors or contractors of one of the processes. Since such firms want to obtain economic benefits from their proprietary process, there is a tendency to compare it with other methods at conditions where the strong points of their process show to the best advantage. Therefore, it is always necessary to take the sources of the information into account when comparing different processes for the same product.

Future developments

The rapid depletion of the world's fossil fuel reserves will most probably have pronounced effects on the chemical industry, among which is the likelihood that alternative processes for nitrogen fixation will become available. Below, a short review is given of the present search for new or improved methods of converting molecular nitrogen into fertilizers or other useful products. A common characteristic of all the methods is that a great deal of energy is required because of the high strength of the bond in the nitrogen molecule. The first question to be answered is if the present-day Haber-Bosch route to ammonia can be appreciably improved energy-wise. Generally speaking, some reduction in the amount of energy required may be possible through the development of highly active ammonia synthesis catalysts. There are some literature reports that heterogeneous catalysts of much improved activity can be made, but these catalysts are very sensitive towards very small quantities of oxygen-containing impurities. Two possible avenues for research are: the development of highly active but more stable catalysts, or of much more efficient and exhaustive purification methods for ammonia synthesis gas.

It should be noted that it is by no means desirable to make ammonia at room temperature and atmospheric pressure. The first problem would then be that the appreciable heat of reaction must be carried off by cooling water, resulting in severe thermal pollution. Furthermore, the equipment would have to be very bulky and ammonia recovery in the liquid form difficult without recourse to refrigeration requiring much power. Rather, catalysts are needed that are active at 200-300 ^{o}C and between 5 and 10 MPa. Even if such catalysts become available, energy savings (though large) will still be marginal in percentage terms.
Any improved process for nitrogen fixation should be able to use nitrogen or air as feed directly, without the prior Production of hydrogen-containing

synthesis gas. Such a process did once, in fact, exist: in the first decade of this century Birkeland and Eyde developed and built a process based on direct oxidation of nitrogen in air in an electric arc. The relevant reaction is

$N_2 + O_2 \rightleftarrows 2\ NO\ (\Delta H_S^o\ =\ +\ 92\ kJ/mole\ NO)$

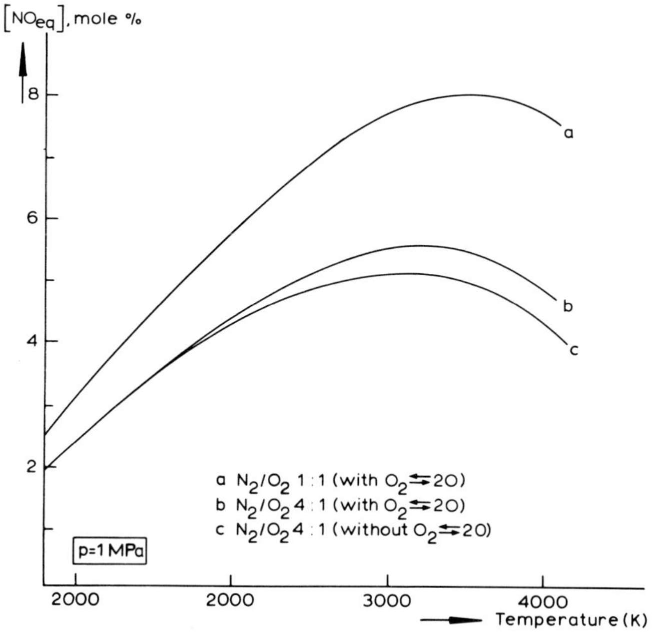

Figure V.5. Calculated equilibrium NO *yields*

Worthwhile conversions to NO are obtained only at 2500-3500 K, where N_2 dissociates into atoms. It is possible to compute equilibrium compositions of the product gas by assuming that noticeable dissociation of oxygen occurs but that the dissociation of nitrogen can be neglected for purposes of calculation. If ideal gas behavior is assumed, the results of Figure V.5. are obtained. It follows that the NO concentration is less than 10 vol%; this is in agreement with experimental observations.

Major difficulties arise in NO recovery. First of all, the reaction mixture should be quenched very rapidly to about 500 ^oC to prevent significant decomposition of NO into N_2 and O_2. This obviously means that the large amount of energy required to reach the very high reaction temperature can be recovered only partly and at a much lower temperature level. Even the best of the many electric arc processes operated commercially or on a pilot plant scale utilized only 3% of the energy input for the chemical reaction. Forty percent of the input was lost to water used as the coolant, thirty percent could be recovered as steam and twenty seven percent was lost via radiation.

Another major obstacle to full-scale development is that the NO must be recovered from a dilute gaseous stream and converted into HNO_3 by oxidation to NO_2 followed by absorption. This requires large reactors made of corrosion-resistant materials. Separation of NO from the dilute reactor product via oxidation and adsorption as NO_2 on acid-resistant molecular sieves, though possible, is also an expensive recovery method.

A comparison of the energy requirement of present-day ammonia synthesis via the reductive route with that of the most optimistic estimate of the oxidative nitrogen fixation shows that the former route is much superior. The energy used for the Haber-Bosch process is equivalent to the heat of combustion of 12 kg CH_4 per 10^3 at N obtained as NH_3, whereas the oxidative route requires at least 21 kg per 10^3 at N. The only possibility for the oxidative synthesis to become feasible in a very distant future would be the application of nuclear process heat at 2500-3500 K. Major obstacles, both as regards safety and materials of construction, make this solution rather unlikely.

Biological and allied methods of nitrogen fixation are more promising; in fact, fertilization via micro-organisms has long been applied in practice (see Chapter IV., introduction). The question arises if these organisms can also be applied industrially. Aerobic as well as anaerobic bacteria known to convert molecular nitrogen to potentially useful products, mostly via a reductive mechanism, should be considered. The reaction is catalyzed by the enzyme nitrogenase. This enzyme consists of several components, at least one of which is oxidized irreversibly during nitrogen fixation. This is the reason why aerobic bacteria with the capacity to convert molecular nitrogen are relatively rare. Some aerobic bacteria contain specially adapted enzyme systems having an oxidation/reduction potential adequate for nitrogen fixation. However, they require relatively large amounts of energy; for example *Azotobacter* shows the highest respiration of all known micro-organisms, which is necessary to maintain the necessary favorable redox potential. The energy consumption is correspondingly high. This energy may be introduced in the form of saccharose; the substrate should also contain important trace metals like Fe, Mo and Mg. At present, the most optimistic estimate of the productivity in terms of energy usage is that about 100 mg nitrogen can be fixed per g of sugar added. If this is converted into energy usage it is found that the process of fixing nitrogen industrially with the aid of *Azotobacter* consumes up to 25 times more energy as the current Haber-Bosch ammonia synthesis. To this energy should be added the amount required to recover the fixed nitrogen from the dilute process liquor. Another complication is that the micro-organisms cannot be used below about 10 ^{o}C (low reaction rates) or above 35 ^{o}C, where the reaction stops. Thus, large volumes of cooling media (water or brine) are required.

REFERENCES

General

1. Skinner, K.J., Chem.Eng. News 54 (41) 22-35 (1976)
2. Brill, W.J., Sci.Am. 236 (3) 68-74, 79-81 (1977)

VI. REMOVAL OF CARBON DIOXIDE AND HYDROGEN SULFIDE FROM INDUSTRIAL GASES

Introduction

Gas purification by some form of sorption by a liquid or solid sorbent is among the most widely applied operations in the chemical and process industries. The sorbent can usually be regenerated but, in a few cases, it is applied in a non-regenerative manner. The interaction between sorbate and sorbent may either be physical in nature or consist of physical sorption followed by chemical reaction. Some well-known examples are listed in Table VI.1.

Table VI.1. Some examples of gas purification processes

Sorbent	Nature of interaction	Regeneration	Examples
Solid	Physical adsorption	yes	Purification of natural gas (H_2S, CO_2); with molecular sieves
		yes	Gas drying operations (cyclic regenerative); mol. sieves
		varies	Odor removal from waste gases (active carbon)
Solid	Chemical reaction	no	H_2S from process gases, with ZnO
		yes	SO_2 from flue gases, with CuO/Al_2O_3
Liquid	Physical adsorption	yes	CO_2 and/or H_2S from hydrocarbon gases; solvents: N-methyl pyrrolidone, propylene carbonate, methanol
Liquid	Absorption + chemical reaction	yes	Many processes for the removal of CO_2 and H_2S from various gases, with solvents like: water + MEA, DEA, DIPA, etc. Agents improving physical solubility may be added (Sulfinol process); H_2S may be recovered as such or oxidized to S
Liquid + Solid	absorption + chemical reaction	varies	Some slurry wash processes for flue gas gas desulfurization

Other gas treatments apply chemical conversion of contaminants to harmless products or to substances which can be removed much more readily than the impurities from which they are derived. Examples of this type of process are methanation of carbon oxides in hydrogen or ammonia synthesis gas, oxygen removal from hydrogen by passage over a Pd catalyst and selective hydrogenation of acetylene in ethylene streams. This group of methods is usually quite specific for the particular application and is not discussed in this chapter. The most important application other than simple gas drying is the removal of CO_2 and/or H_2S from gases like natural gas, synthesis gas, primary products from gasification of coal or heavy fuel oil, and gas streams obtained in the desulfurization of oil fractions. Many different methods have been developed for CO_2 and H_2S removal, some of which are briefly discussed below. Concentrates of H_2S, obtained as by-products of gas desulfurization, are often converted by partial oxidation to elemental sulfur. The Claus Process applied for this purpose is described at the end of this chapter.

Process based on physical adsorption

Adsorption is defined as the concentration of one or more components of a gaseous or liquid mixture at the surface of a solid adsorbent. If the forces of interaction are physical in character, such as Van der Waals' forces, the term physical adsorption is used. The rate of this type of adsorption is so high that equilibrium is usually reached almost instantaneously, attainment of equilibrium being governed solely by diffusional limitations. When chemical interaction between adsorbate and adsorbent occurs, the phenomenon is called chemisorption, i.e. chemical reaction occurring at and limited to the surface of the solid.

Chemisorption shows many characteristics of chemical reactions; it may be irreversible, and its rate is sometimes so low that equilibrium is reached only after a considerable period of time. Since the forces of interaction are usually much stronger in chemisorption, the heat released with physical adsorption (up to 40 kJ/mol sorbate) is often much smaller than with chemisorption, where heat effects up to 200 kJ/mol and higher have been observed. The exothermic nature of many adsorptions also implies that the amount adsorbed at equilibrium decreases with increasing temperature. Adsorbents applied in physical adsorption processes include activated alumina and bauxite, silica gel, active carbon and molecular sieves. The quantity adsorbed by these materials depends on the (internal) surface area, the size of which determines to a large extent the capacity of the adsorbent. However, molecular size is also important, particularly with molecular sieves, adsorbents introduced by Union Carbide Corp. in 1954.

Molecular sieves are crystalline aluminosilicates containing relatively large cavities accessible to adsorbate molecules through rather narrow pore mouths of well-defined dimensions. They have high capacities for small, polar molecules like water, H_2S and CO_2 but do not adsorb larger molecules, which cannot enter the cavities in the crystal structure. Thus, separations are possible based on differences in molecular polarity as well as in size. Molecular sieves are now being applied on a very large scale and have surpassed many of the older type of adsorbents in difficult applications because of their higher selectivity, greater stability and high capacity - even though they are more expensive. Table VI.2. lists some properties of molecular sieve type 4A.

Table VI.2. Properties of Linde molecular sieve

Pore mouth width (nm)	0.4	
Molecules adsorbed	$H_2O, H_2S, CO_2, NH_3, SO_2, C_2H_4, C_2H_6, C_3H_6$	
Molecules excluded	C_3H_8 and higher hydrocarbons	
Capacity at 25 ^{o}C (kg/100 kg)	at p_{eq} = 2.67 kPa	at p_{eq} = 0.13 kPa
H_2O	22	17
H_2S	9.5	4.5
CO_2	11	3.5

Processes based on physical adsorption are usually carried out by passing the gas through fixed beds of adsorbent particles. The beds are mostly operated in a cyclic regenerative manner. The process then consists of the following phases:

adsorption : Impurities are removed from the feed gas at low temperatures until just before breakthrough. If the concentration of impurity is high, the amount of heat released is also high and cooling must be applied. If, on the other hand, small amounts of contaminants are to be removed, adiabatic adsorption is possible.

regeneration : the loaded adsorbent is reactivated by raising the temperature whilst passing through a clean purge gas. Part of the product gas or an auxiliary gas may be used for this purpose. Reduction of the pressure helps to speed up the regeneration. In less exacting separations, where only small amounts of impurities must be removed, untreated feed gas can sometimes be used for regeneration.

cooling : the hot regenerated bed is cooled by passing through cold regeneration gas. This not only prepares the bed for another adsorption phase but also serves to recover part of the heat.

A minimum of two beds is required to make continuous gas purification possible. If three beds are used, improved heat economy can be attained by first passing part of the product gas through an adsorbent bed in the cooling phase and a heater; it is then used for regeneration. This operation, in which some product gas is sacrified to avoid the necessity of using a separate regeneration gas, is feasible only when small amounts of H_2S must be removed. The direction of flow during regeneration is opposite to the flow during adsorption because the highest concentration on the adsorbent then occurs near the outlet of the bed. An example is shown in Figure VI.1. The concentrated stream of impurities obtained can seldom be blown off without an after-treatment, e.g. by a liquid phase absorption process, if it is rich in sulfur compounds.

A different mode of operation uses variation of the system pressure as the only driving force for regeneration (pressure swing operation). A typical application is for the recovery of valuable hydrogen from purge gas of ammonia or methanol synthesis loops. It is also possible to carry out adsorption and regeneration in different vessels; moving or fluid beds are then used to transport the solid from adsorber to regenerator and back. These methods are limited to applications where large volumes of gases are treated at low or moderate system pressures because it is not possible to build large vessels for high-pressure operations at acceptable cost. An important requirement of adsorbents for moving- or fluid-bed processes is that particle wear by attrition, which causes losses and formation of fines, should be low.

Among the advantages of fixed-bed adsorption separations are:

- high product purity can be attained
- automation provides virtually unattended operation
- adsorbent life, particularly with molecular sieves, can be up to 5 years
- within limits, fixed-bed adsorption processes are insensitive to variations in gas load, pressure, temperature and contaminant concentration in the feed
- corrosion problems do not occur.

Disadvantages are:

- investment costs are high
- much energy is required for regeneration and because of pressure drop over the beds
- a separate process must be added for treating the acid gases obtained as a by-product.

The latter disadvantage is particularly important: there are many cases where the combination of fixed-bed adsorption with contaminant aftertreatment is less economic than application of a single wash with a liquid agent.

In the process configuration of Figure VI.1. natural gas containing water, H_2S and CO_2 in excess of the amounts specified for pipeline gas is treated with molecular sieves. Water is adsorbed most strongly and concentrates close to the inlet of the adsorber; this impurity cannot be tolerated in pipeline gas because of the risk of formation of solid gas hydrates at high pressures and relatively low temperatures. H_2S should be removed to prevent air pollution; the maximum CO_2 concentration specified is a few percent. Fortunately, H_2S adsorbs more strongly than CO_2, enabling partial separation between these two impurities.

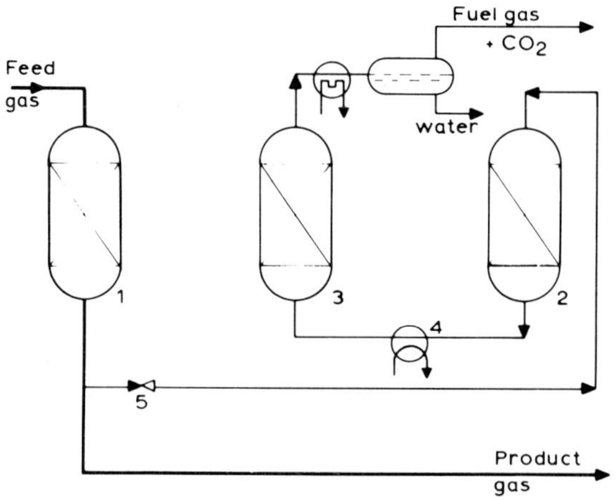

Figure VI.1. Removal of CO_2, H_2S and water from natural gas by adsorption

1. bed in adsorption mode

2. bed in cooling mode

3. bed in regeneration mode

4. throttle valve

Removal of H_2S by chemical reaction with a solid

Hydrated iron oxide has been used to purify coke oven gas since the middle of the 19th century. When gas containing H_2S is reacted with $Fe_2O_3.H_2O$ at temperatures near ambient, the following overall reaction occurs:

$$Fe_2O_3 + 3\ H_2S \rightleftarrows Fe_2S_3 + 3\ H_2O$$

The process is much more complex than is indicated by this equation; for instance, FeS is also formed, particularly at higher temperatures. Any oxygen present reacts to give elemental sulfur:

$$2\ H_2S + O_2 \rightarrow \frac{2}{x} S_x + 2\ H_2O$$

If too much sulfur is formed the process rate diminishes because of diffusion limitation; the same effect is noted when the oxide is overheated and loses water or changes in crystal structure. Thus, careful temperature control is necessary.

A limited number of regenerations is possible; the sulfide is converted to oxide by the addition of small quantities of oxygen to recirculating inert gas:

$$2\ Fe_2S_3 + 3\ O_2 \rightarrow 2\ Fe_2O_3 + \frac{6}{x} S_x$$

If the oxygen concentration is too high, deactivation by overheating as well as formation of SO_2 occur.

This process is applied only in the coal gas industry, both in fixed beds and in a more modern form using fluidized beds at higher pressure. It is competitive only when low concentrations H_2S must be removed from relatively small quantities of gas.

Iron oxide is not suitable for high-temperature H_2S removal. This can be established by equilibrium calculations; the relevant reaction network is:

$$3\ Fe_2O_3 + H_2 \rightleftarrows 2\ Fe_3O_4 + H_2O$$

$$Fe_3O_4 + H_2 + 3\ H_2S \rightleftarrows 3\ FeS + 4\ H_2O$$

$$FeS + H_2 \rightleftarrows Fe + H_2S$$

When using thermodynamic properties of the solids it is found that a concentration of 3-10 ppm H_2S is established at equilibrium in a closed system. In practice, this is not attained because one operates in an open system, where kinetic and mass transport limitations prevent the actual achievement of equilibrium. The reaction which is most responsible for the unsatisfactory equilibrium composition is the reduction to metallic iron.

Zinc oxide is a much better sorbent for H_2S. Less than 1 ppm H_2S can be reached quite easily at temperatures of about 300 ^{o}C by the reaction

$$ZnO + H_2S \rightarrow ZnS + H_2O$$

Thermodynamic calculations indicate that at 300 ^{o}C, atmospheric pressure and a water concentration of 5 mol %, removal of H_2S from hydrogen is possible to a concentration of about 10 ppb. The better performance of ZnO is related to the difficulty of reducing the oxide to Zn. However, a major drawback of ZnO is that in situ regeneration by oxidation of ZnS to ZnO is not possible because the active surface diminishes appreciably by sintering. Also, much of the mechanical strength is lost, and the consequent formation of fines results in a high pressure drop over beds of regenerated sorbents. This low mechanical strength also prevents application of ZnO in regenerative moving bed or fluid-bed applications. As a result, ZnO is applied only to remove small amounts of H_2S in guard beds, e.g. to protect catalysts used in downstream operations such as Cu/ZnO low-temperature shift or methanol production catalysts.

Absorption in liquids - general aspects

The most common way of removing CO_2 and H_2S from industrial gases is by washing with solvents which are selective for these impurities. The separation may either be by physical dissolution or physical dissolution followed by chemical reaction. In some processes H_2S is removed selectively by oxidizing absorbed H_2S to elemental sulfur by aerating the solution.

The distribution of the acid gas components over the gas and liquid phases is, to a first approximation, governed by Henry's law if only physical interactions occur:

$$x_i = p_i/H_i$$

Symbol x stands for mole fraction, p for partial pressure and H for the Henry coefficient; subscript i refers to component i of the gas. If the separation is based solely on differences in H_i, the rate of dissolution is governed by mass transfer from gas to liquid phase. It is possible to obtain lower equilibrium partial pressures by applying solvents in which the acid components react, e.g. with a base. Figure VI.2. shows that at moderate degrees of conversion lower equilibrium pressures of impurities can be obtained.

The two types of absorption processes also behave differently with respect to the operating temperature. With physical absorption the amount of gas dissolved at equilibrium increases with decreasing temperature, with the result that the

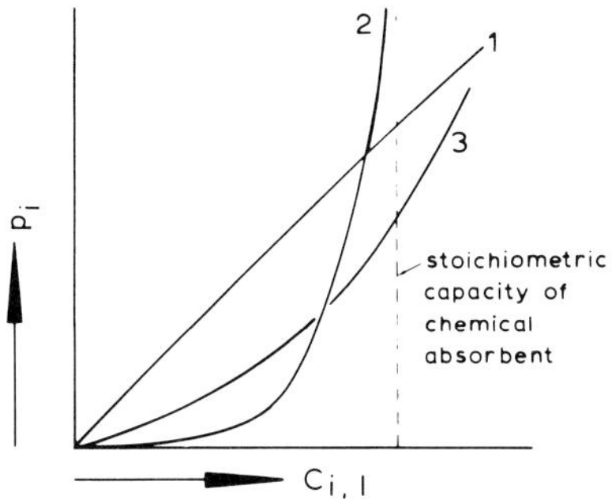

Figure VI.2. Equilibrium relations for absorption in liquids

1. physical absorption

2. chemical absorption

3. absorption with physical and chemical interaction

process becomes more efficient. With chemical absorption, however, a temperature decrease may result in less complete removal of contaminants when the rate of dissolution is largely determined by the rate at which the reaction in the liquid phase proceeds. The rate of dissolution of CO_2 in an aqueous liquid consists of the following steps:

$$CO_2(g) \rightleftarrows CO_2(l)$$

$$CO_2(l) + OH^- \rightleftarrows HCO_3^-$$

followed by further reactions, if these can occur. The second step is generally considered to be so slow that it limits the process rate under unfavorable conditions. This indicates that the absorption can be speeded up by raising the temperature. Although this causes the concentration of CO_2 in the liquid phase to decrease, the effect is overshadowed by the faster attack on CO_2 by hydroxyl ions, resulting in higher rates. A disadvantage is that the equilibrium pressure of CO_2 increases with increasing temperatures, other conditions remaining the same.

A second way of overcoming limitations due to chemical kinetics is to add a solvent in which the physical solubility of CO_2 is much higher. This causes an increase in $CO_2(l)$ and a proportional increase of the rate of bicarbonate formation. Again, the equilibrium partial pressure of CO_2 becomes higher as is shown in Figure VI.2., curve 3.

There are many reports in the literature indicating that additives may be added to speed up CO_2 absorption in chemical absorption processes. Examples are methanol, ethanol, glycols, glycerol, sucrose and dextrose. It is difficult to establish whether these materials act by increasing the physical solubility of CO_2, by catalysis of the reaction of CO_2 to the bicarbonate ion or by both mechanisms. In many cases the increased rate of dissolution seems to be due to both effects.

Absorption of H_2S is invariably fast because mass transfer to the solution is followed by rapid dissociation into H^+ and HS^-. This is utilized in some chemical processes to obtain partial selectivity for H_2S in the presence of CO_2. The temperature must be kept relatively low to ensure that CO_2 absorption is limited to a small fraction. This type of separation is possible with secondary and tertiary amine-containing solvents and is based on kinetic and not on equilibrium phenomena. Absorption of CO_2 and H_2S in solutions is accompanied by a thermal effect. With physical absorption, where the forces of interaction between solute and solvent are weak, the heat of dissolution is usually small, i.e. less than 20 kJ/mol. This indicates that regeneration of the fat solution by raising the temperature is not very effective and that the driving force for regeneration should be obtained mainly by lowering the pressure. Thus, physical absorption is most economic when gases containing high partial pressures of acid components must be treated to products of at best moderate purity. Another consequence is that the amount of heat required for regeneration is relatively small. In the case of absorption by chemical reaction, however, the exothermic heat of absorption is often much higher, exceeding 100 kJ/mol in many processes. In such cases regeneration can be effected by raising the temperature, lowering the pressure, or by both methods. The heat requirements for regeneration of chemically active solvents are much higher than for solvents showing physical interactions only. This does not mean that chemical treatments invariably require more energy than physical absorptions because the amount of solvent circulating in the process is often much smaller for chemical absorption and the energy required for this circulation may offset the energy advantage of physical absorption processes as compared with chemical absorption.

Physical absorption processes

All gas purification plants in which physical absorption is applied have similar basic structures. Any differences are due to solvent properties, product quality requirements, the desired degree of heat economy, which is often greater as plant size increases, and similar process requirements. Another important factor is the extent to which selectivity between CO_2 and H_2S is needed.

Solvents should have a high capacity for acid gas, low tendency to dissolve valuable feed components such as hydrogen and low-molecular weight hydrocarbons, low vapor pressure at operating temperatures to minimize solvent losses, low viscosity, thermal stability, absence of reactivity towards gas components, low tendency towards fouling of equipment and corrosion, as well as acceptable cost. Table VI.3. lists data on solvents applied in commercial units. As is indicated in the table, poly(ethylene glycol dimethyl ether) and N-methyl pyrrolidone are reasonably selective towards H_2S in the presence of CO_2. On the other hand, these solvents are rather expensive, which is important because the solvent adds significantly to the cost of working capital, which includes the initial solvent inventory of a plant.

Table VI.3. Examples of physical absorption processes.

Process	Licenser	Solvent	$C_{eq.H_2S}$ (vol/vol)	$C_{eq.H_2S}/C_{eq.CO_2}$	$P_{solvent}$ (Pa)	Data at T °C
Rectisol®	Lurgi	methanol	92	6	200	-30
Fluor® Solvent	Fluor Corp.	propylene carbonate	12	3.5	< 10	-25
Purisol®	Lurgi	N-methyl 2-pyrrolidone	5.1	12.5	520	+20
Selexol®	Allied Chemical	poly (ethylene glycol dimethyl ether), mol.wt. 280-370	4.8	9	<< 1	+20

$C_{eq.i}$ = equilibrium concentration of dissolved component i at the indicated temperature.

The simplest systems for removal of CO_2 and H_2S contain a single absorption column of the packed or plate type for countercurrent gas/liquid contact at high pressures (up to 80 bar), and a desorption section where the dissolved gases are recovered by two-stage expansion. The gases from the first expansion stage may contain enough valuable components to warrant returning them to the absorber after recompression. If the unit has a large capacity, it is interesting to recover part of the energy by pressure reduction in an expansion turbine. This produces the additional effect of Joule-Kelvin cooling, thus minimizing the energy required to cool the lean solvent fed to the absorber. Such a system is shown in Figure VI.3. In other cases throttle valves are used. The gaseous effluents may still contain traces of solvent and can be purified by water washing to minimize solvent losses.

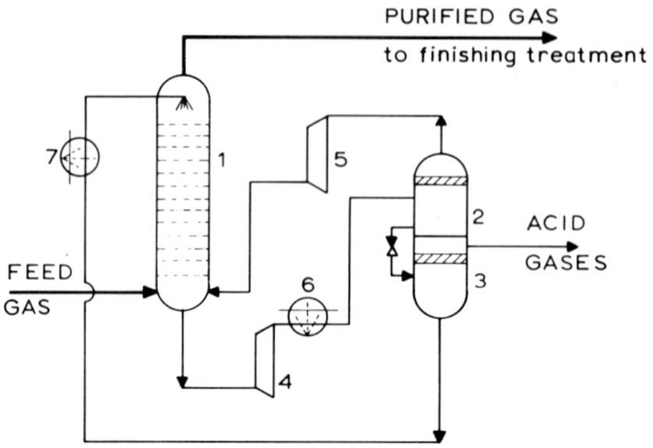

Figure VI.3. Physical absorption process with two-stage expansion regeneration

1. absorber

2. first stage flash vessel

3. second stage flash vessel

4. expansion turbine

5. recycle compressor

6. and 7. heat exchangers

For some applications in which a higher purity of the treated gases is specified, regeneration by expansion alone may not be sufficient. In such cases a thermal solvent regeneration step can be added (see Figure VI.4.). If one wished to regenerate the entire solvent stream by two-stage flashing, much energy would be needed for cooling the solvent to the low temperature required for deep removal of acid gases. Attractive savings on energy consumption are possible by subjecting part of the solvent to thermal regeneration and another part to expansion regeneration. The two solvent streams are taken from and returned to different sections of the absorber, the cleanest solvent being introduced at the top. It is always necessary to weigh the savings in operating costs against the higher investment for such modifications. Another possible variation is regeneration by stripping with an inert gas like nitrogen.

Selectivity between H_2S and CO_2 removal can be accomplished in two ways, viz. by carrying out both the absorption and the regeneration in two steps, such as is shown in Figure VI.4., or by absorbing in one step and regenerating in two stages. Since H_2S is more soluble than CO_2, the acid gases obtained by regenerating solvent from the bottom part of the absorber are relatively rich in H_2S. The same applies to gases released in the second regeneration step. The partial separation between H_2S and CO_2 can be utilized in two ways:

- virtually complete removal of H_2S is possible even when only partial CO_2 removal is required, such as is customary when producing pipeline quality gas from sour natural gases;
- when a gas contains much CO_2 relative to its H_2S content it may still be possible to concentrate the H_2S in such a manner that application of the split-stream type of Claus processing (see below in this chapter) is feasible. In other words, increased processing flexibility is obtained.

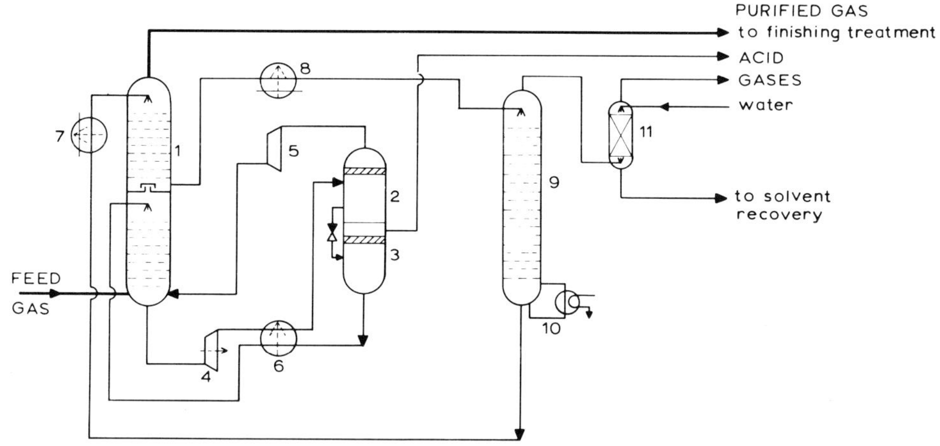

Figure VI.4. Physical process with two-stage expansion and thermal regeneration

1. through 7. see Figure VI.3.

8. fat solvent heater

9. regenerator

10. reboiler

11. acid gas wash column

Chemical gas/liquid absorption processes

In general, solutions of weak bases (15-40% w) in water are applied for the removal of CO_2 and H_2S from gases. Strong bases are inapplicable because they would give irreversible absorption. In most processes absorption is effected between 20 and 40 ^{o}C in countercurrent operation in a column, preferably at the highest possible pressure. The amount of non-acid gas components is so small that the quantity released upon pressure reduction is too small to return them to the absorber; moreover, expansion turbines cannot be applied here. The regeneration is carried out at atmospheric pressure by applying heat; since much water is removed from the solution along with desorbed H_2S and CO_2 it is customary to cool the top product of the regenerator with a cooler/condenser and to return the water to the column as reflux. A simple two-column flowchart is shown in Figure VI.5.

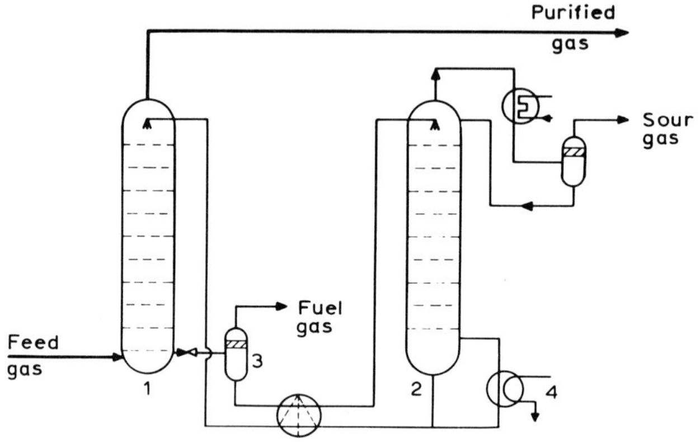

Figure VI.5. Chemical absorption process

1. absorber

2. regenerator

3. flash vessel

4. reboiler

Many processes have been developed which make use of alkanolamines as the basic solvent component, notably monoethanolamine (MEA, $NH_2CH_2CH_2OH$), diethanolamine (DEA, $NH(CH_2CH_2OH)_2$) and diisopropanolamine (DIPA). These bases are produced by reaction of ammonia with ethene oxide and propene oxide respectively. They react with H_2S by a simple acid-base reaction:

$$R_2NH + H_2S \rightleftarrows R_2NH_2^+ + HS^-$$

The reaction is extremely fast, the absorption of H_2S being limited by mass transfer. This is not so for CO_2, as has been discussed earlier. This absorption proceeds according to:

$$CO_2 + OH^- \rightleftarrows HCO_3^-$$
$$HCO_3^- \rightleftarrows CO_3^{--} + H^+$$
$$HCO_3^- + R_2NH \rightleftarrows R_2NCOO^- + H_2O$$

The formation of the (bi)carbonate and carbamate ions is responsible to a large extent for the favorably low equilibrium partial pressure of CO_2 over the solution. Regeneration by heating the solution leads to virtually complete desorption of CO_2 and H_2S. A comparison between MEA, DEA and DIPA shows that MEA is the cheapest of the three but shows the highest heat of reaction and corrosion, whereas the reverse is true for DIPA.

Another important group of processes employs carbonate solutions as the absorbent, mostly K_2CO_3 solutions in water. In this case CO_2 reacts to give bicarbonate according to the overall reaction:

$$CO_3^{--} + CO_2 + H_2O \rightleftarrows 2\ HCO_3^{-}$$

The heat of reaction is lower than with alkanolamine absorption; moreover, the reaction between dissolved CO_2 and solution components is slow at low temperatures. Therefore, the absorbers of the first generation of hot carbonate processes operated close to 100 ^{o}C and the regeneration was effected mainly by pressure reduction, with only a slight rise in temperature. Since complete regeneration, necessary to obtain low lartial pressures of acid gases above the regenerated solution, consumes much energy when applied to the entire volume of fat solution, *split stream regeneration* (Figure VI.6.) was often applied. In such plants, the fully regenerated part of the solution was introduced at the top of the absorber to obtain high gas purity. More modern processes usually operate with an activator (see the previous discussion on process principles earlier in this chapter). This either permits operation at lower temperatures or a reduction in size of the absorber when the temperature is kept high.

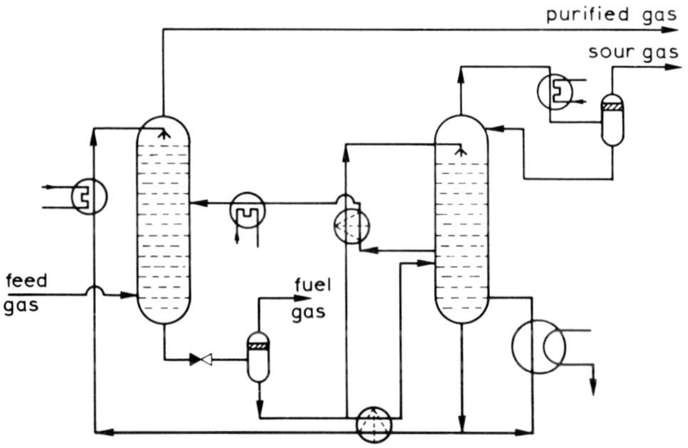

Figure VI.6. Chemical absorption process with split-stream regeneration

Other processes employ oxidation of H_2S to elemental sulfur by atmospheric aeration of the solution coming from the bottom of the absorber. Figure VI.7. shows an example, the Stretford Process. The rate of oxidation to sulfur is accelerated by the addition of a catalyst combination consisting of disodium salt of anthraquinone disulfonic acid and sodium vanadate. The sulfur is separated by froth flotation and then upgraded by purification, i.e. filtration or centrifugation, if necessary combined with heating to produce molten sulfur. Advantages of this process are that relatively little CO_2 is removed and that a separate Claus plant is not required, but the quality of the sulfur obtained is sometimes poor.

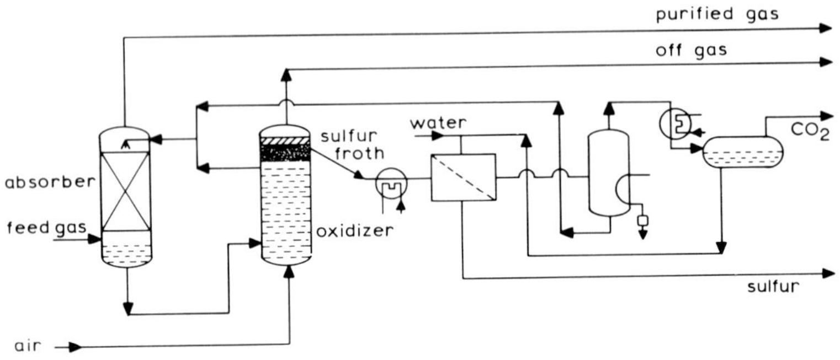

Figure VI.7. Stretford process

Table VI.4. lists a number of chemical absorption processes applied in industry. One of these, the Sulfinol ^{R} process, is a special case in that a solvent showing high physical solubility for CO_2 and COS has been added to an alkanolamine solution. This speeds up the dissolution of CO_2 and also raises the rate at which COS is hydrolyzed. These advantages are to some extent offset by the higher solubility of other gas components in the absorption liquid. In other words, the Sulfinol process can be considered as a hybrid between a physical and a chemical absorption process.

When applying chemical absorption processes to gases obtained by partial oxidation of heavy fuels or by coal pyrolysis or gasification care should be taken that HCN is not present in the feed to the absorber because it would hydrolyze to formic acid and thus permanently neutralize the base. This implies that HCN should be removed beforehand, e.g. by water wash or by catalytic conversion.

General requirements of absorption processes

Table VI.5. gives a survey of the four main types of processes used for the removal of H_2S and CO_2 from gases. Apart from the process properties mentioned in the table the following general requirements apply:

- high product gas purity should be attainable

Table VI.4. Examples of chemical absorption processes

Process:	Absorption agent:	Remarks:
Alkazid®	alkali + amino acids in water	BASF
ADIP®	diisopropanolamine in water	Shell
Sulfinol®	diisopropanolamine in water and sulfolane	Shell
Benfield	K_2CO_3 in water (also with an activator	Benfield
Catacarb®	K_2CO_3 + amine/borate activator in water	Eickmeyer
Giammarco-Vetrocoke®	K_2CO_3 + As_2O_3 (or glycine) in water (sometimes also an alkanolamine	Giammarco-Vetrocoke oxidation to sulfur possible
Stretford®	alkaline solution containing anthraquinone disodium sulfonate + sodium vanadate, in water	North-Western Gas Board; oxidative regeneration to sulfur
Seabord®	Na_2CO_3 in water	Koppers

- selectivity between acid gases and other gas components shoud be acceptable
- it should be possible to remove H_2S in preference to CO_2
- the H_2S-rich gas produced should be easy to process in a Claus unit
- if sulfur is the end product it should be sufficiently pure
- COS and CS_2 should be removed almost completely
- the consumption of energy and chemicals should be low
- corrosion should be low
- the plant should be as compact as possible, require minimal investment and be easy to operate
- the amount of solvent in recirculation per unit mass of H_2S and CO_2 should be acceptable
- environmental pollution should be within acceptable limits.

In view of the wide variety of processes available for gas purification it is always necessary to make a careful study of a number of methods, particularly when the H_2S/CO_2 removal process is incorporated in a new chemical process or applied in new situations. Optimum integration in the entire plant should be the final objective.

Table VI.5. Summary of processes for removing H_2S from process gases

	Chemical adsorption	Physical adsorption	Chemical absorption	Physical absorption
Equilibrium relation	almost irreversible adsorption:	very fast, reversible physical adsorption:	equilibrium reaction:	physical equilibrium (Henry's law):
	$p_{i,g}$ 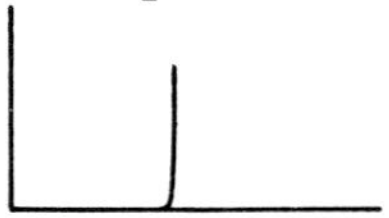 $c_{i,s}$	$p_{i,g}$ 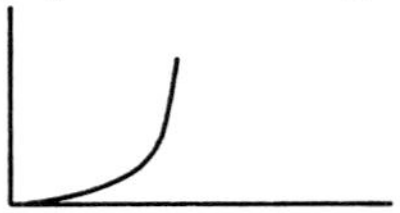 $c_{i,s}$	$p_{i,g}$ 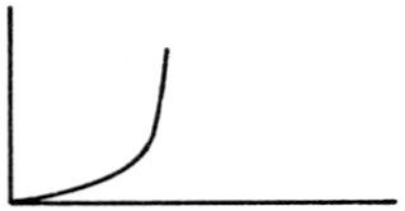 $c_{i,l}$	$p_{i,g}$ 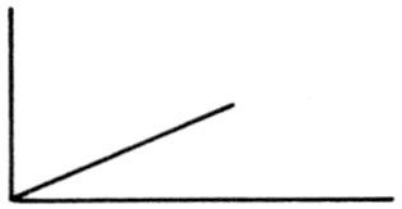 $c_{i,l}$
Maximum load	limited by stoichiometry	limited by surface area of adsorbent	limited by stoichiometry	no clear limit observed
Product purity	extremely high	high	high	high only when low temperatures are applied
H_2S/CO_2 selectivity	very selective	almost non-selective	partial selectivity obtainable	partial selectivity obtainable
H_2S/hydrocarbon selectivity	very selective	almost non-selective	selective	partially selective
Energy requirements				
- for desorption	high (if regenerative)	moderately high	high	low
- for other operations	low	rather low	low	moderate to high
- total	--	moderately high	rather high	moderate
Feed requirements	low partial pressure of contaminants	not feasible with very high partial pressure of contaminants	no limitations	high partial pressure of contaminants required
Mode of operation	mostly batch; sometimes cyclic regenerative	cyclic regenerative	continuous	continuous
Main applications	guard beds	aftertreatment	generally applicable	bulk removal of contaminants

Conversion of hydrogen sulfide into sulfur

The best way of disposing of the H_2S produced in a more or less concentrated form is to convert it into sulfur by means of the *Claus process*. The relevant reactions are:

$$2\ H_2S + 3\ O_2 \rightleftarrows 2\ SO_2 + 2\ H_2O$$

$$2\ H_2S + SO_2 \rightleftarrows \frac{3}{x}\ S_x + 2\ H_2O$$

The overall reaction,

$$2\ H_2S + O_2 \rightarrow \frac{2}{x}\ S_x + 2\ H_2O$$

can be carried out by reacting the H_2S-containing gas directly with air in a burner reactor if the gas can be burnt with a stable flame. Other equilibria which should be taken into account are the formation of sulfur dimer, hexamer and octamer (S_2, S_6 and S_8) as well as the dissociation of hydrogen sulfide:

$$H_2S \rightleftarrows H_2 + \frac{1}{x}\ S_x$$

Moreover, COS and even CS_2 may be formed, particularly when the gas is burnt with less than stoichiometric amounts of air in the presence of hydrocarbon impurities or large amounts of CO_2.

Equilibrium data on the reaction between H_2S and SO_2 indicate that the equilibrium conversion is almost 100% at relatively low temperatures and that it diminishes at first at higher temperatures, in accordance with the exothermic nature of the reaction (see Figure VI.8.). It also follows from the figure that a further rise in temperature causes the equilibrium conversion to increase again. This is a consequence of the dissociation of the sulfur polymers mentioned above into monoatomic S.

Catalysis by alumina is necessary to obtain good equilibrium conversions; the thermal Claus reaction is fast only above 500 ^{o}C. There is also a lower temperature limit which is not caused by low rates but by sulfur condensation in the catalyst pores and consequent deactivation of the catalyst. The lower limit at which satisfactory operation is still possible depends on the pore size and pore size distribution of the catalyst; with aluminas having wide pores the conversion still proceeds satisfactorily at about 200 ^{o}C.

In all Claus process configurations several conversion steps in adiabatic, i.e. cheap, reactors are used, with intermittent and final condensation of the sulfur produced. There are three main process forms, depending on the concentration of H_2S and other sulfur compounds in the gas to be converted, viz. the

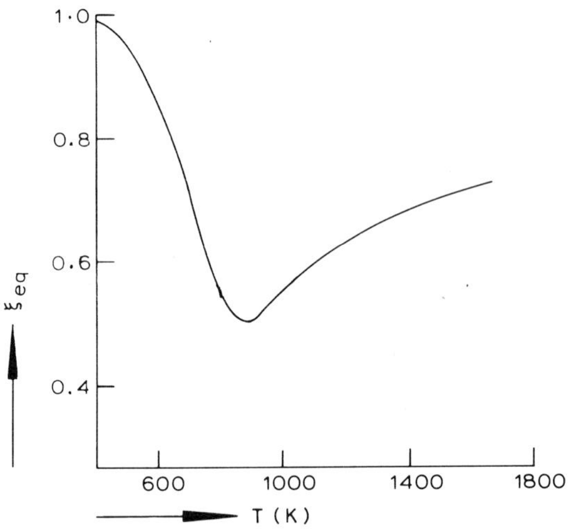

Figure VI.8. Equilibrium conversion of Claus reaction

straight-through, the split flow and the direct oxidation processes.

The straight through process (Figure VI.9.) applies when the feed contains more than 50 vol % H_2S. Feed gases of this type can be burnt with the stoichiometric amount of air to give sulfur, without the risk of the flame becoming unstable. The combustion reactor is followed by a combined waste heat boiler and sulfur condenser from which liquid sulfur and steam are obtained. The gases are then reheated by in-line fuel combustion to the temperature of the first catalytic convertor, which is usually kept at about 350 ^oC to decompose any COS and CS_2 formed in the combustion step. Subsequent steps are cooling with steam production and sulfur condensation, another line burner, a second catalytic convertor operating at as low a temperature as possible to obtain high final conversions, and a final condensation. This set-up gives higher flexibility than application of heat exchange between reactor feeds and products, another possible means of optimizing heat economy. Claus plants employing line burners are net steam producers.

Care should be taken to avoid condensation of an aqueous phase in the system because of the extreme corrosivity of the liquid phase produced. Another operating problem is concerned with mist formation, a phenomenon which occurs very readily when condensing sulfur. Demisters are installed to prevent this; if this is done, wellrun plants reach overall yields of 96%. Residual sulfur is converted to SO_2 by incineration of the tail gas from the last condenser, to prevent emission of other sulfur compounds and to dilute the effluent to reduce

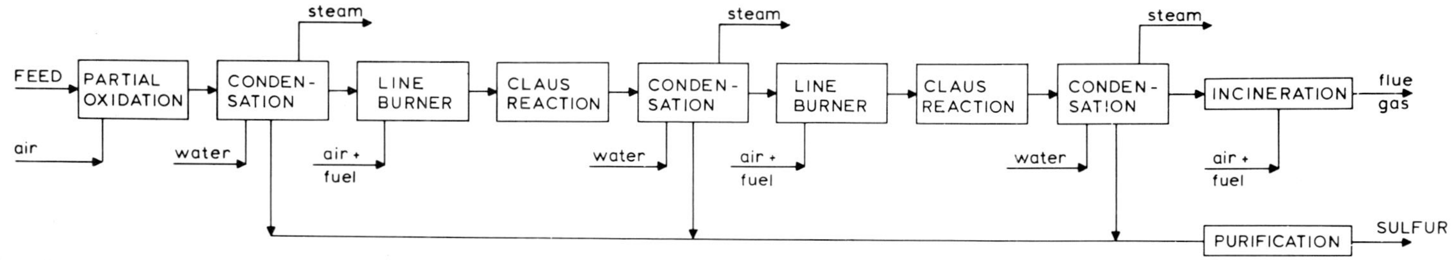

Figure VI.9. Straight-through Claus process

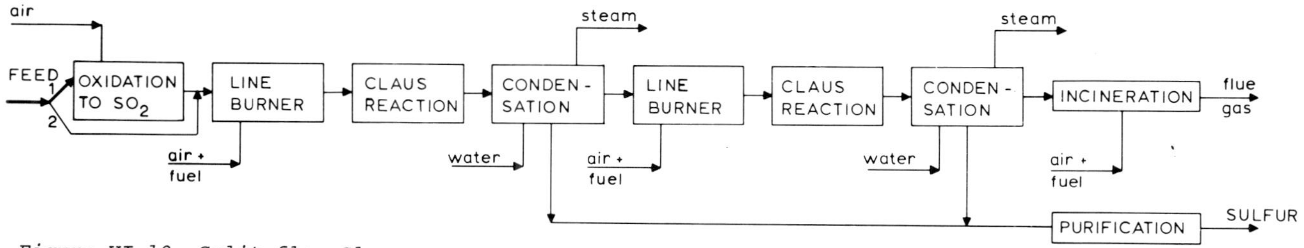

Figure VI.10. Split-flow Claus process

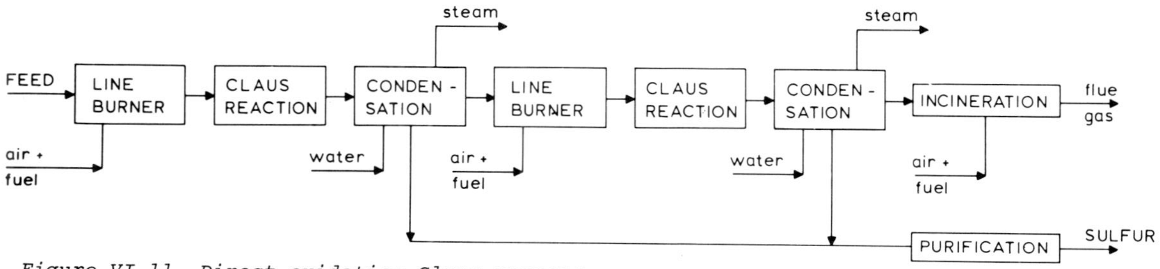

Figure VI.11. Direct oxidation Claus process.

ground level SO_2 concentrations. The sulfur produced is corrosive and still contains some H_2S, which reacts to polysulfides imparting an objectionable odor to the product. A remedy is to spray the sulfur in a closed vessel while adding some ammonia to decompose polysulfides. This also reduces the risk of explosions.

The split-flow configuration (Figure VI.10.) is used whenever the H_2S content of the feed is between 15 and 50 vol % H_2S. Such gases cannot be burned with a stable flame with the stoichiometric quantity of air needed to give sulfur, but stoichiometric combustion to SO_2 is quite feasible. The feed is now split into two flows with a volume ratio of 2 : 1; the smaller flow is converted to SO_2. The gases are recombined, reheated if necessary, and then fed to the first catalytic convertor. This reactor now operates at low temperatures because COS formation is small when burning to SO_2. In this case only two convertors need be used to give reasonably high final sulfur yields. High quality sulfur is produced only when the hydrocarbon content of the feed is sufficiently low. The same applies to feeds converted by the direct-oxidation process, in which stable combustion of the feed is not possible without the addition of combustible gas during its partial oxidation (Figure VI.11.). The overall sulfur yield of this type of Claus plant is rather low. Therefore it is important to ensure that a feed gas is produced in the previous absorption process which contains enough H_2S to apply the split-flow process.

Although modern Claus plants can achieve high conversions and thus give relatively low SO_2 emissions, local legal requirements are often so strict that reduction of the SO_2 concentration in the off-gas is mandatory. This has led to the development of a fairly large number of methods for desulfurizing these gases. The main methods used are based on the following principles:

(a) a final Claus conversion step in the liquid phase; an example is the IFP process developed for this purpose;

(b) a final Claus reaction at low temperature using activated alumina catalyst; the temperature is now so low that the sulfur formed is taken up almost entirely in the catalyst by capillary condensation. This last stage is operated in a cyclic regenerative manner, i.e. the sulfur is removed from the catalyst bed by heating. Several absorption beds must be used to ensure that the process is continuous. This type of operation is also practiced in the Lurgi Sulfreen® process, in which active carbon is used as absorber.

(c) hydrogenation of the sulfur in the Claus tail gas followed by removal of the H_2S produced by a selective alkanolamine wash, returning the H_2S rich gases to the Claus plant feed. The hydrogen present in the Claus off-gas is usually sufficient to achieve complete conversion when Co/Mo-sulfides on alumina catalysts are used. This type of process, of which the Shell Scot® process is an example, can increase the overall sulfur yield to more than 99.8%. Another possibility is to hydrogenate and then to remove the H_2S formed by means of the Stretford process (see Figure VI.7.); this combination is known as the Beavon Sulfur Removal® process.

REFERENCES

General

Kohl, A.L., F.C. Riesenfeld, 'Gas Purification', McGraw-Hill, New York 1960.

Special

Ginneken, A.J.J. van, J.P. Klein, Proc. World Petr. Congr. 9th, Vol. 6, 107-116 (1975).
Table VI.5. is reproduced from this article with permission.

VII. MANUFACTURE OF INORGANIC ACIDS

Sulfuric acid

Sulfuric acid is one of the most important industrial acids and used in the processing of minerals, production of fertilizers, nylon 6, rayon, titanium dioxide, detergents etc. Its function is mainly as a carrier of 2 protons, the sulfate group being often non-functional. In those cases the sulfate group is finally discarded as a waste product in a form such as $CaSO_4$, or recovered as $(NH_4)_2SO_4$, which is a fertilizer but is economically uninteresting. Hence several alternative processes have been developed in which HNO_3 takes over the function of H_2SO_4.

It is much easier to recover nitrogen oxides from nitrate-containing products which can be recycled to the HNO_3 plant than to recover SO_2 from $CaSO_4$. The production is based on the oxidation of sulfur or sulfides to SO_2, catalytic conversion of SO_2 to SO_3 and hydration of the latter to H_2SO_4.

Production of SO_2

Sulfur from mining operations or produced by the Claus process is burned with air according to:

$$S + O_2 \rightarrow SO_2(g) \qquad \Delta H^o_s = -297 \text{ kJ/mole}$$

Because air is also needed to oxidize SO_2 in the second step of the process a corresponding excess of air is used to burn S. Hence the concentration of SO_2 is only 8 - 10 vol%. This has the advantage of completely oxidizing the sulfur. The large amount of reaction heat, which leads to gas temperatures in the region of 1000 ^oC, is recovered in a boiler. In the first step a very small amount of SO_3 is formed which nevertheless can be the cause of severe corrosion if any condensation with water vapor from the air occurs in the boiler.

If sulfides such as ZnS (important in the production of zinc) and FeS_2 are used the oxidation is carried out as a 'roasting' process. ZnS is converted to ZnO:

$$ZnS + \frac{3}{2} O_2 \rightarrow ZnO + SO_2; \qquad \Delta H^o_s = -443 \text{ kJ/mole}^{-1}$$

The roasting of pyrite is more complicated because 5 reactions can take place:

	ΔH^o_s
$FeS_2 \rightarrow FeS + S$	82 kJ/mole FeS_2
$S + O_2 \rightarrow SO_2$	-297 kJ/mole SO_2 formed
$2\ FeS + 3\ O_2 \rightarrow 2\ FeO + 2\ SO_2$	-1227 kJ/mole Fe_2O_3 formed
$2\ FeO + \frac{1}{2}\ O_2 \rightarrow Fe_2O_3$	
$3\ FeO + \frac{1}{2}\ O_2 \rightarrow Fe_3O_4$	-108 kJ/mole FeO

The main parameters determining the end result are temperature and air-ore ratio. Because of the strongly exothermic effect high temperatures are easily reached resulting in SO_2 and Fe_2O_3 as the final products. Care must be taken, however, not to surpass 940 ^oC. Above this level eutectic melts with FeS are formed which hinder the diffusion of oxygen into the sulfide core. Then the roasting stops and low conversions are obtained.

Technically, the most attractive possibility of carrying out the reaction is in a reactor having a large capacity, thus making it possible to recover the correspondingly large heat production, viz. in a fluid bed. Broken and classified ore of about 6 mm particle size is continuously fed to the reactor bed, which consists mainly of iron oxide. Steam is produced in coils within the reactor as well as by heat exchange with the hot SO_2-containing gas. It is possible to keep the residence time of the gases so short that SO_3 formation is at a very low level.

The product gas is freed of entrained solid particles by cyclones and cleaned further with an electrostatic precipitator (Cottrell). The iron oxide can be used in the steel industry if it conforms to a certain specification. Figure VII.1. gives a simplified flow sheet.

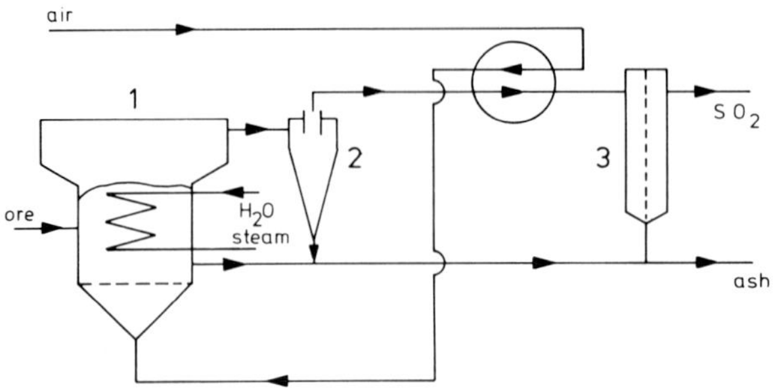

Figure VII.1. Fluid bed roasting

1. fluid bed reactor

2. cyclone

3. electrostatic precipitator

Since calcium sulfate is available in large quantities as a by-product of many operations it would be interesting to use it in making SO_2. This is possible, using carbon or CO as reducing agent:

$$2\ CaSO_4 + C \rightarrow 2\ CaO + 2\ SO_2 + CO_2 \quad (\Delta H_s^o = +304\ kJ/mole\ CaSO_4),\ or$$

$$CaSO_4 + CO \rightarrow CaO + SO_2 + CO_2 \quad (\Delta H_s^o = +217\ kJ/mole\ CaSO_4)$$

Both process routes require much energy and a temperature of at least 1400 ^oC. This makes calcium sulfate reduction economically unattractive even if the sulfate is available at low cost. Only when combined with large-scale cement production is the reduction of $CaSO_4$ worthwhile:

$$8\ CaSO_4 + 4\ C + 2\ Al_2O_3 + SiO_2 \rightarrow 4\ CO_2 + 8\ SO_2 + 2(3\ CaO.SiO_2) + 2(CaO.Al_2O_3)$$

In this process, which has been operated for some time in the UK by ICI, the silica and alumina reactants are added as clay and sand. The advantage of this manner of producing cement lies in the direct use of CaO; otherwise, it must be added to a cement kiln as $CaCO_3$. As in cement manufacture, a rotating kiln is used.

Oxidation of SO_2 to SO_3

It is necessary to use an oxygen transfer agent to oxidize SO_2 to SO_3. At first, nitrogen oxides were used for this purpose in the so-called lead chamber process.

Nowadays a heterogeneous catalyst is applied for the direct conversion according to:

$$2\ SO_2 + O_2 \rightleftarrows 2\ SO_3 \qquad (\Delta H^{o}_{s} = -98.5\ \text{kJ/mole}\ SO_2)$$

Initially platinum was used as a catalyst because it is active from 400 ^{o}C upwards. It was possible to reach a conversion of well over 95% by using a number of adiabatic reactors in series, with interstage cooling. This is necessary because equilibrium is established; since the reaction is exothermic the equilibrium conversion is less favorable at higher temperatures. It is necessary to attain as high a conversion as possible because unconverted SO_2 is an undesirable atmospheric pollutant. The use of excess air is desirable for obtaining high yields of SO_3.

Because Pt is expensive and very sensitive to poisons such as As (often present in pyrite), V_2O_5 is the only catalyst in current use. The oxide is supported on kieselguhr or zeolites and promoted with K_2SO_4. It is less active than Pt, requiring a temperature of 430 ^{o}C or higher, and acts through a redox mechanism in which oxygen is transferred from the surface to SO_2, followed by reoxidation of the surface by the oxygen in the feed. The optimum degree of conversion when using air is 0.98, which would result in losses of SO_2 unless precautions are taken. Since moreover, environmental laws do not allow gaseous emissions of relatively high SO_2 content (normally a SO_2 concentration in the stack gas of less than 500 ppmv is required), there is only one feasible solution, apart from the rather expensive expedient of stack gas desulfurization: most of the SO_3 formed must be removed by interstage cooling and absorption followed by conversion of nearly all the residual SO_2 after reheating.

The reactors used are generally fixed beds with interstage cooling, either by raising steam in internal cooling coils (Figure VII.2.) or in externally placed heat exchangers. The changes of the conversion in a multi-bed reactor of this type are shown in Figure VII.3. The principle is to convert most of the SO_2 in the first bed where the temperature rises from 430 ^{o}C to 600 ^{o}C. At that temperature, about 4 - 5% of the SO_2 is still unconverted; therefore the gases are cooled to about 450 ^{o}C and further converted in subsequent beds, with steadily decreasing outlet temperatures. Suitably optimized reactors make it possible to convert 5 tons SO_2 per day per m^3 catalyst. Catalyst life is quite long: 5 years for the catalyst operating at the higher temperature and up to 1 year at lower temperature.

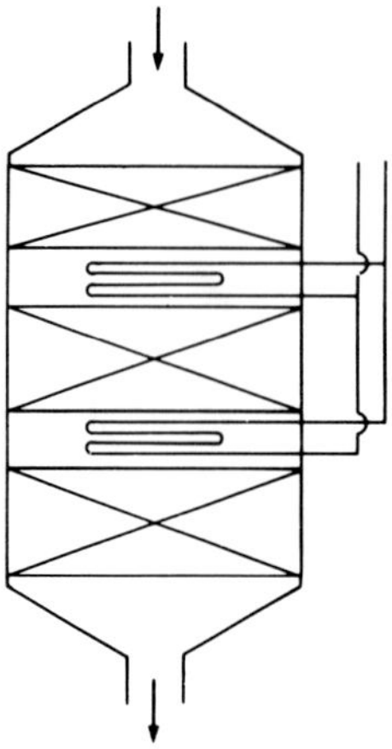

Figure VII.2. Reactor with interstage cooling

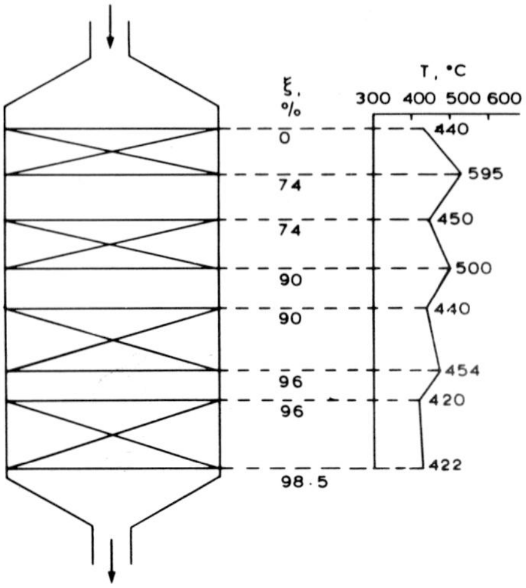

Figure VII.3. Temperature profile in a multi-bed SO_2 convertor

The overall process rate is determined both by chemical kinetics and by mass transfer in the catalyst pores. At 420 °C the process rate is governed entirely by chemical conversion; at 450 °C resistance against pore diffusion and chemical reaction is about equal, whereas at 490 °C the process rate is largely governed by pore diffusion. In other words: most of the conversion then proceeds in the outer layer of the catalyst particles.

Absorption of SO_3

The reactor gas is cooled to 100 °C during which time oleum is separated. The gas phase is treated with 98 - 98,5% sulfuric acid (produced from oleum) in a packed scrubber made of steel and clad with ceramic material. The sulfuric acid concentration is very critical because a more diluted acid gives rise to a persistant mist which is difficult to absorb, and a more concentrated acid does not absorb all the SO_3. To make sure that no mist is carried away the gas is passed through a mist eliminator. A simplified flow chart of a sulfuric acid plant in which interstage SO_3 removal is applied as shown in Figure VII.4.

Nitric acid

The production of nitric acid occurs in three main steps:
- oxidation of NH_3 to NO
- oxidation of NO to NO_2
- absorption of NO_2 in H_2O

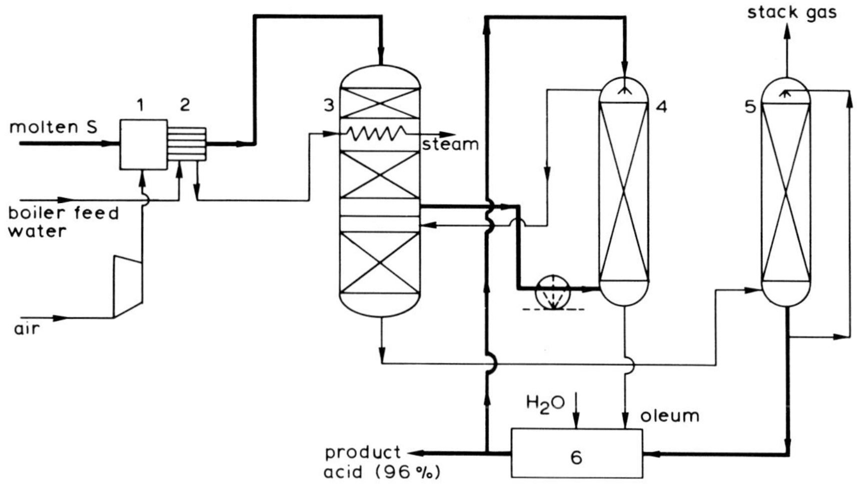

Figure VII.4. Production of sulfuric acid (dual absorption process)

1. *burner*
2. *boiler*
3. *convertor*
4. *interabsorber*
5. *final absorber*
6. *dilution system*

Oxidation of ammonia

Several parallel reactions are possible:

	ΔH^o_s/mole NH_3	reaction
$2\ NH_3 + \frac{3}{2}\ O_2 \rightarrow N_2 + 3\ H_2O$	$-$ 315 kJ	1
$2\ NH_3 + 2\ O_2 \rightarrow N_2O + 3\ H_2O$	$-$ 276 kJ	2
$2\ NH_3 + \frac{5}{2}\ O_2 \rightarrow 2\ NO + 3\ H_2O$	$-$ 226 kJ	3
$2\ NH_3 + \frac{7}{2}\ O_2 \rightarrow 2\ NO_2 + 3\ H_2O$		4
$2\ NH_3 \rightarrow N_2 + 3\ H_2$		5
$2\ NH_3 + NO \rightarrow \frac{5}{2}\ N_2 + 2\ H_2O$		6

The equilibrium constants of the first five reactions which are the most important are presented in Table VII.1. The related ΔG-values are given in

Table VII.1. Equilibrium constants for the oxidation of NH_3

p = 0.1 MPa

T(K)	reaction number				
	1	2	3	4	5
300	$7.3x10^{56}$	$7.3x10^{47}$	$6.4x10^{41}$	$7.7x10^{47}$	$1.7x10^{-3}$
500	$7.1x10^{34}$	$4.4x10^{28}$	$1.1x10^{26}$	$1.4x10^{28}$	3.3
700	$2.6x10^{25}$	$2.7x10^{20}$	$2.1x10^{19}$	$5.0x10^{19}$	$1.1x10^{2}$
900	$1.5x10^{20}$	$7.4x10^{15}$	$3.8x10^{15}$	$9.9x10^{14}$	$8.5x10^{2}$
1100	$6.7x10^{16}$	$9.1x10^{12}$	$1.5x10^{11}$		$3.2x10^{3}$
1300	$3.2x10^{14}$	$8.9x10^{10}$	$3.4x10^{11}$		$8.1x10^{3}$
1500	$6.2x10^{12}$	$3.0x10^{9}$	$2.0x10^{10}$		$1.6x10^{4}$

Figure VII.5. as a function of temperature. It is clear that the system has a strong tendency to preferential formation of N_2. This can be overcome, however, by using a specific catalyst and short residence times at a high temperature. Noble metals suit this purpose as, to a lesser degree, do some oxides. Pt is widely used at a temperature of 880 ^oC, where a selectivity of 96% is reached. The residence time is extremely short (10^{-4} to 10^{-3} s). The temperature is a weak function of the pressure (100 ^oC shift from 0.1 MPa to 0.8 MPa). Under these circumstances the reaction velocity is only dependent on the diffusion of NH_3 to the oxygen-covered Pt surface. Hence a large catalyst surface area is not required and a reactor containing layers of a woven catalyst net can be

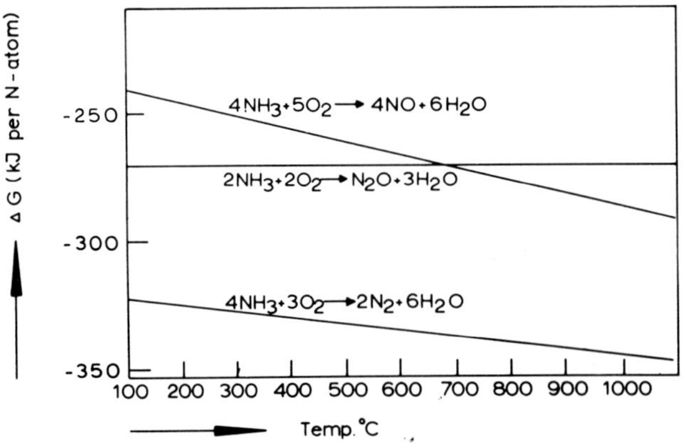

Figure VII.5. ΔG-values in the oxidation of ammonia

applied. The Pt is strenghtened by alloying with 10% Rh. Due to recrystallization and erosion at the high gas velocity employed, some Pt is lost with the product gas (0,18 g/ton HNO_3); and this is sometimes recovered by filters. Catalyst life is about two years. Because the absorption of NO_x is a pressure process (see later) the oxidation is also often carried out under pressure. This, however, lowers the oxidation efficiency, which is 97% at 0.1 MPa, 95% at 0.5 MPa and 92% at 1.2 MPa. Another drawback is an increasing loss of catalyst with higher pressures (0.36 g/ton HNO_3 at 1 MPa). This has partly been solved by adding a gold-palladium net to the Pt-gauzes to absorb Pt carried away by the gas stream. The ammonia reactor is of the adiabatic type: all reaction heat is carried away as sensible heat in the product gas. It is recovered to a large extent in a boiler. Because of strong radiation the reactant gas quickly assumes the reaction temperature. The system has a peculiar property which can be understood from Figure VII.6. The dotted line represents the heat produced and the full line the heat carried away, both as a function of temperature. The almost horizontal part of the dotted line is the range where the process rate is governed by diffusion, which is not very dependent on temperature. At lower temperatures the reaction is kinetically controlled, the relationship between heat production and the temperature being exponential (Arrhenius). The point of intersection between the two curves represents the catalyst temperature. The effect on this temperature of the gas throughput is very significant: if the space velocity is too low the flame detaches from the net and shifts forward, resulting in a greatly increased production of N_2 by homogeneous gas-phase oxidation of ammonia. If the air/NH_3 ratio is too high the net is 'blown cold': heat production by reaction then is insufficient to preheat the reactants to

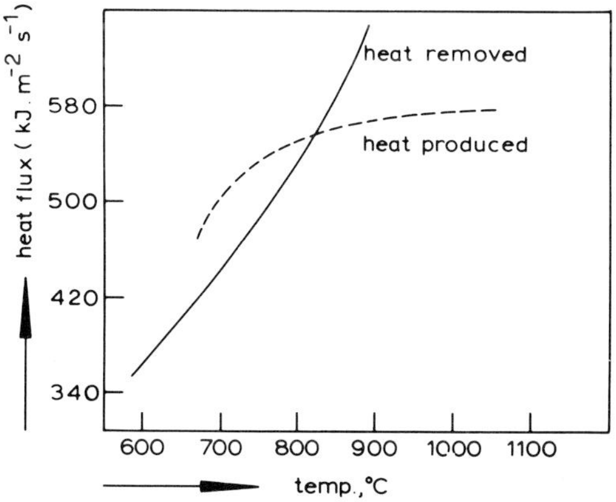

Figure VII.6. Stability of an ammonia burner

the high temperature required for selective oxidation. With increasing gas velocity the heat transfer coefficient increases proportionally to the mass transfer coefficient causing the temperature to remain virtually constant. The figure also illustrates what would happen if the reactants are preheated: only the amount of heat carried away by the gas flow diminishes, i.e. the full curve shifts downward and the net temperature rises. If the ammonia concentration is increased, a similar temperature effect occurs. However, the NH_3 concentration should not be higher than about 14 vol% at 0.1 MPa because ammonia/air mixtures are explosive above this limit. The limit concentration shifts downwards by a few vol% at higher pressures. Excess air is therefore used to keep the concentration in the region 10-12½ vol%, depending on reactor pressure.

Oxidation of NO

The homogeneous gas phase oxidation of NO with oxygen is a famous reaction because it seems to be one of the few third order reactions known:

$$R = k.p^2_{NO}.p_{O_2}$$

Remarkably, k has a negative temperature coefficient (Figure VII.7.). These data suggest a stepwise mechanism. The reaction is exothermic:

$$2\ NO + O_2 \rightleftarrows 2\ NO_2 \qquad \Delta H^o_s = -113.8\ kJ$$

The equilibrium composition depends of course on the temperature. The reaction velocity is relatively low, even after cooling the nitrogen oxide from the ammonia burner down to 50 ^oC. In the cooler a weak acid of 40% condenses. At very short residence times this concentration is lower. The non-condensed NO_x leaves the cooler to be mixed with secondary air.

Absorption of nitrogen oxides

It is found that the absorption is quite complex. Several reactions can occur, in the liquid phase as well as in the gas phase. Usually these are the main steps:

$$2\ NO_2 \rightleftarrows N_2O_4$$

$$N_2O_4 + H_2O \rightleftarrows HNO_3 + HNO_2$$

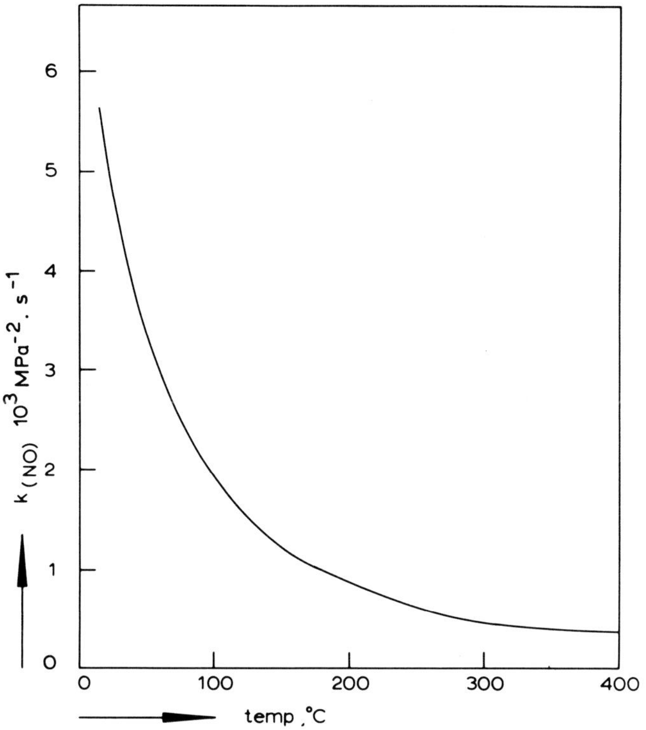

Figure VII.7. Reaction rate constant of the homogeneous NO-oxidation

$2\ HNO_2 \rightarrow NO + NO_2 + H_2O$

About 90% of the HNO_3 is formed in the liquid phase. The overall reaction can be written as:

$3\ NO_2 + H_2O \rightleftarrows NO + 2\ HNO_3$

NO is partially formed back and must be reoxidized, requiring a large reaction volume (low rate) which is even more serious at low NO concentrations (2nd order in NO). Technically the problem is to construct a large gas-liquid contact area alternating with oxidation sections. This can either be done in series of towers or in a single large tower partitioned by trays. The heat of reaction is removed by cooling the trays internally. Figure VII.8. presents a flowchart with a bubble tray column.

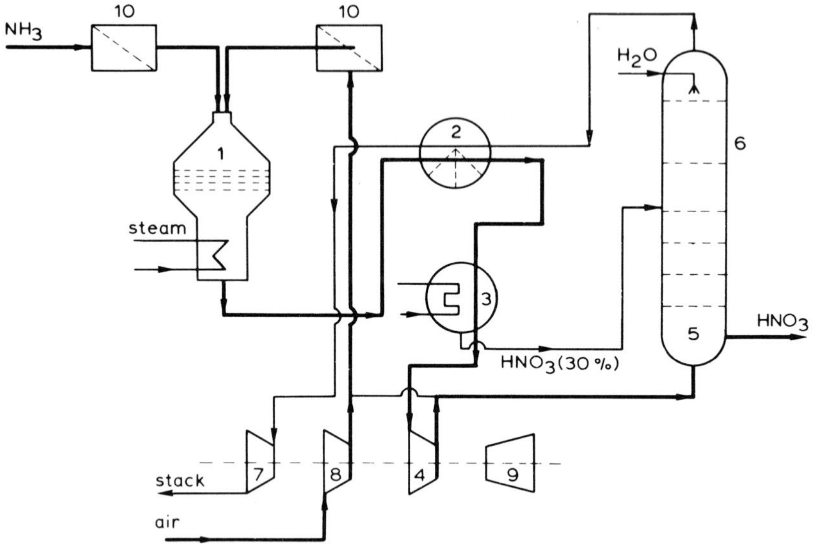

Figure VII.8. Production of nitric acid

1. *ammonia burner*
2. *gas cooler*
3. *cooler - condenser*
4. *NO_x compressor*
5. *oxidation*
6. *absorption tower*
7. *expansion turbine*
8. *air compressor*
9. *steam turbine*
10. *filters*

To improve the performance of this column it is usual to run the absorber under pressure. An absorption pressure of about 0.8 MPa has been economical for a long time. The nitric acid produced has a concentration of at most 67% and the exit gas then contains 2000 - 4000 ppm NO. In several countries, however, a concentration of only 200 ppm is allowed in the effluent gases because of the poisonous nature of nitrogen oxides and the danger of photochemical smog. This makes it necessary to cool the nitrogen oxides and the acid in the absorption section to rather low values, to increase the number of trays, and to raise the pressure to at least 1 MPa. If the ammonia oxidation is carried out at a lower pressure the NO_x has to be compressed in a special stainless steel compressor. To reduce costs the compression can be carried out before the oxidation of NH_3 but then the disadvantages of a lower selectivity and higher catalyst loss still occur. Such a solution is probably more attractive than cleaning the stack gas. Two processes have been developed for the latter; both reduce nitrogen oxides. Reducing agents such as CH_4 and H_2 convert NO_x into N_2 if noble metal catalysts are used. Unfortunately the excess oxygen fed to the oxidation process is preferentially reduced. This means a danger of overheating and a relatively large amount of reducing agent. Ammonia is the only reducing agent wich attacks NO_x first, but it is also more expensive than CH_4:

$$3\ NO + 2\ NH_3 + \frac{5}{2}\ N_2 + 3\ H_2O$$

Because a nitric acid plant is a large steam producer the steam can be used in a turbine to drive the compressor. Conditions in which a maximum of 200 ppm NO_x in the stack gas is specified require a high absorption pressure at which nearly all the steam is used in the plant. Otherwise a net production of steam results. Because the stack gas must be expanded to 0.6 MPa it is useful to apply an expansion turbine. Expansion turbine, steam turbine and compressor are coupled (one shaft). The largest outlet for nitric acid is in the fertilizer industry. Other applications are the use of HNO_3 as an oxidizer (e.g. in the production of adipic acid) and as a nitrating agent (dyestuffs and explosives).

Phosphoric acid

Two processes have been developed for the production of phosphoric acid, a so-called 'wet' process and a 'dry' process. Both are based on the conversion of the mineral fluorapatite, $Ca_{10}F_2(PO_4)_6$. In the wet process the mineral is dissolved in sulfuric acid:

$$Ca_{10}F_2(PO_4)_6 + 10\ H_2SO_4 \rightarrow 2\ HF + 6\ H_3PO_4 + 10\ CaSO_4$$

Depending on concentrations and temperature the calcium sulfate can be crystallized as gypsum ($CaSO_4 \cdot 2\ H_2O$), hemihydrate ($CaSO_4 \cdot \frac{1}{2}\ H_2O$) or anhydrite ($CaSO_4$). Sulfate is separated by filtration, hence the conditions of precipitation must be chosen in such a way that the crystals are well-formed and do not contain phosphate. This depends not only on concentration and temperature but also to a large extent on fluorine and impurities content as well as on residence time.

Figure VII.9. presents a flowchart. The product is used in the fertilizer industry to produce monocalcium phosphate and ammonium phosphates.

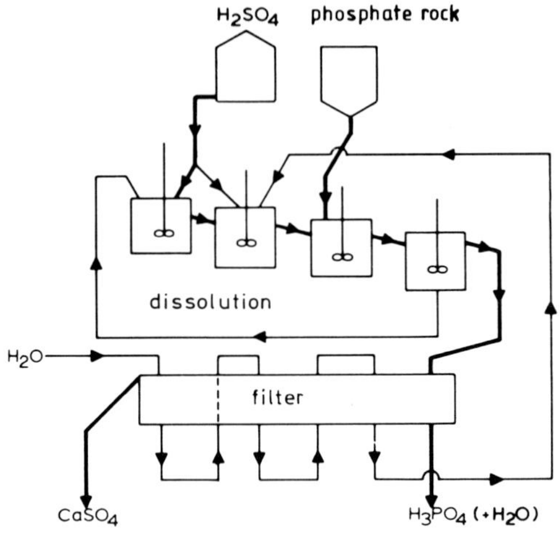

Figure VII.9. Production of phosphoric acid

Calcium sulfate coproduction is a problem, although mineral calcium sulfate in the form of gypsum is mined to produce plasterboard, building elements etc. Chemical calcium sulfate, however, contains impurities and is wet. To produce building quality material extra steps must be added to the H_3PO_4 process and extra energy input is necessary. The amount of discarded gypsum in the world is at least 60 Mton per year.

In the dry variant the apatite is reduced by carbon under addition of sand:

$$6\ Ca_{10}F_2(PO_4)_6 + 43\ SiO_2 + 90\ C \rightarrow 3\ SiF_4 + 90\ CO + 20(3\ CaO\ .\ 2\ SiO_2) + 9\ P_4$$

Phosphorus is condensed, burned to P_2O_5 and absorbed in water. The process requires a lot of energy of high cost because the reaction temperature is kept at 1500 ^{o}C electrically (14.300 Kwh per ton P). A highly concentrated pure phosphoric acid is produced suitable to make phosphate detergents (sodium phosphate, pyrophosphate and metaphosphate).

REFERENCES

General

1. Waggaman, Wm.H., 'Phosphoric Acid, Phosphates and Phosphatic Fertilizers", A.C.S. Monograph, Reinhold Publ. Co., New-York 1952

2. Slack, Editor A.V., 'Phosphoric Acid', Vol. 1-2, Marcel Dekker Inc., New-York 1968.

3. Moore, Editor A.I., 'Fertilizer Acids', Proc. 3rd Conf. on Fertilizers, The British Sulphur Co., London 1979.

Special

1. Ketjen. G.Th., Proc.Fert.Soc. 170, 1-25 (1978)

2. Yamaguchi, M., K. Matsushita, K. Takami, Hydrocarbon Process. 55 (8), 101-106 (1976).

3. Hoftijzer, P.J., Chem.Weekbl., 52, 71-77 (1956).
Figures VII.5. and VII.7. are reproduced from this article with permission.

4. Oele, A.P., Chem.Eng.Sci., 8, 146-157 (1958).
Figure VII.6. is reproduced from this article with permission.

VIII. PRODUCTION OF FERTILIZERS

Introduction

The largest share of the World's total food is produced agriculturally. Since the world population is still increasing, agricultural production should also grow. Although the application of irrigation can extend the area for agriculture, costs are very high. Clearly the conclusion is, that in old as well as in new areas the intensification of agriculture is unavoidable.
Plants need light, water and minerals and normally the latter is the determining factor in production. Only in the Netherlands and Japan is the use of fertilizers as source of minerals shifting into the region of deminishing additional returns. As example of the effect of fertilizers, consider that the production of wheat in the Netherlands has risen from 1000 kg/ha in 1800 to 5000 kg/ha in 1970. All developed countries show this effect for a large number of products. Developing countries in Africa, Asia and Latin America, where productivity per unit area is much lower, are undoubtedly regions where the response to essential minerals is very large and also very necessary. Compared to the use of 300 kg nitrogen per ha per year in the Netherlands the mean application is 10 - 15 kg world-wide.
Essential as plant food are nitrogen, phosphorus and potassium. A regular supply of nitrogen is particularly important because all soils lose nitrogen much more readily than phosphate and potassium by leaching and microbiological activity. These losses are especially high in tropical zones.
Nearly all artifical nitrogen fertilizers have their origin in the Haber-Bosch ammonia synthesis (Chapter IV) in which nitrogen from the air is 'fixed'. Normally the resulting ammonia is converted again in another process to a salt which can be handled more easily than NH_3 itself. However, in certain irrigation cultures ammonia can be injected directly into the water system. Other areas permit the use of ammonia which is injected into the soil, as is the case for cotton and corn in the U.S.A.
Historically, ammonium sulfate has been an important fertilizer because sulfuric acid production techniques were rather simple and widely known in the fertilizer

industry (superphosphate). Neutralization of acid with ammonia followed by crystallization results in a product which even in the tropics has good storage qualities because it is not hygroscopic and does not recrystallize into other modifications. However, the sulfate group normally is ballast (only a few areas in the world need sulfur, e.g. Australia) and the nitrogen content is only 20%, resulting in extra transportation costs. From this point of view nitric acid is a more attractive acid. It can be produced from NH_3 and converted to ammonium nitrate in which not only the ammonium ion has an agricultural value but the nitrate ion as well. The product contains 35% N. Ammonium nitrate has a disadvantage, however, of being an explosive substance under certain conditions. Therefore it is usual in many countries to add fillers like calcium carbonate or dolomite to render the product harmless. The carbonate helps in counteracting the tendency of soils to acidify as a result of the nitrification process. The advantage of a higher N-concentration is then partly lost (max. 26% N). Since less attractive storage properties also somewhat diminish the quality of product, it is only applied in moderate climates.

Nitrophosphates

Much better plant value than in $CaCO_3$ is found in calcium and ammonium phosphates, which also have the capacity to decrease the explosive properties of NH_4NO_3 markedly. Such phosphates are always produced from natural minerals. Orginally the necessity for production of phosphate fertilizers led to the superphosphate industry. Mineral phosphates are converted by sulfuric acid into monocalcium-phosphate or phosphoric acid. In this way the extremely insoluble phosphate minerals, generally with the formula $Ca_{10}F_2(PO_4)_6$ for the main component, are converted to water soluble $Ca(H_2PO_4)_2$:

$$Ca_{10}F_2(PO_4)_6 + 7\ H_2SO_4 \rightarrow 3\ Ca(H_2PO_4)_2 + 2\ HF + 7\ CaSO_4$$

This process has the disadvantage that the product contains a large amount of useless $CaSO_4$. Although this may be overcome by using phosphoric acid according to $Ca_{10}F_2(PO_4)_6 + 14\ H_3PO_4 \rightarrow 10\ Ca(H_2PO_4)_2 + 2HF$ it must be realized that the production of H_3PO_4 produces a large amount of $CaSO_4$ as by-product.
The virtues of the NH_4NO_3 process can be combined with the superphosphate process and the disadvantages of both avoided. If H_2SO_4 is replaced by HNO_3 a mixed fertilizer results containing two essential plant food components. The process can be divided into 4 steps (Figure VIII.1.):

1. dissolution of phosphate
2. adjustment of composition, i.e. reduction of the Ca/P ratio

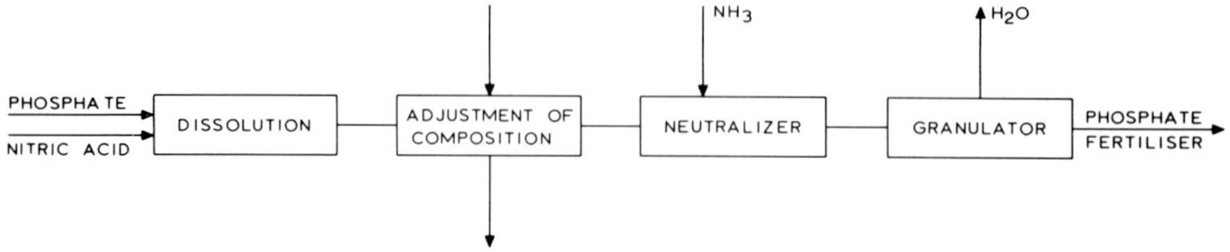

Figure VIII.1. Production of nitrophosphate

3. neutralization with ammonia
4. granulation and drying

The first step is digestion of the solid mineral with nitric acid:

$$Ca_{10}F_2(PO_4)_6 + 20\ HNO_3 \rightarrow 6\ H_3PO_4 + 10\ Ca(NO_3)_2 + 2\ HF$$

With industrial nitric acid (56 wt% and higher) the reaction does not lead to $Ca(H_2PO_4)_2$ under practical conditions and a solution results consisting of phosphoric acid and calcium nitrate (and part of the fluoride). This strongly acid material can be neutralized by ammonia but then two problems may arise: if the neutralization is complete the fluorapatite $Ca_{10}F_2(PO_4)_6$ is reformed and if the reaction with NH_3 is not complete the final product contains $Ca(NO_3)_2$. This is a very hygroscopic substance which makes the final product unsuitable for transport and storage. An attractive solution to these problems is to reduce the calcium content of the acid solution before neutralization. One variation is to crystallize a part of the $Ca(NO_3)_2$ by cooling. If, for example, 3 moles are crystallized, the neutralization occurs according to:

$$7\ Ca(NO_3)_2 + 6\ H_3PO_4 + 2\ HF \xrightarrow{14\ NH_3} 6\ CaHPO_4 + 14\ NH_4NO_3 + CaF_2$$

Dicalcium phosphate is barely soluble in water, but is soluble in organic acids such as are present in soils and plants. If certain soils and crops prefer water-soluble phosphate, more $Ca(NO_3)_2$ can be crystallized, e.g., 6 $Ca(NO_3)_2$:

$$4\ Ca(NO_3)_2 + 6\ H_3PO_4 + 2\ HF \xrightarrow{11\ NH_3}$$

$$\xrightarrow{11\ NH_3} 3\ CaHPO_4 + 3\ NH_4H_2PO_4 + 8\ NH_4NO_3 + CaF_2$$

A second variant is to precipitate part of the calcium by adding the proper amounts of sulfuric acid, sulfate or phosphoric acid. In both variants the Ca/P-ratio in the acid solution is adjusted in such a way that the subsequent neutralization does not precipitate calcium-rich phosphates.

Figure VIII.2. presents schematically the production of nitrophosphates. After neutralization the slurry is fed to an evaporator and finally granulated and dried.

The final product composition depends on the amount of nitric acid used in the first step (not necessarily stoichiometric) and the final Ca/P-ratio in the second step. A very well-known composition is N-P 23-23 (23% N, 23% P_2O_5). Such a fertilizer combines two nutrients, is concentrated and saves not only transportation costs but also costs of handling by the farmer. From the above it can also be concluded that for adjustment of the Ca/P-ratio extra sulfuric acid or phosphoric acid is necessary, or else calcium nitrate is produced as a

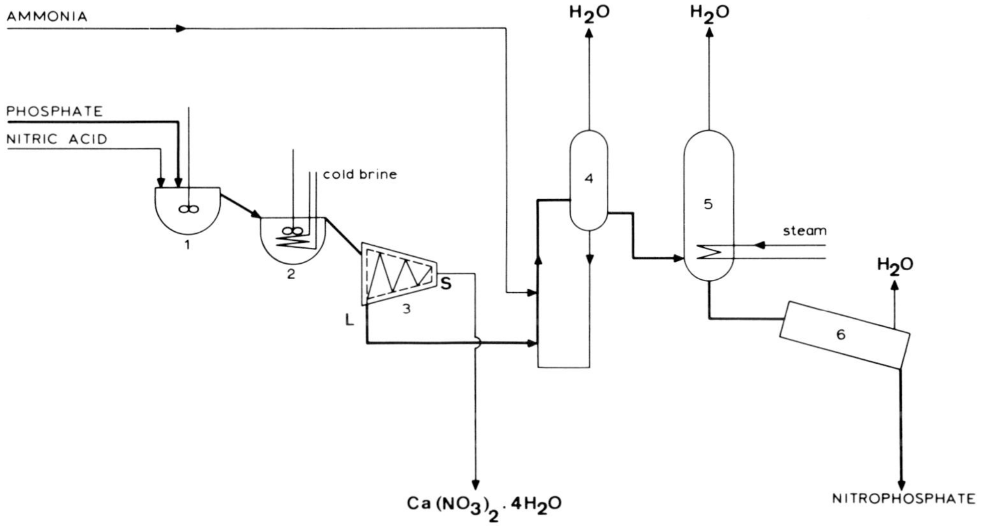

Figure VIII.2. Production of nitrophosphate

1. dissolution

2. crystallizer

3. centrifuge

4. neutralizer

5. evaporator

6. granulator

by-product. The latter may be converted to a double salt with ammonium nitrate or converted directly to ammonium nitrate itself, but its production is in fact a less attractive aspect of the nitrophosphate process. Therefore a process has been developed in which the by-product $Ca(NO_3)_2$ is thermally dissociated at 700 °C in a fluid bed. The endothermic reaction produces valuable nitrogen oxide for recycle to the absorption unit in the nitric acid plant:

$$Ca(NO_3)_2 \rightarrow CaO + 2\ NO_2 + \tfrac{1}{2}\ O_2$$

By-production of calcium oxide is not a problem because it can be used in agriculture or the building industry. The integration of a plant for N-P production with recycle of nitrogen oxide is represented schematically in Figure VIII.3 and the dissociation in Figure VIII.4.
Heat and mass balances are given as Sankey diagrams (Figures VIII.5. and VIII.6.).

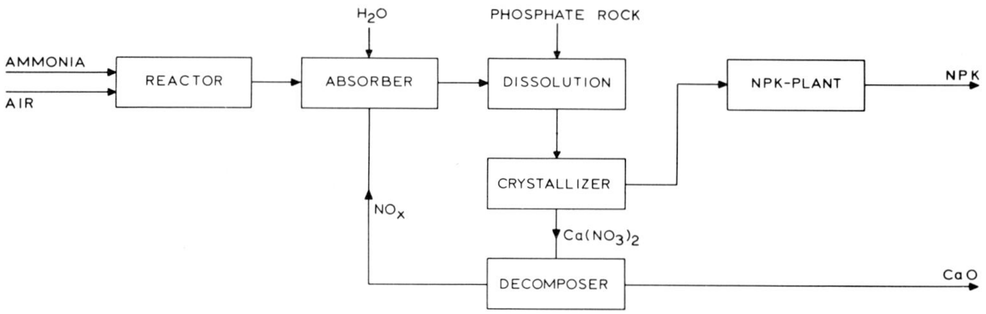

Figure VIII.3. Production of NPK fertilizer with thermal decomposition of by-product $Ca(NO_3)_2$

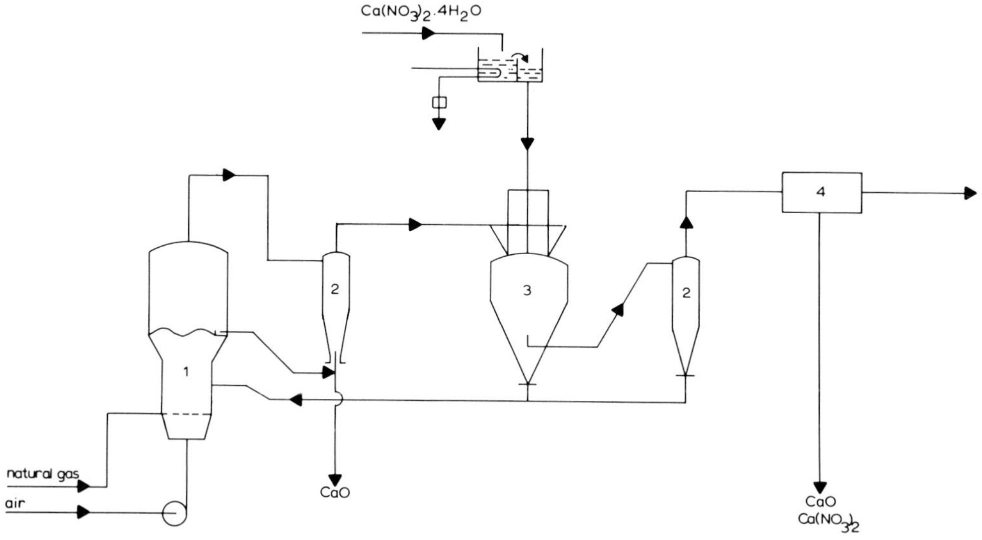

Figure VIII.4. Decomposition of $Ca(NO_3)_2$ in a fluid-bed reactor

1. *fluid-bed*
2. *cyclone*
3. *spray dryer*
4. *filter*

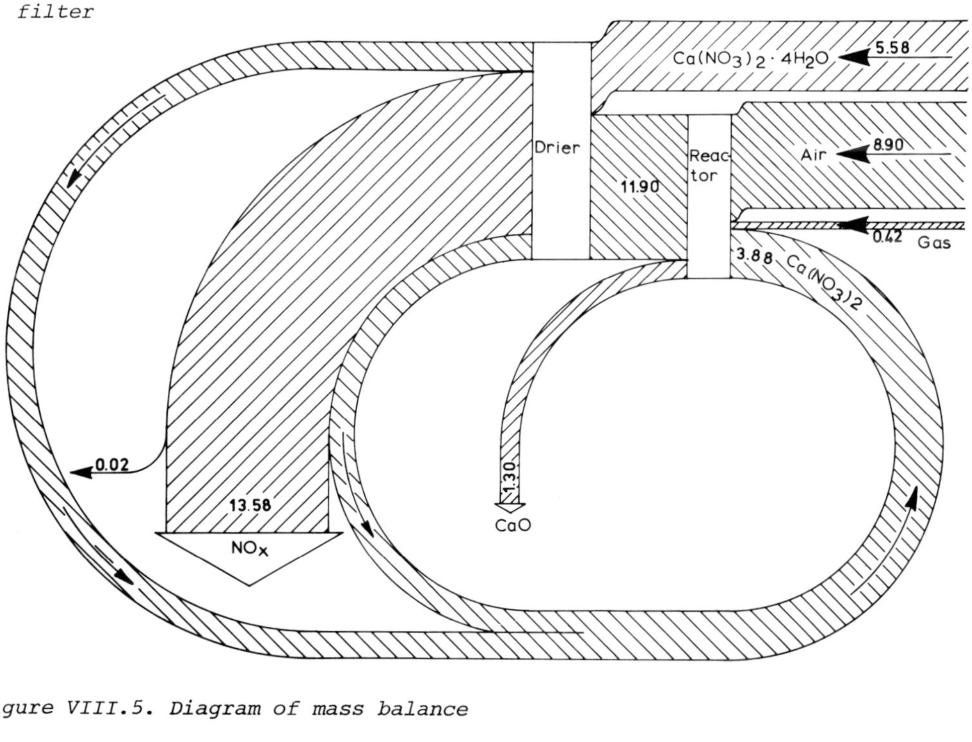

Figure VIII.5. Diagram of mass balance

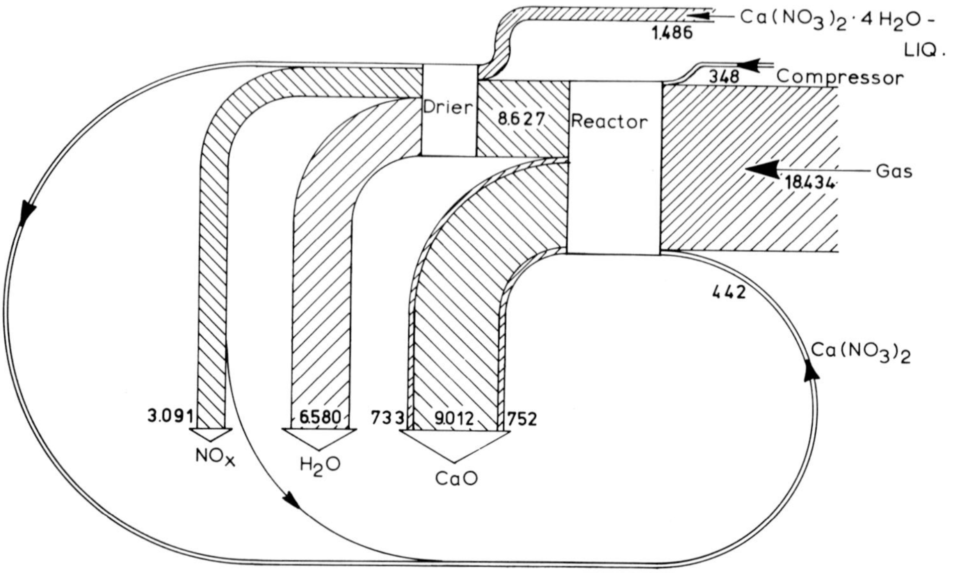

Figure VIII.6. Diagram of heat balance of the decomposition of calciumnitrate in a fluid bed reactor.

Urea

After the introduction of the nitrate group instead of the sulfate group in ammonium fertilizers a third acid has been the source of much innovation: carbonic acid. It is possible to synthesize urea from NH_3 and CO_2 in the following overall reaction:

$$2\ NH_3 + CO_2 \rightleftarrows NH_2CONH_2 + H_2O$$

This is an attractive possibility for producers of ammonia because carbon dioxide is produced in the hydrogen section where CO is converted to CO_2 and H_2.

Urea is a fertilizer with 46.6% nitrogen which is advantageous from the view of transport costs. Because urea has a higher agricultural efficiency in tropical zones than in moderate ones, it is no wonder that since its introduction in 1950 there has been a tremendous increase in production capacity. Because the synthesis is technically rather complex much process development work has been carried out to find optimal conditions. This work is a good demonstration of the use of technological principles leading to different solutions for a given problem.

The synthesis of urea consists of two equilibrium reaction steps:

$$2\ (NH_3)_G + (CO_2)_G \rightleftarrows (NH_2COONH_4)_L \quad \Delta H_r = -117.2\ \text{kJ}$$

$$(NH_2COONH_4)_L \rightleftarrows (NH_2CONH_2)_L + (H_2O)_L \quad \Delta H_r = 15.5\ \text{kJ}$$

The first reaction is fast and results in the formation of ammonium carbamate. Carbamate dehydrates to urea at temperatures above 150 $^{\circ}C$, but only in the liquid phase. Because the system NH_3-CO_2-H_2O-carbamate forms such a condensed phase only at pressures of 13 MPa and more (at T $\geq$ 150 $^{\circ}C$) the urea synthesis is a high pressure process. The rate of the dehydration is low and, even at the above temperatures, residence times of 30 min and more are used. A catalyst is not applied, because the reaction mixture is highly corrosive and the possible decrease in reactor volume is more than offset by technical difficulties of catalyst recovery.

Although earlier processes had one reactor for each reaction step, it is more economical nowadays to combine them. This makes it also easier to use the heat of carbamate formation for dehydration to urea. Summarizing, a reactor is needed working at temperatures and pressures around 180 $^{\circ}C$ and $\geq$ 135 atm. Ammonia is fed as a liquid and is easily brought to the right pressure with a pump. Carbon dioxide is a gas which must be compressed. To minimize the volume of the autoclave, baffles or sieve plates are installed, which serve to diminish backmixing.

The urea synthesis is an equilibrium reaction: under technically realizable conditions the degree of conversion, ξ, is between 0.5 and 0.7. To separate urea from unconverted carbamate profit can be gained from the large difference in kinetics between urea reaction and carbamate formation. It is possible to lower the pressure after the reactor (expansion valve) which shifts the first reaction step to the left: the gas phase will contain mainly ammonia and carbon dioxide (and some water vapor). The composition of the gas phase depends on the final expansion pressure: the higher this pressure is, the more ammonia it contains (and the more carbon dioxide relative to NH_3 stays in solution). This phenomenon is more pronounced at higher NH_3-CO_2 ratios in the reactor feed (above stoichiometry). The liquid phase contains nearly all the urea and the bulk of the water; if the residence time at a high temperature ($\geq$ 130 $^{\circ}C$) is short hardly any urea will hydrolyze.

The separation of a gas phase containing excess ammonia and carbon dioxide does not mean that this unconverted material can be recycled to the reactor easily. The mixture should be compressed again to bring it back to synthesis pressure. Then, however, a condensed phase containing liquid carbamate (or solid carbamate at lower temperatures) may be formed. No compressors exist which can handle droplets or crystals. In the early days of urea manufacture one did not bother with this problem (Figures VIII.7. and VIII.8.). The gases were used to produce some by-product, e.g. NH_4NO_3. Neutralization of the valuable NH_3 is possible with HNO_3 and CO_2 is sacrificed. With large productions this results in a weight ratio of products which will seldom reflect the needs of the market. Integration of a number of processes for different products can be useful in a certain location and at a certain time, but also troublesome. From the solutions to the recycle problem which now follow, it is concluded that these determine the character of the process to a large degree.

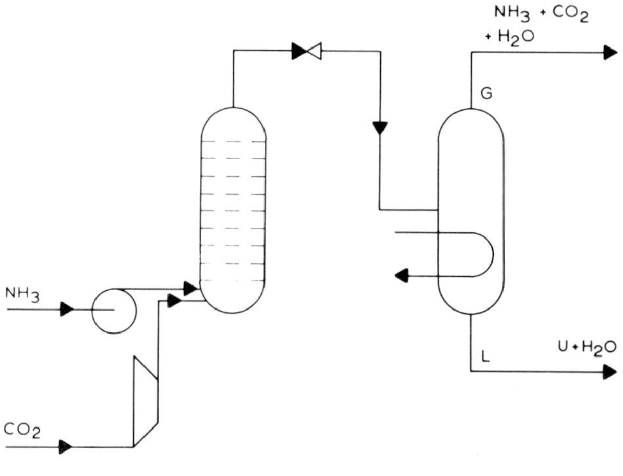

Figure VIII.7. Once through process for urea

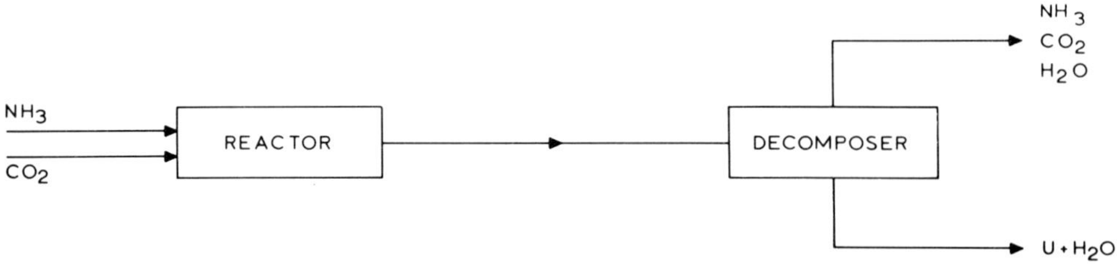

Figure VIII.8. Once through process for urea

a. Recycle as a solution in water

In this case the gaseous phase after expansion is dissolved in water and recycled by pumping (Dupont). Technically it is simple, although corrosion has to be overcome. Technologically, however, it has a disadvantage because water shifts the urea synthesis equilibrium back towards the starting materials. The consequences are lower conversion, larger reactor volume, higher energy costs, more water to evaporate. This can be compensated by using much higher NH_3/CO_2-ratios in the reactor.

But, if such high ratios are used advantage can be taken of the fact that partial expansion gives an NH_3-rich phase. Accordingly D.S.M. and Mitsui Toatsu have developed the system shown in Figures VIII.9. and VIII.10.. The first expansion step is from synthesis pressure to about 1,8 MPa a second step to 0,3 MPa and finally to 1 atm. The gas-phase originating from this last step is dissolved in a small amount of water and pumped back to the 3 atm. stage. The solution is used to dissolved the gas-phase in that stage, pumped to 18 atm. etc..

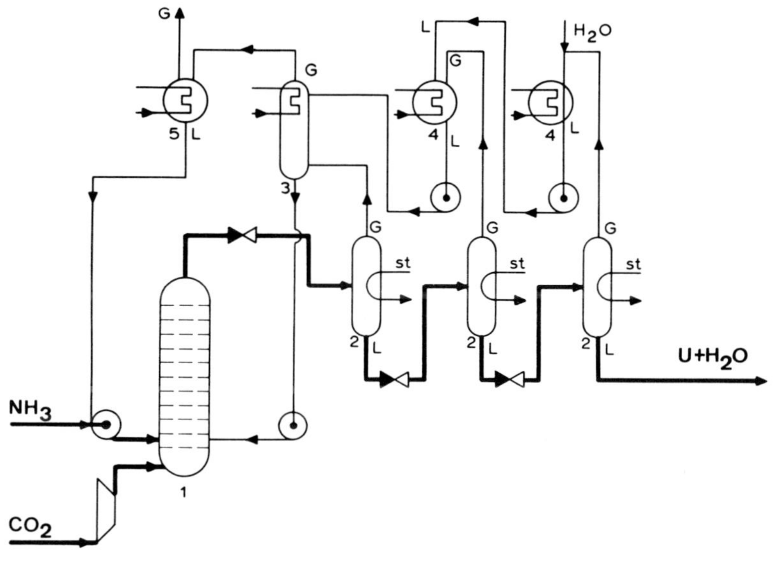

Figure VIII.9. Total recycle process for urea

1. reactor

2. separators

3. rectifier

4. condenser

5. condenser

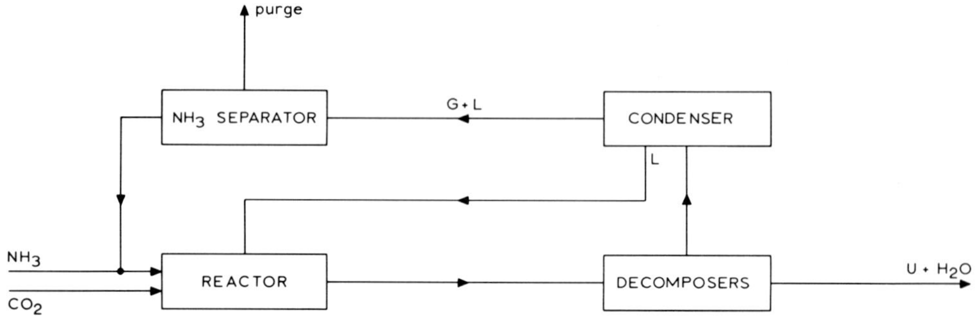

Figure VIII.10. Total recycle process for urea

The final result is that a small amount of water is used throughout the whole process. The gas phase in the first expansion stage is so rich in ammonia that it can be recycled to the reactor via a rectifier and NH_3-condenser.
The choice of expansion pressures and temperatures is determined by amounts and composition of gas phase in relation to liquid phase and the heat necessary for carbamate decomposition. The higher efficiency of such a process clearly necessitates extra investments.

b. Recycle with an auxiliary agent

Instead of water oil may be used to recycle ammonium carbamate. One advantage is that special measures to prevent attack by the strongly corrosive carbamate solution are unnecessary. Péchiney demonstrated (Figures VIII.11. and VIII.12.) that a suspension of carbamate in oil can be pumped.
After expansion the liquid contains urea, oil and water. This mixture separates into two layers. Urea and water flow to the evaporating section, oil is used in the recycle of carbamate which originates from condensation of the gas phase. Unfortunately the process has not been a success because oil deteriorates at the high synthesis temperature. This means poor separation of urea solution and oil, leading to extra costs due to losses and contamination.

c. Recycle of separated reactants

Several companies decided not to recycle NH_3 and CO_2 together but to separate them first. This can be done by selective solvents. Acid solvents like NH_4NO_3

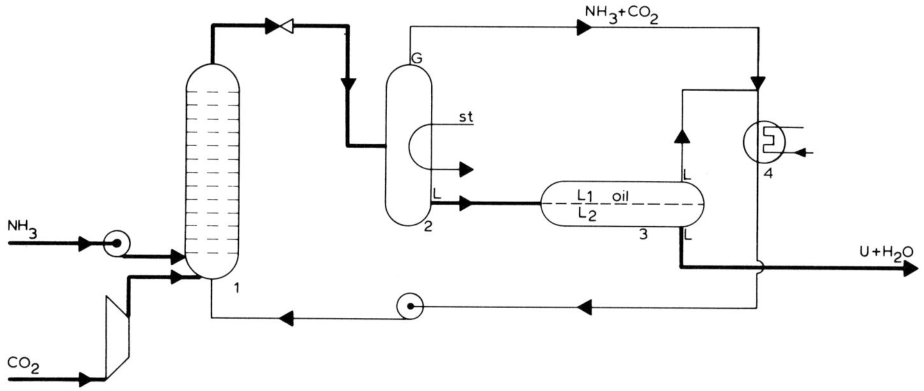

Figure VIII.11. Oil recycle process for urea

1. reactor

2. separator

3. separator

4. condenser

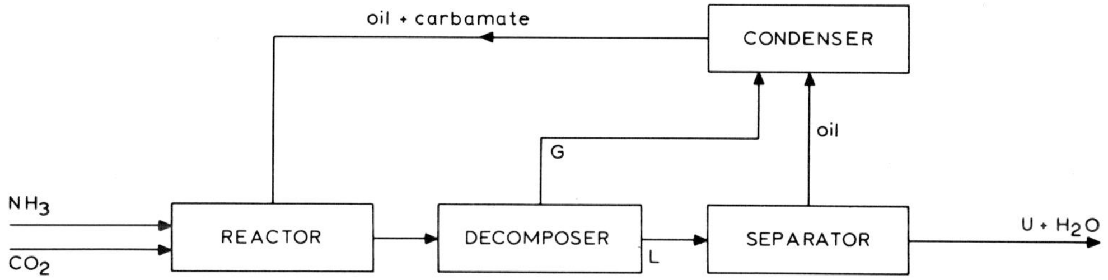

Figure VIII.12. Oil recycle process for urea

solution absorb ammonia but not carbon dioxide, which is recompressed and recycled. Ammonia is recovered in a second column by heating the solution and also recycled. Alkaline solvents use the same principle in absorbing one reactant (CO_2). These systems are energetically expensive because absorption means cooling and desorption heating. The principle is demonstrated in Figures VIII.13 and VIII.14.

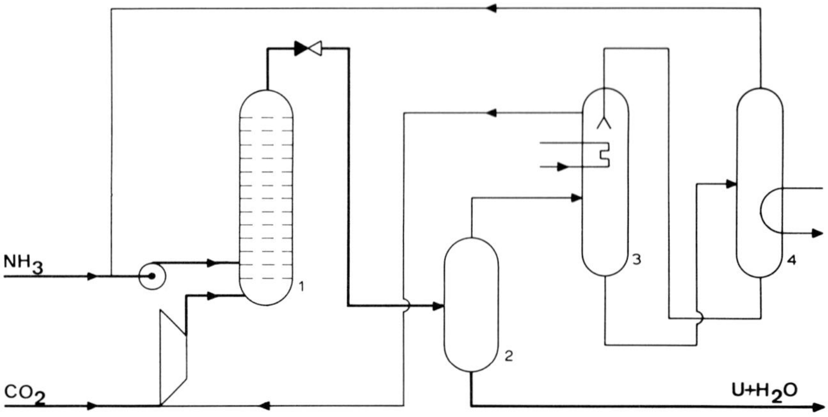

Figure VIII.13. Urea process with recirculation of separated NH_3 and CO_2

1. *Reactor*
2. *Separator*
3. *Absorber*
4. *Desorber*

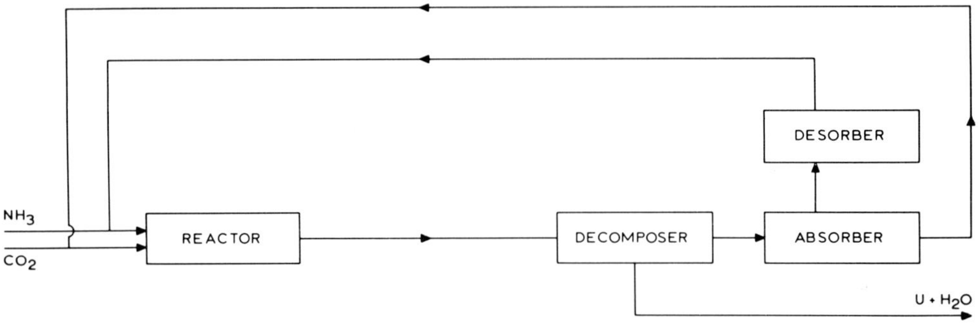

Figure VIII.14. Urea process with recirculation of separated NH_3 and CO_2

d. Recycle by stripping

It took some time to notice that it is not at all necessary to separate urea from reactants and carbamate by expansion. D.S.M. and S.N.A.M. developed processes in which the synthesis solution is stripped by a gas. If this is done at reactor pressure recompression is superfluous. Although a gas such as nitrogen can expel NH_3 and CO_2 from the urea solution, it is much more attractive technologically to use one of the reactants. D.S.M. chose carbon dioxide and S.N.A.M. developed stripping with ammonia. Such stripping agents lower the vapor pressure of the other reactant with the consequence that carbamate

decomposes. Figures VIII.15. and VIII.16. show the mode of operation.
The stripping agent contacts the synthesis solution, heat is supplied at the same time. The stripped solution is expanded to release dissolved stripping agent. The gases leaving the stripper flow to the reactor where the NH_3/CO_2-

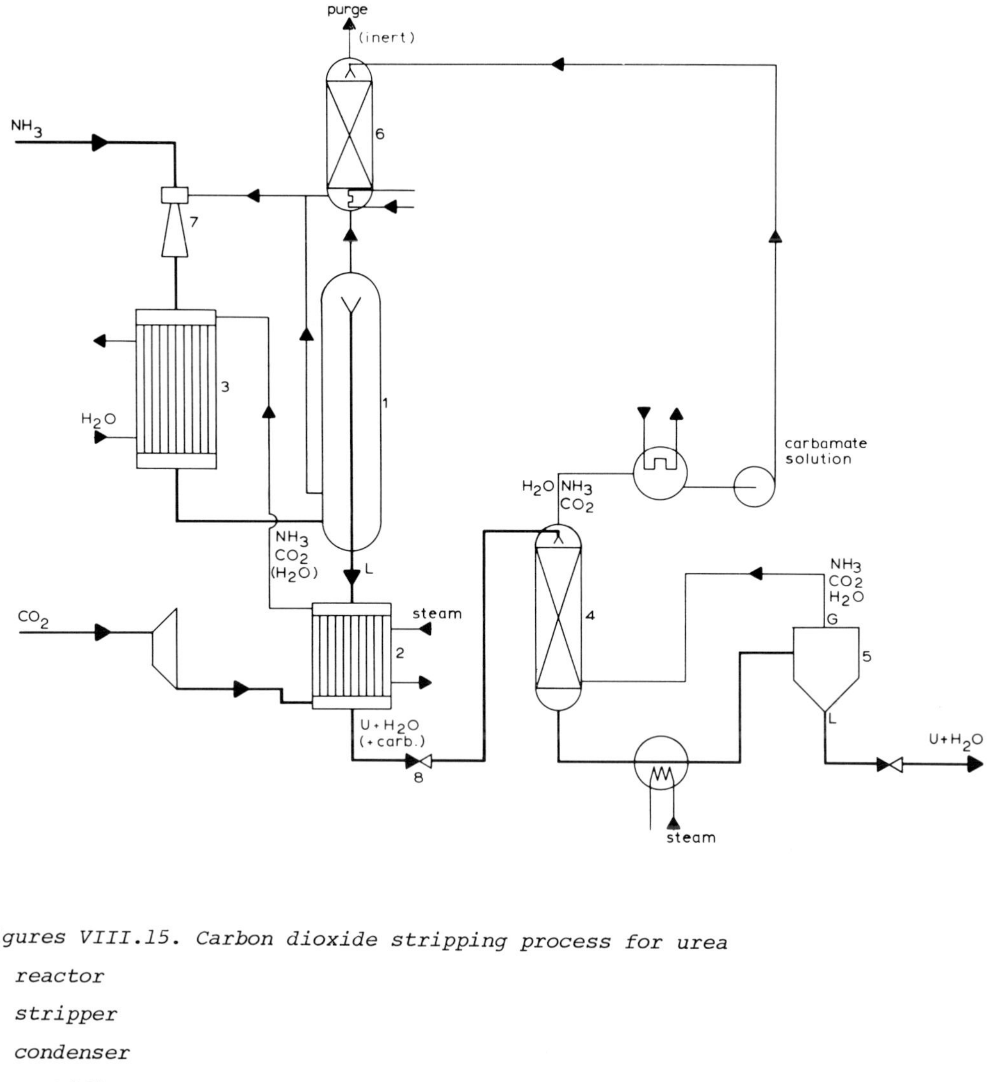

Figures VIII.15. Carbon dioxide stripping process for urea

1. reactor

2. stripper

3. condenser

4. rectifier

5. separator

6. scrubber

7. ejector

8. expansion valve

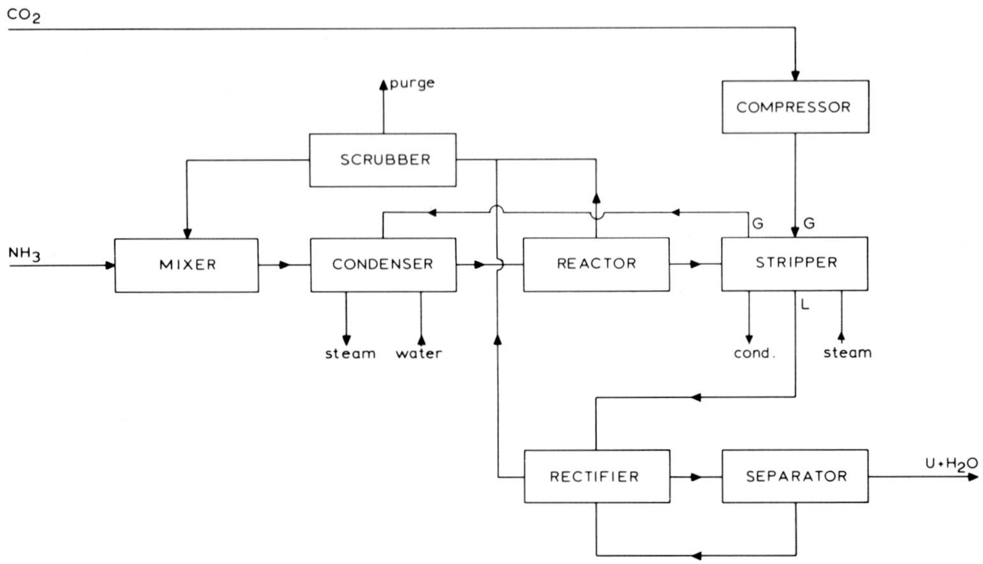

Figure VIII.16. Carbon dioxide stripping process for urea

ratio is brought to the right value by supplying the second reactant. During stripping with CO_2 biuret may be formed according to the reaction:

$$2\ NH_2CONH_2 \rightleftarrows NH_2CONHCONH_2 + NH_3$$

This can be minimized by assuring that the residence time in the stripper is short. Hence a film reactor (bundle of tubes) is chosen in which mass and heat transfer are fast. In the SNAM-process this problem does not exist but another problem does arise since ammonia is very soluble in the urea solution: its recovery requires more energy.

e. Recycle by hot compression

Recycle of NH_3 and CO_2 by compression may easily result in a condensed phase (liquid and solid), unless the compression occurs at temperatures above about 230 °C. Recently, this old idea of I.G. Farbenindustrie has been studied again. Extensive investigation of the phase system NH_3-CO_2-H_2O-urea have been made to demonstrate at which process conditions recycle by direct compression would be feasible. The enormous corrosion problems could be overcome, making a 'hot' compressor possible. These studies and pilot plant work proved that such a process can be built and run. The economics are equal to the D.S.M.-stripping process. Figures VIII.17. and VIII.18. represent the process schematically. As to the future, further research and development is expected, because the importance of urea as a fertilizer is still increasing. There is a growing trend towards regarding NH_3 as an intermediate product. It could be worthwhile to integrate ammonia plants with urea installations, with the following advantages:

1. CO_2 must be absorbed from synthesis gas after carbon monoxide conversion. Instead of the usual alkaline absorbents a urea synthesis solution may be used;
2. gases from the hydrogen plant and NH_3 synthesis may be used as stripping agents.

There are two disadvantages to such an extensive integration. The first is an increased sensitivity to technical troubles (the stream factor of urea and ammonia plant is 0.9). The second is that the optimal size of single-train ammonia plants is about 1800 tons/day and this implies 3200 tons/day of urea. Production of this amount in a single urea plant has not yet been demonstrated.

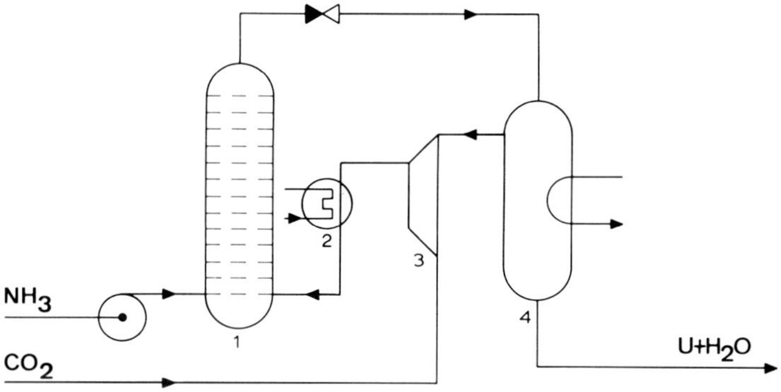

Figure VIII.17. Hot gas recirculation process for urea

1. reactor

2. condenser

3. compressor

4. separator

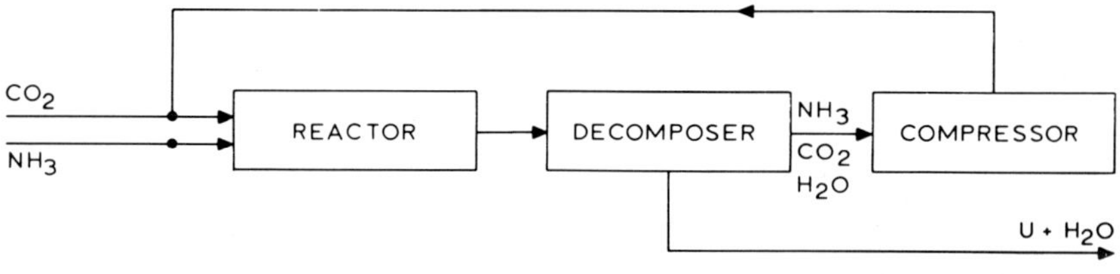

Figure VIII.18. Hot gas recirculation process for urea

REFERENCES

General

'Granular Fertilizers and their Production', The British Sulphur Co., London 1977.

Special

1. Berg, P.J. van den, A.C. Smit, H. Hellmer, A.G. Montfoort, Delft Progr. Rep., Ser. A, 1, 103-106 (1975).
   Figures VIII.3., VIII.4. VIII.5. and VIII.6. are reproduced from this article with permission.

2. Berg, P.J. van den, Isr.Chem.Eng. To be published in 1980.
   Figures VIII.7., VIII.9., VIII.11., VIII.13. and VIII.15 are reproduced from this article with permission.

3. Figure VIII.16. is based on British Patent 1365707.

IX. SOME CHEMICALS BASED ON SODIUM CHLORIDE

The main products made from sodium chloride are caustic soda (NaOH), soda ash (Na_2CO_3) and chlorine. The raw material, NaCl, is obtained by mining (as rock salt), from sea water by solar evaporation, or as brine by injecting water into underground salt deposits, analogous to the Frasch method for recovery of sulfur. Depending on the application, (re)crystallization after partial evaporation of the water and/or purification is necessary. Usual impurities are calcium, magnesium, heavy metal ions and sulfate. Ca and Mg are usually present as sulfates and are eliminated by a treatment with milk of lime and soda ash:

$$Mg^{++} + Ca(OH)_2 \rightarrow Mg(OH)_2\downarrow + Ca^{++}$$

$$Ca^{++} + Na_2CO_3 \rightarrow CaCO_3\downarrow + 2Na^{+}$$

Heavy metal ions should not be present in salt used in making chlorine and caustic soda by electrolysis in mercury cells (see below). If present, most of the ions are removed in the crystallization of NaCl and in the precipitation step with Na_2CO_3 or, less frequently, NaOH. Sulfate is removed by precipitation as $BaSO_4$. All precipitates are removed by filtration; filterability depends strongly on residence time, temperature and amount of reactants added.
After removal of precipitate the purified brine is concentrated by evaporation. This is usually done in a multiple-effect evaporator to save energy, a method which is also used in e.g. concentrating caustic soda, in the sugar and paper industries and for sea water distillation. The system shown in Figure IX.1. is essentially a multi-stage evaporation, in which the evaporation temperature is lowered in successsive stages by reducing the pressure.
The first stage is fed with fresh steam and subsequent stages by steam from the previous stage. Steam consumption is lower as the number of stages increases, but investment costs and electric power consumption increase with the number of stages applied. The economic optimum is usually between 3 and 5 stages.
In some designs more elaborate and complex flow schemes are used, depending

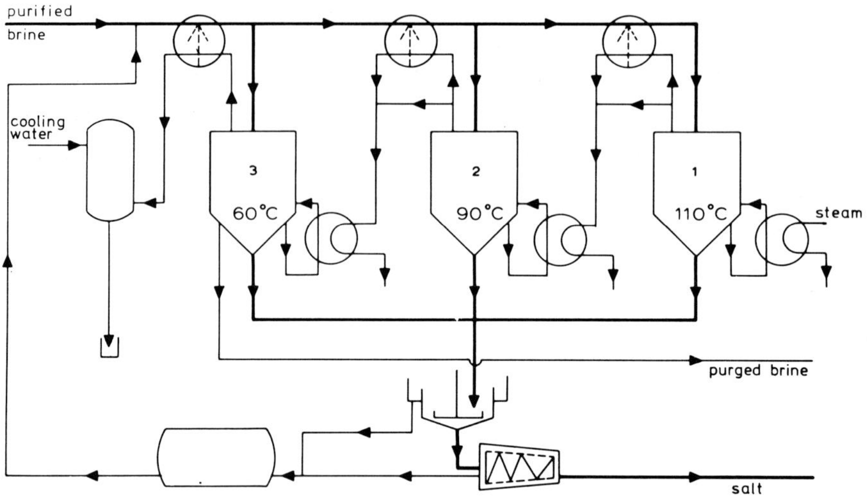

Figure IX.1. Multiple-effect evaporation

1. crystallizer, first effect

2. crystallizer, second effect

3. crystallizer, third effect

on (i) the necessity of integrating purification or crystallization with evaporation, (ii) the prevailing ratio between energy and investment costs, (iii), the nature of the required product, (iv) the size of the unit and several other factors.

Table IX.1. Steam requirements of multiple effect evaporators

Number of stages	kg water evaporated per kg fresh steam
1	0.9
2	1.7
3	2.6
4	3.3
5	3.9

Soda ash by the Solvay process

The process for making Na_2CO_3 developed by Ernest Solvay more than a century ago is still used today. Figure IX.2. shows a block diagram of the main process steps. Brine is purified, used to scrub off-gases and then to absorb ammonia. In the subsequent carbonatation the ammonia is reacted with carbon dioxide to give ammonium bicarbonate:

$$NaCl + H_2O + NH_3 \rightarrow NaCl + NH_4OH$$

$$NaCl + NH_4OH + CO_2 \rightarrow NaCl + NH_4HCO_3$$

Upon cooling, sodium bicarbonate precipitates and is separated by filtration or centrifuging and then calcined to soda ash:

$$NaCl + NH_4HCO_3 \rightarrow NaHCO_3 + NH_4Cl$$

$$2\ NaHCO_3 \rightarrow Na_2CO_3 + CO_2 + H_2O$$

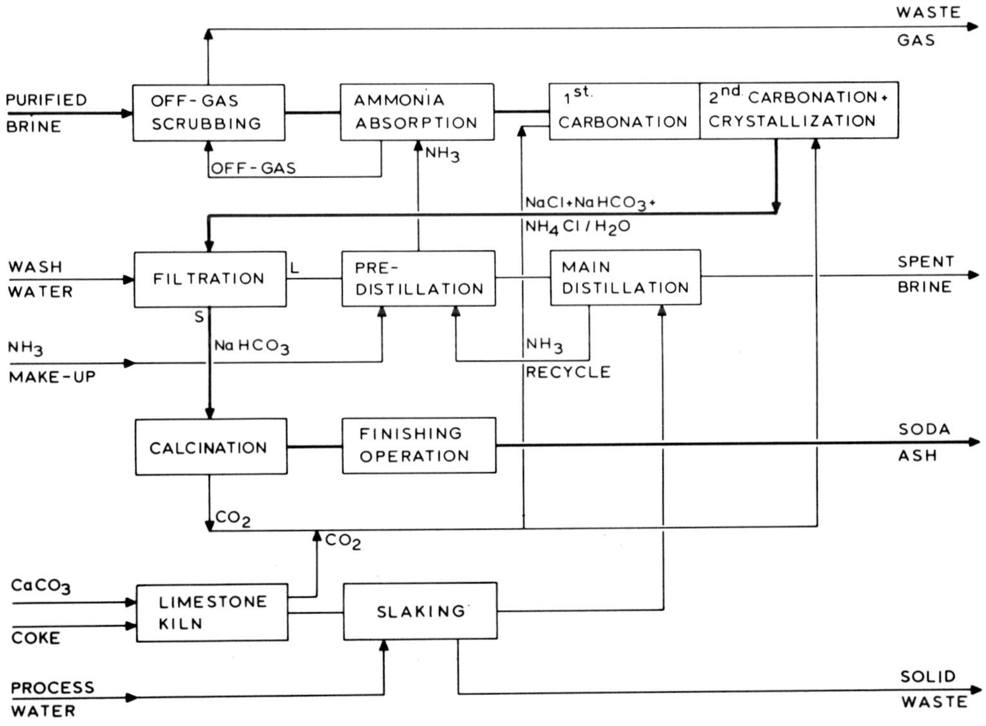

Figure IX.2. Block diagram of Solvay process

The ammonia used in above sequence of reactions must be recovered nearly quantitatively if the process is to be economic. This is done in the distillation step where the filtrate is treated with lime; the ammonia set free is driven off by heating:

$$2\ NH_4Cl\ +\ Ca(OH)_2\ \rightarrow\ CaCl_2\ +\ 2\ NH_3\ +\ 2\ H_2O$$

The ammonia is recycled and the calcium chloride/NaCl-solution discarded. The $Ca(OH)_2$ needed for distillation is produced from CaO obtained by burning limestone; this also produces the required carbon dioxide:

$$CaCO_3\ \rightarrow\ CaO\ +\ CO_2$$

The flow chart corresponding to block-diagram IX.2. is shown in figure IX.3.. The brine fed to the Solvay process should be substantially free of Ca and Mg-ions because they would give rise to insoluble carbonates in the carbonation step, e.g. Na-Mg-chlorocarbonate, NaCl. $MgCO_3$. Na_2CO_3 would cause deposit formation in piping and on heat exchanger surfaces. Most of the sulfate in the feed solution leaves the system in the spent $CaCl_2$/NaCl-solution, but sulfates also tend to give sparingly soluble Ca-sulfate deposits in the distillation equipment. Periodic cleaning is needed, even with careful control of temperatures and concentrations.

The equipment used in the scrubbing and ammonia absorption operations can be either packed towers or tray absorbers. The scrubber serves to recover small amounts of ammonia from the off-gases; it is generally possible to recycle more than 99½% of the NH_3.

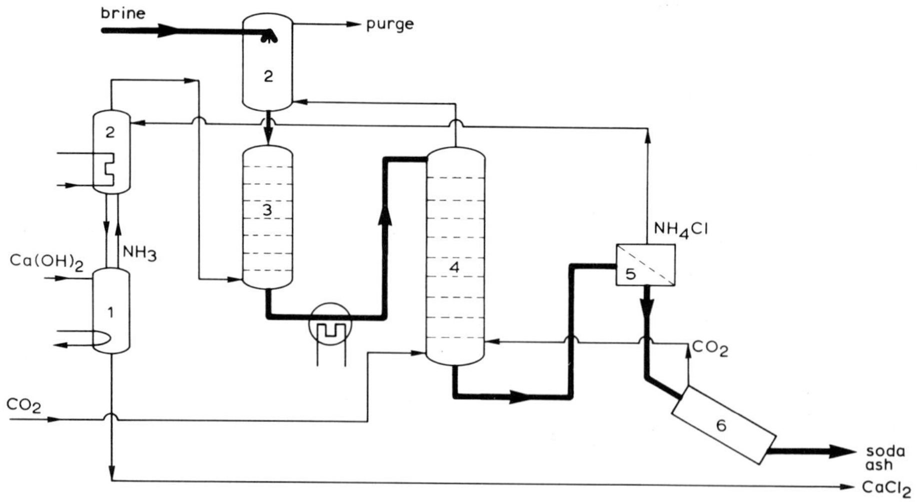

Figure IX.3. Flowchart of the Solvay process

Since CO_2 is almost insoluble in brine unless ammonia is present, any excess CO_2 entering the scrubber is lost. This is not serious since the limestone calciner produces enough CO_2; losses of NH_3 should however be prevented because these are an important economic factor in the Solvay process.

Precipitation is effected in a tower which is usually operated countercurrently. The top section is intended to remove the CO_2 from the carbonizing gas feed by absorption. The CO_2 can be introduced at two positions: at the bottom, below the cooling section which serves to induce crystallization of $NaHCO_3$, the high grade CO_2 recovered from the calcining operation is introduced. Make-up gas from the calciner is fed in just above the cooling section.
Since the process of forming bicarbonate is fairly exothermic, large amounts of heat must be removed. If the precipitation and carbonation operations are carefully controlled, it is possible to convert up to 80% of the NaCl into $NaHCO_3$. The NaCl conversion is maximal if

- the solution resulting after NH_3 absorption is almost saturated relative to NaCl
- the molar ratio NH_3/NaCl is between 1.10 and 1.12
- the CO_2 contains as low a quantity as inerts as possible to minimize stripping of NH_3
- precipitation temperature is 30-40 ^{o}C.

The reaction system of this process step is quite complex and has been the subject of numerous theoretical studies. Apart from bicarbonate it is also possible to form ammonium carbonate and ammonium carbamate, whilst the solubility of NaCl decreases with increasing NH_3 concentration.
The crude bicarbonate obtained by filtration or centrifuging is washed with a small amount of water and then sent to the calciner, a rotary kiln in which water is evaporated, bicarbonate decomposed and residual ammonium chloride converted into sodium chloride and NH_3. The off-gas is freed of small soda ash particles by passage through cyclones. The product is a light, fluffy powder ('light caustic ash') which is not suitable in many applications and should be transformed into a more compact product. This is usually accomplished by recrystallization as the monohydrate or by a briquetting operation.
Distillation is usually performed in two towers. In the first tower labile NH_3/CO_2 compounds are decomposed, and in the second the almost carbonate-free liquor from the first distillation is reacted with lime to give a concentrated stream of NH_3. Stripping with steam is necessary here to allow a sufficiently complete ammonia recovery. Here again, packed or plate-type columns may be

applied. The liquid residue is clarified, neutralized if necessary, and discarded in a suitable way, e.g. at sea.
Limestone burning is usually carried out in a moving-bed reactor; rotary kilns and fluid-bed reactors are used less frequently. The dissociation of $CaCO_3$ being an endothermic process, heat must be supplied by burning coke:

$$CaCO_3 \rightarrow CaO + CO_2 \qquad (\Delta H_s^o = 176 \text{ kJ/mole})$$

$$C + O_2 \rightarrow CO_2 \qquad (\Delta H_s^o = -396 \text{ kJ/mole})$$

Figure IX.4. shows a simplified sketch of a vertical shaft kiln. A stationary reaction zone is maintained between a preheating and a cooling zone. In some cases fuel gas is used instead of coke. The particle size distribution of the feed, the fuel/limestone ratio and the temperature profile in the kiln should be carefully controlled to minimize the risks of low conversion by 'overburning' (sintering), which would lead to high limestone consumption and low product reactivity.

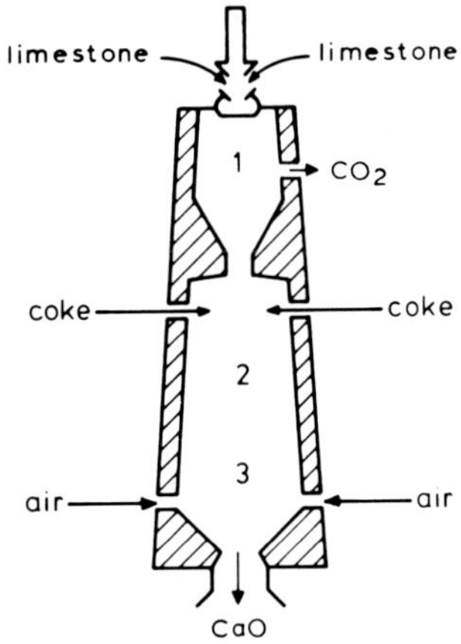

Figure IX.4. Lime kiln
1. preheating section
2. reaction zone
3. cooling zone

Chlorine and caustic soda by electrolysis

General aspects

Sodium hydroxide and chlorine are produced by electrolysis of a concentrated NaCl solution according to the overall reaction.

$$NaCl + H_2O \rightarrow \tfrac{1}{2}Cl_2 + NaOH + \tfrac{1}{2}H_2 \quad (\Delta H_s^o = 218 \text{ kJ/mole})$$

Important discharge reactions at the electrodes of ions present in NaCl solutions are:

$$2\,H^+ + 2\,e \rightarrow H_2 \quad (E_o = 0{,}000 \text{ V, by definition})$$

$$2\,Cl^- \rightarrow Cl_2 + 2\,e \quad (E_o = 1.358 \text{ V})$$

$$4\,OH^- \rightarrow O_2 + H_2O + 4\,e \quad (E_o = 0.401 \text{ V})$$

$$Na^+ + e \rightarrow Na \quad (E_o = 2.711 \text{ V})$$

The standard potentials, E_o, of the above reaction equations apply to reversible electrode reactions, i.e. are measured at infinitely small currents, at 25 ^oC, concentrations of 1 mole per litre and atmospheric product gas partial pressures, against the standard hydrogen electrode, which is assigned the value 0.000 V. At other concentrations and temperatures the Nernst equation can be applied to calculate the electrode potential:

$$E = E_o + \frac{RT}{nF} \ln C_{ion}$$

in which R is the gas constant, T the absolute temperature, n the charge of the ion and F Faraday's constant. (1 F = 96500 A.s = 26.8 Ah).
However, modern cells operate at *high current density* (the cell current per area of electrode surface), mainly to reduce investment costs. Therefore, actual conditions are far from equilibrium and the cell potential differs from the potential calculated by the Nernst equation. The difference is caused by (i) the electrical resistance of the system and (ii) the *overvoltage*. The overvoltage or overpotential represents the difference between the observed electrode potential and the corresponding calculated value; it is due to

- depletion of ions in the boundary layer because diffusional transport from the bulk of the electrolyte to the electrode surface is relatively slow (a phenomenon called *concentration polarization*),
- resistance to ion discharge caused by gas bubbles on the electrode surface
- if mercury is applied as the cathode: slow diffusion of Na in the amalgam formed by discharge of Na^+

The above causes of the overvoltage explain why it depends on the following process variables:

1. nature and surface structure of the electrode
2. nature of the ions
3. composition and purity of the electrolyte; some impurities may be discharged preferentially, thereby changing the surface structure and composition of the electrode
4. temperature, and
5. current density

To a first approximation, the overvoltage increases with the logarithm of the current density. Some practical values shown in Figure IX.5.. It follows that the overvoltage for OH^- discharge on graphite is much larger than for Cl^-.

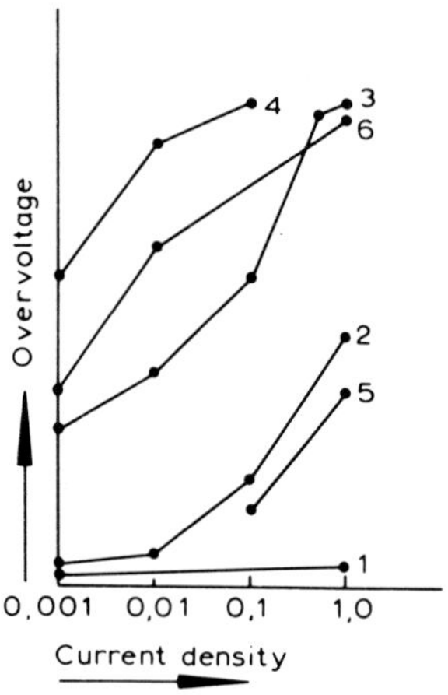

Figure IX.5. Overvoltages on different surfaces

1. H_2/Pt-black

2. H_2/Pt, smooth

3. H_2/Fe

4. H_2/H_g

5. Cl_2/ graphite

6. O_2/ graphite

Since, moreover, the Cl^- concentration during brine electrolysis is many decades higher than the concentration of hydroxyl ions near the anode, Cl_2 is formed in preference to oxygen, although the standard potential for oxygen evolution is appreciably lower. Another technically significant example is the difference between iron and mercury cathodes. On Fe, hydrogen is discharged in preference to sodium because of the relatively low hydrogen overvoltage on Fe, but on Hg sodium is discharged with the formation of sodium amalgam, the overvoltage for hydrogen discharge on the very smooth Hg surface being very high. Apart from the desired discharge reactions some parallel reactions occur, with the result that the *current efficiency* in modern cells is between 0.95 and 0.97, depending on cell concstruction and operating conditions. For example, the following may occur:

- impurities in or on an electrode, such as in graphite anodes, and mixing processes in the cell whereby the OH^- concentration near the anode is raised, cause some oxygen formation due to OH^- discharge
- subsequent reactions between chlorine and hydroxyl ions may occur:

$$2\,OH^- + Cl_2 \rightarrow OCl^- + Cl^- + H_2O$$

This implies that contact between the chlorine formed at the anode with the strongly alkaline liquid at the cathode must be prevented if chlorine is to be produced. At very low pH and intense mixing it is even possible to obtain chlorate:

$$3\,OCl^- \rightarrow ClO_3^- + 2\,Cl^-$$

$$6\,OCl^- + 3H_2O \rightarrow 2\,ClO_3^- + 4\,Cl^- + 1\tfrac{1}{2}\,O_2 + 6\,H^+ + 6\,e$$

In practice, three types of electrolytic cell are applied, viz. the mercury cell, the diaphragm cell and the membrane cell. Processes based on these cells are briefly outlined below.

Diaphragm cells.

The principle of this type of cell is shown in figure IX.6. At the electrodes the following reactions occur:

cathode: $2\,H^+ + 2\,e \rightarrow H_2$ and anode: $2\,Cl^- \rightarrow Cl_2 + 2\,e$

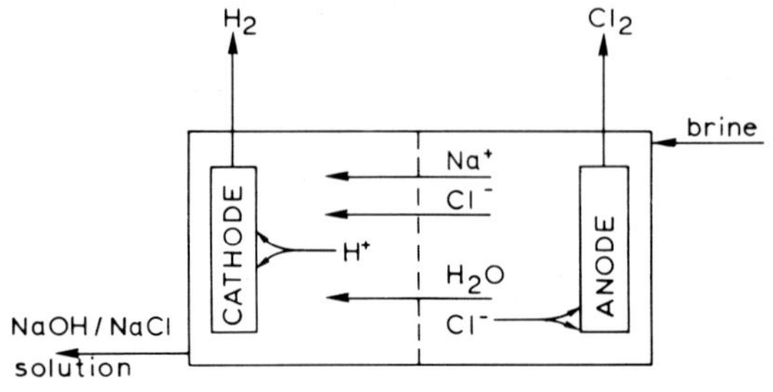

Figure IX.6. Principle of diaphragm cell

The diaphragm separates the chlorine evolving at the anode from the NaCl-containing caustic soda solution at the cathode which would otherwise react to give hypochlorite or, at higher conversion, chlorate. Moreover, chlorine and hydrogen should not mix because these gases form explosive mixtures. A net flow of liquid is maintained from the anode to the cathode compartments through the asbestos diaphragm, which in modern cells is supported by the steel cathode. To that end, a somewhat higher liquid level is maintained in the anode chamber. This minimizes migration of hydroxyl ions in the direction opposite to the flow and allows NaCl conversions of up to 80%. A disadvantage of the diaphragm is that aging and plugging by solid particles (e.g. of graphite if this material is used for the anode) reduce the permeability of the diaphragm, thus necessitating its periodic replacement. Until about 1970, graphite was widely applied as the anode material, but this has two main disadvantages: discharge of OCl^- and OH^- not only releases oxygen, which contaminates the chlorine, but also produces CO_2 by attack on the graphite. As a result, the anode/cathode distance must be adjusted periodically and new anodes must be installed more frequently. A further result of graphite electrode consumption is that the overvoltage rises somewhat, causing an increase in power consumption.
These disadvantages are not shown by the newer dimensionally stable electrodes made of titanium coated with ruthenium oxide, a material that can be worked into various and even complex electrode shapes relatively easily, in contrast to graphite. Although these electrodes are more expensive there is an overall saving in costs because labor costs and loss of production during installation of new electrodes are much reduced. Energy consumption is also lower because the resistivity of the material is less than that of graphite, and the overvoltage is appreciably below that of the older electrode materials.

Optimal operation with a brine feed of about 400 g/l NaCl leads to a product stream containing about 12%w NaOH and about 15%w NaCl, i.e. a conversion of 50-55%. Purification is therefore necessary: this is done by first concentrating the caustic/NaCl solution in a multiple-effect evaporator and then crystallizing the NaCl which is much less soluble than NaOH. The resulting solution contains 50%w NaOH along with 1%w NaCl and is applicable as industrial caustic soda solution.

Solid NaOH can be made by further evaporation, solidification and flaking or pelleting. However, in some applications like rayon fibre manufacture, a purer caustic with less NaCl contamination is needed; methods developed for further purification are based on fractional crystallization of hydrates of NaOH after dilution and cooling, addition of Na_2SO_4 to precipitate NaCl as the triple salt NaOH. Na_2SO_4. NaCl thus reducing the NaCl content to 0.3%w, and selective extraction of NaCl by liquid ammonia. Traces of heavy metals can, if necessary, be removed electrolytically.

An example of a diaphragm cell is shown in Figure IX.7.

Mercury cells

Mercury cells contain the same type of anodes as diaphragm cells, but the cathode is liquid mercury, which is converted to amalgam during electrolysis:

$$Na^+ + e + x\,H_g \rightarrow NaHg_x$$

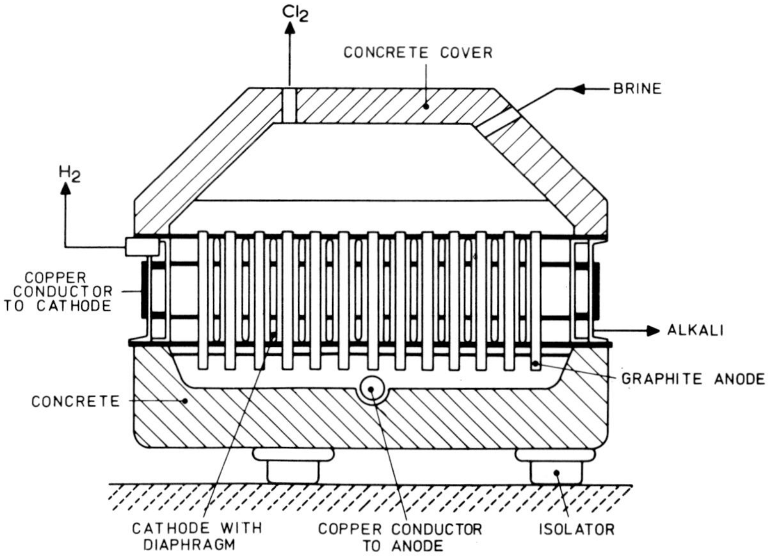

Figure IX.7. Hooker diaphragm cell

The anode reactions are similar to those in diaphragm cells. As can be seen from Figure IX.5., the hydrogen overvoltage on mercury is quite high; moreover, the concentration of Na^+ is very much higher than that of H^+ just above the mercury where the electrolyte is fairly strongly alkaline. As a result, Na^+ is discharged in preference to H^+, although a small amount of hydrogen is still formed. The concentration of Na in the amalgam is normally about 0.2%w; at higher concentrations the viscosity of the amalgam rises and at 0.8%w Na solidification sets in. Moreover, the electrode potential of the $NaHg_x$/NaOH, water electrode rises as the Na content of the amalgam increases. Overall, the energy consumption for the electrolysis is higher than for the diaphragm cell because of the higher cell potential. For a simplified flow sheet the principle is shown in Figure IX.8.. A simplified flow sheet of the mercury cell process is shown in Figure IX.9.. The cell, a rectangular box, is slightly inclined so that the mercury flows over its bottom in a thin layer under a series of anodes made of graphite or titanium/ruthenium oxide. The sides of the cell and the cell cover are internally coated with a chlorine and alkali-resistant material to prevent corrosion. Mercury and virtually saturated brine flow cocurrently through the cell; brine conversion is about 15%. At the lower end the amalgam is separated from the NaCl solution, washed with some water and then sent to the decomposer, where it is short-circuited with graphite in the presence of aqueous alkali to give NaCl-free caustic soda solution according to

$$NaHg_x + H_2O \rightarrow \tfrac{1}{2} H_2 + x\,Hg$$

The decomposer may either be built horizontally, like the electrolytic cell, or vertically, in the form of a packed bed of graphite over which mercury and aqueous phase flow countercurrently. The hydrogen evolved is cooled, washed and sent to storage and the mercury returned to the electrolyzer by a centrifugal pump.

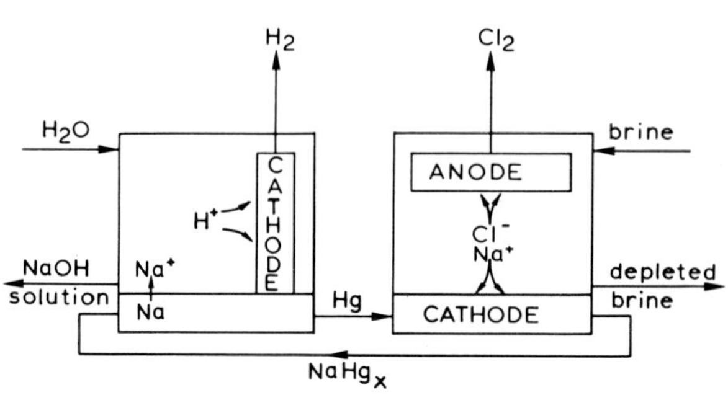

Figure IX.8. Principle of amalgam cell

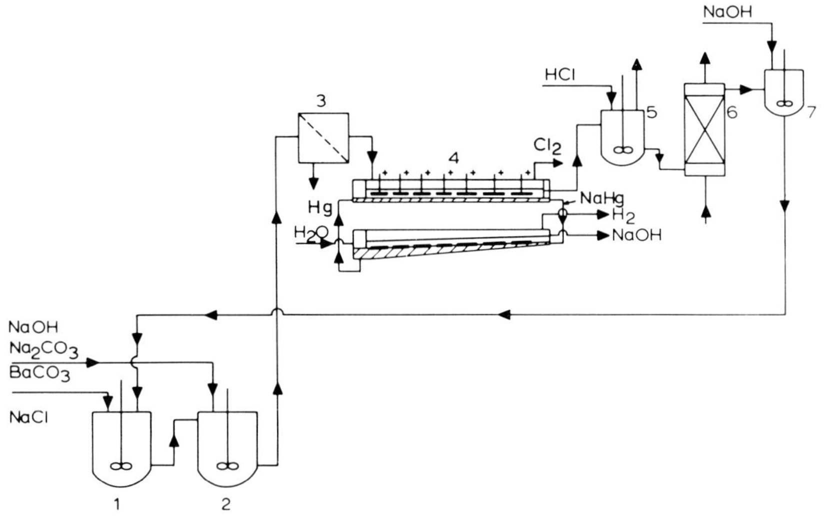

Figure IX.9. Mercury process

1. *dissolution*
2. *purification*
3. *filter*
4. *mercury cell*
5. *removal of chlorine*
6. *removal of last chlorine*
7. *recirculation of brine*

Sodium chloride is fed to the process as solid salt. The partly depleted brine from the electrolyzer is freed of chlorine by stripping with air after addition of some HCl to lower its pH, and then sent to a dissolution vessel where salt is added. Contaminants (Ca, Mg, heavy metal ions, sulfate) are removed by precipitation and filtration before the reinforced brine is returned to the process. Product chlorine is cooled, freed of entrained brine droplets, dried, filtered and liquefied by compression and cooling. Care should be taken to avoid the formation of explosive mixtures of chlorine and the small amounts of hydrogen formed as a by-product.

A disadvantage of the mercury cell process is that a small amount of mercury is lost with the process streams, gases and liquids. At first mercury losses were a significant factor in process economy, but in recent years the environmental aspects of these losses have become more important than economics. Hg and its compounds can now be removed so exhaustively from the effluents that losses are extremely small; it has become possible to meet very severe pollution limits, e.g. 0.1 - 0.3 g Hg/ton product as specified in the USA. Mercury removal can be accomplished by oxidation to Hg^{++} followed by adsorption on an ion exchanger containing SH-groups. Regeneration of the ion-exchanger is possible.

Membrane cells

Membrane cells (for principle see Figure IX.10.) are a relatively recent development. In this case an ion-selective membrane between cathode and anode prevents contact between Cl_2 and H_2, and also the passage of Cl^- ions from the anode to the cathode compartment. Only positive ions (H^+, Na^+) can pass through the membrane. Chlorine is produced at the anode and hydrogen along with NaCl-free caustic solution at the cathode. Brine is introduced into the anode chamber and withdrawn from it after partial conversion; water fed to the cathode compartment is drawn off as caustic soda solution. Thus, this type of cell has important advantages over each of the two other cells:

- like the mercury cell it produces chloride-free caustic without the need for the extensive and energy-intensive purification necessary in the diaphragm process.
- its cell potential and, consequently, the consumption of electrical energy, is about equal to that of the diaphragm cell and lower than in the mercury process. Moreover, membrane cells can be constructed as compactly as diaphragm cells.

Future Outlook

Uses of products. Caustic soda, soda ash and chlorine are used in the manufacture of a great variety of secondary products.

NaOH	is consumed in the detergent, paper, steel, petroleum and food industries and for making many chemicals
Na_2CO_3	has similar outlets as caustic soda and is used in large amounts by the glass industry
Chlorine	is converted to many organic and inorganic chemicals. The most important organics are the lower-molecular weight chlorinated aliphatics, in particular vinyl chloride. It also serves as disinfectant in water treatment.

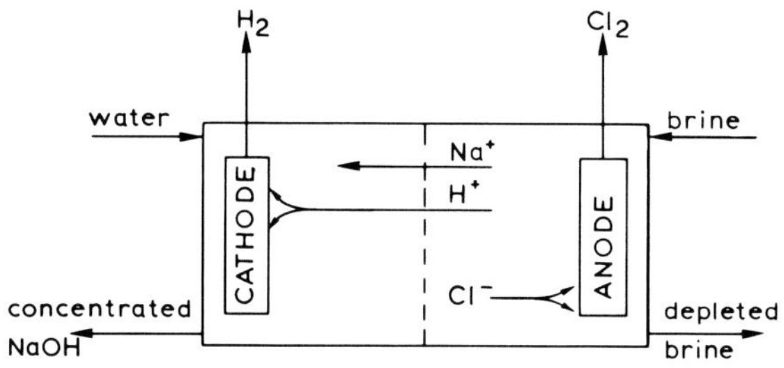

Figure IX. 10. Principle of membrane cell

Economics. In principle, the production costs of chlorine and caustic soda can be divided over the two products. However, since the markets for them are quite different, the question arises whether the production and demand ratios are similar. Changes in demand ratio should be assessed carefully if new investments in process facilities are contemplated. Before World War II the demand for caustic soda was much higher than that of Cl_2. For that reason, the non-electrolytic route to NaOH based on natural or synthetic sodium carbonate was applied at that time to balance product demand:

$$Na_2CO_3 \; + \; Ca(OH)_2 \; \rightarrow 2\ NaOH \; + \; CaCO_3$$

Since then the demand for chlorine has risen much more rapidly than for caustic soda, causing the lime-soda route to be abandoned. In some highly industrialized countries a surplus of caustic soda has even developed over the past decade. However, several possible developments make predictions of what may happen uncertain:

- it may become more common to produce additional chlorine by electrolysis of HCl solutions or by catalytic oxidation over a Cu-chloride on silica catalyst promoted with rare earth chlorides (the Shell chlorine process):

$$4\ HCl + O_2 \rightarrow 2\ Cl + 2\ H_2O \quad (\Delta H_s^o = -220\ kJ/mole\ HCl)$$

- by-product HCl produced in chlorinating organic compounds can be converted into chlorinated hydrocarbons by oxychlorination over copper-containing catalysts, e.g.:

$$C_6H_6 \; + \; HCl \; + \tfrac{1}{2}\ O_2 \; \rightarrow \; C_6H_5Cl + H_2O$$

$$C_2H_4 + 2\,HCl + \tfrac{1}{2}O_2 \rightarrow C_2H_4Cl_2 + H_2O$$

The latter reaction is one of the process steps applied in making vinyl chloride.

- in some applications caustic soda may replace soda ash
- some of the commercial outlets for chlorine are under disfavor because of environmental or health hazards: the use of a number or chlorine-containing insecticides is no longer permitted, less chlorine/bromine-containing scavengers are required because the content of TEL-type anti knock-agents in gasoline is being limited for reasons of pollution control, and the outlook for vinyl chloride-containing polymers is somewhat uncertain in view of reported carcinogenic properties of the monomer.
- processes in which much chlorine has traditionally been consumed (e.g. for propylene oxide, or epichlorohydrin) are being replaced by other methods in which Cl_2 is not used to the same extent
- some products made by processes in which much caustic soda is consumed, e.g. regenerated cellulose fibers, are under severe competitive pressure from other products (e.g. synthetic fibers).

Of the above factors, the outlook for PVC is probably the most important.

Process developments. It is expected that the market share of the membrane cell will grow considerably at the expense of the other two processes, mainly because of the lower energy requirements. The outlook for the mercury process relative to the diaphragm process is more difficult to establish: on the one hand, most of the environmental problems concerned with the use of mercury have now been solved, but on the other hand new materials for diaphragms are being developed by which the latter cell type could become even more competitive.

REFERENCES

General

1. Dahl, S., Chem.Eng., 82 (17), 60 (1975).

2. Sconce, J.S. 'Chlorine, its manufacture, properties and uses', Reinhold Publ. Co., New-York 1962.

Special

Figure IX.7. is published with permission of Hooker Chemical Corporation, Niagara Falls, N.J.

X. UNSATURATED HYDROCARBON INTERMEDIATES

Introduction

Ethylene*, propylene*, butadiene and, to a lesser extent isoprene and butenes, are all important building blocks for more complex products; nearly all plastics and rubbers produced on a large scale are derived from these intermediates. Although acetylene may also be used for organic syntheses its production has decreased for quite some time because it is relatively expensive and because storage and transportation under pressure are impossible in view of the explosion hazard: acetylene can decompose explosively into its elements below about 1000 $^{\circ}C$ since it is thermodynamically unstable. Besides, many products formerly based on acetylene can now be made from ethylene as well, although the corresponding reactions of acetylene are generally easier to carry out.

At first ethylene and propylene were recovered from coke oven gas, which contains 1-2 vol% ethylene and even less propylene. The increased demand for these olefins was a stimulus to develop processes based on petroleum. As a result, thermal cracking, i.e. pyrolysis at high temperature, short residence time and substantially atmospheric pressure is now widely used, ethane, propane, refinery gas, naphtha and light gas oil serving as feedstocks. Moreover, work has been done to produce C_2-C_4 olefins from even heavier oil fractions.

The reason that catalytic dehydrogenation cannot be applied follows from Figure X.1.: at equilibrium very low conversions are obtained except at very high temperatures. Then, however, C-C bonds are broken and a very complex product mixture is obtained, requiring costly separation. Also, coke may be deposited on the catalyst. Improvement of the equilibrium yield by lowering the partial pressure is possible but not very attractive: the equipment must be large, energy is needed to maintain the low pressure and leakage may occur. Alternatively,

* The systematic names ethene and propene are not often used in the industry.

an inert component may be added to keep the pressure atmospheric but products and inerts must then be separated after the reaction.

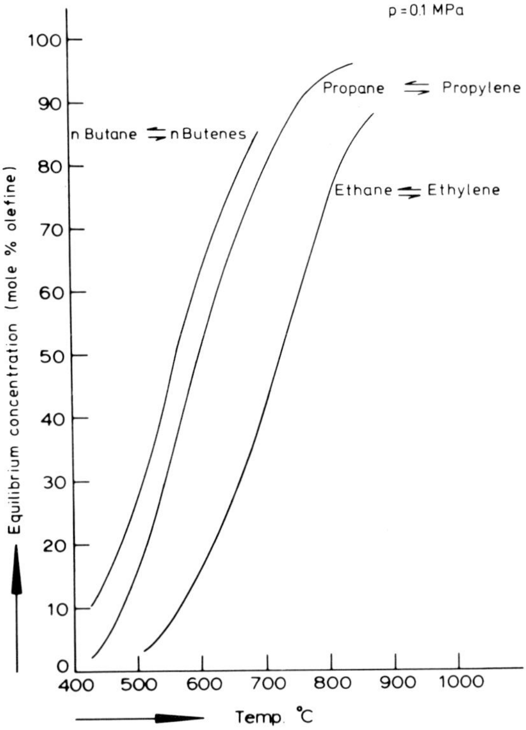

Figure X.1. Equilibrium position of the reaction paraffin ⇄ olefin + H_2

Mechanism and prccess conditions

The thermal cracking of hydrocarbons is an endothermic process with a complex reaction path based on a free radical mechanism with the usual phases: initiation, propagation and termination. The reactions may be represented schematically as outlined below.

Initiation $R_1R_2 \rightarrow R_1\cdot + R_2\cdot$ and $R_3H \rightarrow R_3\cdot + H\cdot$

he second reaction, abstraction of a hydrogen atom, becomes more important at higher temperatures. Initiation is followed by propagation, a type of reaction in which the number of radicals does not change. Possible reactions are:

Propagation	Hydrogen abstraction	$R_1\cdot + R_4H \rightarrow R_1H + R_4\cdot$
	Addition	$R_3\cdot + R_5H \rightarrow R_3R_5H\cdot$
	Isomerization	$R_4CCCC\cdot \rightarrow R_4CC\dot{C}C$
	Scission	$R_{n+m} \rightarrow R_n\cdot + \overset{=}{R}_m\cdot$

Addition reactions occur with olefins and aromatics. The fact that isomerization of radicals is generally very slow, combined with the relative frequency of scission reactions of a carbon-carbon bond in β-position with respect to the unpaired electron accounts for the high concentration of ethylene in the reaction products:

R-CCCCCC. → RCCC. + C=C or → RCC=C + CC. (β-scission)

The radical RCCC. may undergo further scission, whereas the ethyl radical decomposes according to

C-C. → C=C + H.

Obviously, a large amount of ethylene cannot be expected if the primary radicals would isomerize rapidly:

if e.g. R-CCCCC. → R-CCCĊC then the products of β-scission would be

R-CC. + CC=C and R-C=C + CĊC

The high proportions of ethylene in the reaction products clearly is a kinetic phenomenon because, in general, the stability of paraffinic radicals increases in the order primary < secondary < tertiary. However, a necessary condition is that the feedstock should be rich in n-paraffinic chains, the more so because the ease of abstraction of hydrogen atoms also increases in the order primary < secondary < tertiary. A further advantage of cracking feedstocks rich in paraffins is that aromatics generally show more addition reactions which, if they proceed much further, eventually lead to coke formation. Since higher boiling oil fractions generally are richer in aromatics, the tendency for coke formation rises with increasing average molecular weight.

The free radical chains are broken by

Termination e.g. R_2. + R_7. → R_2R_7 and R_8. + H. → R_8H

The importance of β-scission combined with the slowness of radical isomerization reactions can be illustrated by comparing the composition of the cracked products obtained from n-butane and isobutane (see Table X.1.). These hydrocarbons essentially react via

CCCC + R. → CCCC. + RH and CC(C)C + R. → CĊ(C)C + RH

The primary H-atoms are abstracted less easily than the secondary or tertiary H's. Of the resulting radicals, only the primary butyl radicals can undergo β-scission to ethylene + an ethyl radical. Propylene and a methyl radical or methane and a propyl radical are formed from the secondary butyl radical. This type of scission is impossible with the isobutene radical. Ethane cracking is often called thermal dehydrogenation because a high percentage of ethylene is formed. In this case initiation starts by CC-bond cleavage:

Table X.1. Butane cracking (650 °C, 0.1 MPa)

Composition (mole %)	by pyrolysis of	
	n-C_4	i-C_4
H_2	3	35
CH_4	34	14
C_2H_4	15	1
C_2H_6	14	1
C_3H_6	34	14
i-C_4H_8	0	35
Activation energy (kJ/mole)	310	277

$CH_3CH_3 \rightarrow 2CH_3\cdot$

followed by the propagation steps

$CH_3CH_3 + CH_3\cdot \rightarrow CH_4 + CH_3CH_2\cdot$

$CH_3CH_2\cdot \rightarrow CH_2{=}CH_2 + H\cdot$

$CH_3CH_3 + H \rightarrow CH_3CH_2\cdot + H_2$

The chain terminates by combination of two radicals. Because the initiation is rather difficult and the activation energy of ethylene formation high as compared with other reaction steps the cracking of ethane must be carried out at a somewhat higher temperature than is the case with naphtha.

From the above mechanism it can be understood that the following process conditions are favorable:

a. very short residence times at high temperature to avoid consecutive reactions such as additions. Older types of reactor have residence times of 1-2 s, but in modern plants 0.05-0.2 s is used. A higher temperature must then be applied if the same conversion level is to be attained;
b. very fast cooling after reaction to prevent consecutive reactions;
c. the highest possible temperatures to obtain high conversions; successive β-scissions then lead to maximal ethylene production;
d. low hydrocarbon concentration, hence gas-phase cracking with steam dilution at total pressures near 0.1 MPa, again to avoid addition reactions which would ultimately lead to coke. Since heavier hydrocarbons start to decompose at lower temperatures they would give rise to too high a radical concentration.

(and, therefore, more addition and recombination reactions) if the temperature is not reduced somewhat relative to that needed for ethane and propane, and if insufficient steam is added.

The most crucial problem is how to supply the heat required for the rapid rise in temperature and the endothermic reactions at the very short residence times encountered here. A further difficulty is that the heat content of the product stream must be utilized, e.g. by raising steam, if adequate heat economy is to be obtained. The treating, separation and purification of product streams also require careful design to achieve economic production.

Equipment

Externally heated tubular reactors are applied in nearly all modern ethylene plants. A block diagram is shown in Figure X.2. and a simplified flowchart in Figure X.3. The feed is preheated in the convection section of the cracking furnace, mixed with superheated steam and fed to the tubular reactor at 600-650 $^{\circ}C$. The temperature is raised to about 850 $^{\circ}C$ and the product gases formed are quenched in a high-performance cooler to 400 $^{\circ}C$. The term 'quenching' is often used for very rapid cooling, generally by direct injection of a cooling liquid but, in this case, it also applies to cooling by heat exchange over a considerable temperature range in a very short time. After the quench cooler an oil quench is applied by injection of cold by-product oil obtained as the residue in distilling the products. This ensures uniform cooling in a short time with minimum pressure drop. The oil quench is followed by a water quench in which a condensate rich in benzene and its homologs is obtained. The above sequence applies to naphtha and gas oil cracking; if ethane is used as feed the oil quench can be omitted.

Plant design is further characterized by a low pressure drop (0.1 MPa max.) and maximum heat recovery. To that end, hot oil from the quenching section is heat-exchanged with fresh feed and boiler feed water and used to raise low-pressure steam.

In spite of the many precautions, coke deposits in the tubes are formed after some time and the heat transfer diminishes. The tubes are cleaned periodically (every two or three months) with a mixture of steam and air at 750-800 $^{\circ}C$. Deposition of coke is faster under more severe operating conditions, with higher-boiling feedstocks and increasing roughness of the tube walls.

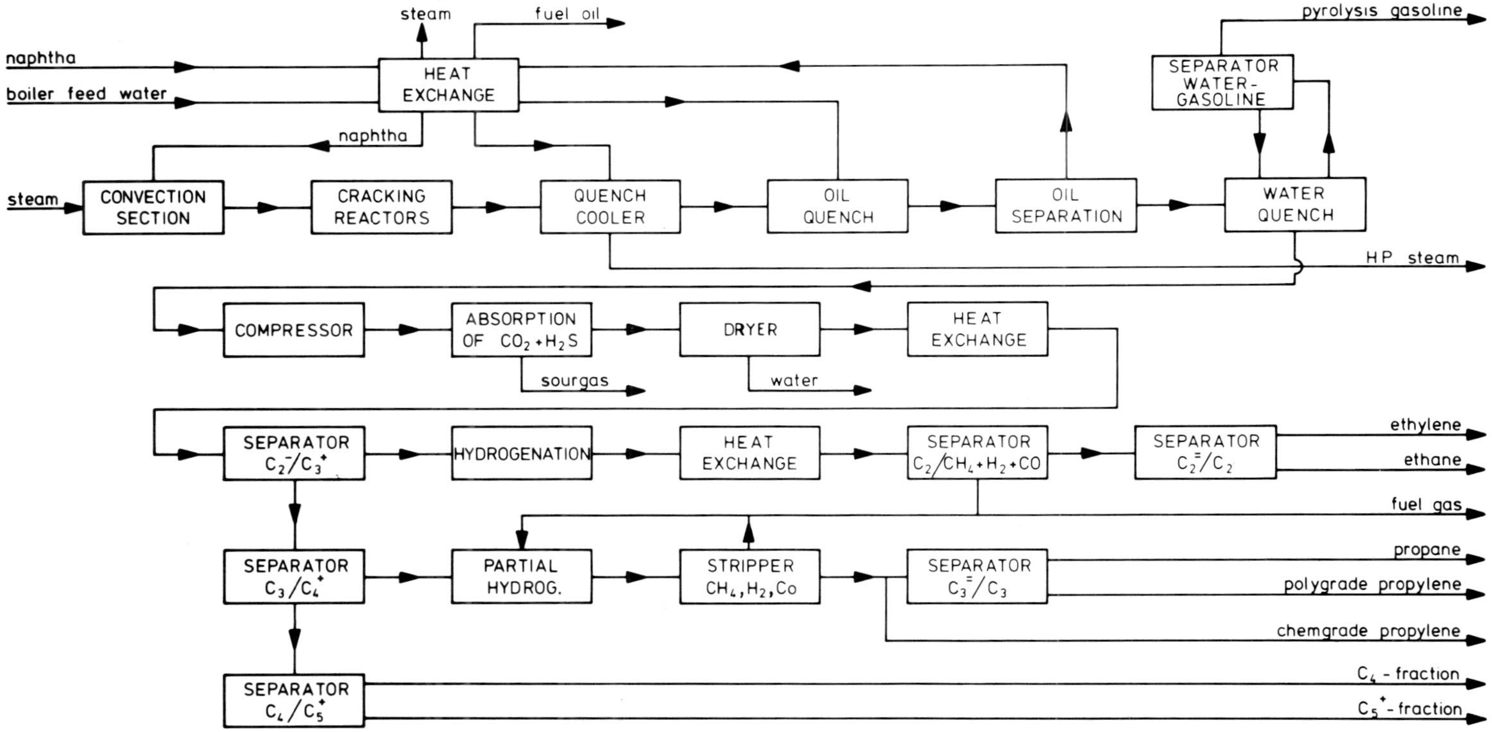

Figure X.2. Olefins by steam cracking of naphtha.

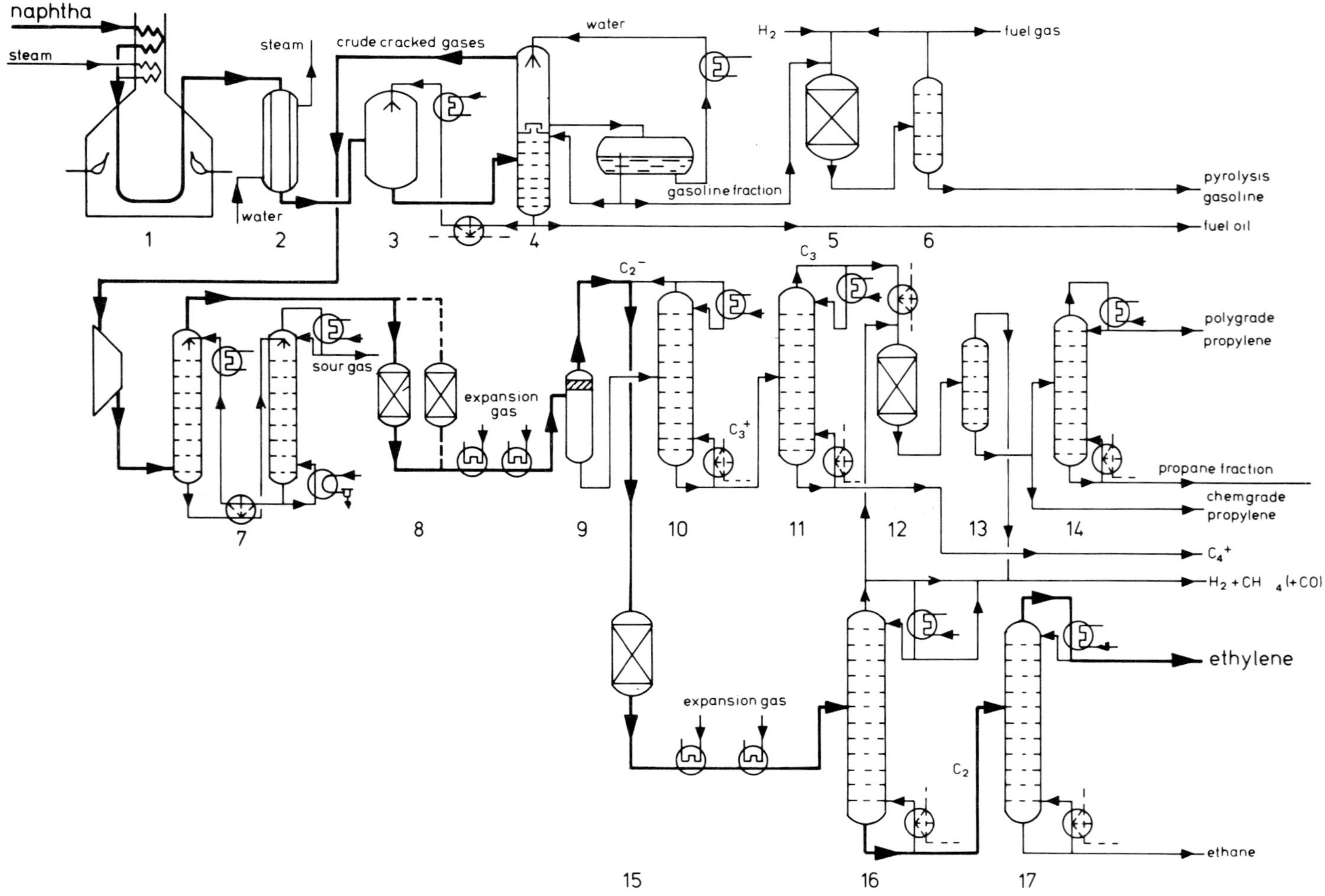

Figure X.3. Production of olefins by steam cracking of naphtha

Another common precaution against rapid coking is the use of a feed containing 25-50 ppm sulfur in the form of organic sulfur compounds. These deactivate the tube wall by decreasing the catalytic effect of nickel in the deposition of carbon.

Recent developments

All the improvements introduced in recent years have been aimed at maximum olefin production. Because the need for ethylene is mostly greater than for propylene this also implied a shift towards higher C_2/C_3 ratios. The modifications consisted of increasing the reactor outlet temperature and decreasing the residence time or, in other words, increasing the severity of cracking. The favorable effect of raising the temperature is shown in Table X.2.

Table X.2. Effect of cracking temperature on product composition (naphtha with 70% alkanes, τ = 0.5 sec., constant space veolcity)

T_{exit} in ^{o}C	Ethylene (wt %)	Propylene (wt %)
790	21	16
815	23	16
825	25	16
840	27	15
860	28	14
885	30	abt 10

To reach high reactor outlet temperatures the quality of the cracking tubes and the furnace construction had to be improved. Reactor tubes made of a Cr-Ni alloy (25-20) by centrifugal casting, a technique which is also used in making steam reformer tubes, are now generally applied.

More advanced materials are also being introduced. e.g. Cr-Ni 50/50 with 1½%w Nb, which permit skin temperatures above 1050 ^{o}C. A complication of such high temperatures is the increased formation of acetylene and its homologs from 900 ^{o}C upwards. The acetylenes are usually converted to the corresponding olefins or removed by washing with a selective solvent and used as fuel. It may become of interest to recover acetylene as a by-product.

In early plants cracking tubes of more than 100 m length were installed, but modern practice demands U-shaped tubes of 40 m length with a diameter of 10 cm and a wall thickness of 1 cm. Per tube about 10 000 tons ethylene are produced anually; for productions of 300 000 tons per year, a common capacity for new units, groups of 4-8 tubes are placed in parallel in each oven. Every oven is equipped with its own quench cooler. In this case the capacity is enlarged by multiplication of production units rather than scale-up even though the saving on capital is, of course, much less. This does not apply to the separation train, which is a single unit for the entire plant.

Alternative cracking methods

In the past, the difficulties of fast heat transfer, short residence times and carbon deposition have been solved in other ways than those described above. A more recent incentive for process innovations stems from the high and still rising cost of the preferred feeds naphtha, ethane and LPG. Some of the alternatives are briefly mentioned below.

1. Direct heat transfer can be achieved by cracking with the aid of superheated steam (900 ^{o}C) or by partial oxidation of the feed. Disadvantages are that separation between cracking products and steam or oxidation products is needed.
2. It is possible to apply recuperative ovens in a cyclic process. The cycle consists of cracking - inerts flush - oxidation of carbon formed with preheated air - inerts flush, etc. The carbon formed during the cracking phase is deposited on the ceramic packing of the reactor, which also acts as a heat sink by taking up the heat produced during the subsequent oxidation phase. Disadvantages of this type of process are the frequent valve switches at high temperatures, the decreasing degree of conversion (severity) during the cracking periods and the fluctuating composition of the reactor product gases which complicates the separation section.
3. An improvement relative to the cyclic process is the use of a moving bed. In such processes solids like coke particles, sand or pebbles are heated by carbon burn-off in a regenerator and cracking in a separate reactor. The solid is circulated from reactor to regenerator and back. This mode of operation in fact utilizes the carbon formed as the source of heat needed for cracking. Nevertheless, capital costs are relatively high, maintenance is expensive because of erosion by hot solids, solid/gas separations are needed and some fines are formed by solids attrition.
4. Fluid beds may also be used; in this case, there is less difficulty in transporting the solid and the rate of heat and mass transfer is very much

higher than in the moving bed systems mentioned above. Small particles of sand and/or coke are kept in a fluid-like state by the upflowing gas resulting in very uniform temperature in the reactor. Gas/solid separation is by cyclones. There are two ways to carry out this process:

- cracking and heat generation can be carried out in a single fluidized bed if carbon formed is burnt just above the distributor plate, generating hot gases and heating the coke to a sufficiently high temperature for the cracking reactions to proceed in the top section of the bed. A disadvantage is that oxidation products and cracking products must be separated. This may be facilitated by using oxygen as the regeneration gas but at increased cost;
- cracking and regeneration (heating) may be carried out in separate fluid beds, the solid being circulated from reactor to regenerator and back. In this case there is no problem of separating CO and CO_2 from cracked products and air can be used as the oxidant.

Both these fluidized bed processes have been developed by BASF up to and including the pilot plant stage; moreover, there has also been development activity in Japan.

Feedstocks

One of the most significant economic factors in thermal cracking for olefins production is the type of feedstock used. Depending on the location of the cracker and the prices of the various raw materials, the most attractive feedstock may be light hydrocarbons, i.e. the C_2-C_4 fraction, often in the form of refinery gases, naphtha or gas oil. Feedstock choice has a very significant effect on process conditions and product composition, as follows from Table X.3. In some locations, notably the USA and more recently the UK, ethane is widely used. Since this preferred feed is unavailable in Continental Western Europe, naphtha and gas oil are used in that area. In the latter case much by-product butadiene and to a certain degree also benzene, toluene and xylenes is obtained. As a result, butadiene is available in sufficient amounts to make other processes for its manufacture, e.g. by (oxidative) dehydrogenation of n-butane or n-butene, superfluous - unlike the situation in the USA.

However, the picture for the future is not quite clear: on the one hand ethane from natural gas is tending towards shortage of supply, whereas the demand for naphtha in making hydrogen or high-octane gasoline components is rising, resulting in price increases for naphtha. This is the reason why new large-scale crackers

Table X.3. Product distribution by steam cracking

Feed	Ethane %w, C_2	Propane %w, C_3	n-butane %w, C_4	Naphtha 76% w alkanes 4% aromatics 20% naphthenes	Gasoil boiling range 230-340 °C
T_{exit}, °C	830	840	820	820	820
steam/feed w/w	0.25	0.40	0.33	0.60	0.90
Product %w					
CH_4	10.9	22.4	24.7	14.6	12.5
C_2H_2	0.3	0.1	0.4	0.3	0.3
C_2H_4	46.0	30.3	29.7	26.1	19.0
C_2H_6**	38.3	3.8	7.4	4.0	3.5
C_3H_4+C_3H_6	1.9	20.5	17.4	16.5	14.3
C_3H_8	0.3	19.4	0.5	0.5	0.5
C_4		3.6	8.7	12.4	12.3
C_5-200 °C	2.3*	-	11.2	25.6	21.5
residue	-	-	-	-	16.1

* C_4 inclusive

** C_2H_6 is often recirculated, increasing the C_2H_4 output to 70%

are designed to run on naphtha as well as gas oil, the lower yield from gas oil being compensated for by the lower cost of the feed. The 10% higher investment for multi-feed plants must be earned back by savings on feedstock and higher prices for by-products. The tendency to build ethylene plants that can run on various heavier feeds is now spreading to the USA, an additional reason being the rising demand for propylene.

Separation of products

The main method used for separation of the cracked products is distillation at low temperature and high pressure. For some components (e.g. acetylenes) absorption in a selective solvent may be applied. The presence of impurities like water, CO_2, H_2S, COS and propadiene must be taken into account; the same applies to conjugated dienes which make it necessary to keep the temperature below 120 °C to avoid unwanted polymerization reactions which would cause fouling of the equipment.

The crude product mixture freed from pyrolysis gasoline components is compressed and then scrubbed in a regenerative absorption process, e.g. using alkanolamines, to remove H_2S and CO_2 (see Chapter VI.). The hydrocarbon gases are then dried exhaustively to prevent ice formation in the low-temperature section; this can be done by drying the gases between compression steps by means of a glycol wash followed by a final drying step with molecular sieves. Alternatively, molecular sieves may be used as the only drying agent. The low-temperature distillation can be carried out in various ways. The flowchart of Figure X.3. begins with a separation of fractions boiling higher than ethane and ethylene. The top product of this separation also contains acetylene; although it is, in principle, possible to recover this compound it is usually hydrogenated selectively to ethylene over a Pd catalyst at about 100 ^{o}C. The light products are then cooled further to separate the C_2-fraction from hydrogen and methane; the latter are partly used to hydrogenate higher acetylenes and propadiene. The C_2-fraction is separated in a third column into ethylene and ethane, which is often recycled to the cracking process and preferably cracked in separate reactors because of the higher temperature needed to obtain high yields of ethylene from it.
The C_3^+-fraction is separated by distillation in a C_3-cut and heavier hydrocarbons. After partial hydrogenation of the C_3-stream a propylene concentrate is produced which is often directly used for chemical conversions ('chemgrade propylene'), but in some applications, such as for making polypropylene a higher purity is necessary and further distillation to give pure propylene is needed.
The C_4^+-fraction is distilled into a butadiene concentrate and a butenes fraction; the former is the feed to an extractive distillation for the production of high-grade butadiene. From the C_5-fraction isoprene and cyclopentadiene can be isolated.

The pyrolysis gasoline contains a high percentage of aromatics, in particular benzene, if naphtha is used as feedstock. The aromatics are separated by liquid/liquid extraction after hydrogenation of dienes, in particular conjugated dienes and acetylene homologs, to the corresponding mono-olefins over NiS on a carrier, a selective catalyst in this application. The raffinate from the extraction then is sufficiently stable to be added to motor gasoline. If this raffinate is to be recycled to the thermal cracker all olefins must be hydrogenated because otherwise enhanced coke formation would occur.

The fuel oil fraction recovered from the thermally cracked materials is not suitable for storage for long periods of time because reactive hydrocarbons (conjugated dienes, styrenes and the like) are easily oxidized to give insoluble resins, even at room temperature.

An impression of the product spectrum obtained by naphtha cracking is shown in Table X.4. If necessary process conditions can be adapted to make a product package corresponding to the continually changing market demands.

Table X.4. By-products of naphtha cracking

Per 1000 tons of ethylene the following products can be recovered:

	Severity	
	low naphtha 40-110 °C	high naphtha 50-170 °C
Propylene	640	540
Butadiene	160	180
Benzene	190	270
Toluene	80	160

Product uses

The main outlets for ethylene are, in decreasing order of importance: high-density polyethylene, vinyl chloride, low-density polyethylene, ethylene oxide, styrene, acetaldehyde and ethanol.
The most important propylene uses include: polypropylene, solvents (acetone, isopropanol), diacetone alcohol and its derivatives, acrylic acid and derivatives, methacrylic acid and derivatives and, particularly in the USA, high-octane motor gasoline components made by alkylation (see Chapter XIII.).

Butadiene manufacture

Recovery from cracked products by extractive distillation

The butadiene concentrate obtained in the separation section of an ethylene plant contains on an average 40%w butadiene, the balance being butanes, alkynes and allenes. Separation by distillation is much more difficult than for the C_3 fraction. A first problem is that selective hydrogenation of alkynes is not possible because butadiene would be hydrogenated as well. Moreover, the boiling points of the C_4-hydrocarbons are very close together, as is shown in Table X.5. This problem can be solved by applying extractive distillation with a solvent causing sufficiently large shifts in the relative volatilities of the various hydrocarbons, such as furfural, N-methylpyrrolidone and acetonitrile. Relative volatility data for furfural containing 4% water are shown in Table X.5. The relative volatility is defined here as follows:

$$\alpha_R = \frac{p_{component}}{p_{butadiene}}$$

and is given in the Table at infinite dilution. Apparently, butadiene dissolves better than the mono-olefins mentioned in Table X.5. and much better than the butanes. In practical processes, however, high solute concentrations are an advantage and it is necessary to know the effect of concentration on the relative volatilities.

Table X.5. Relative volatilities V_R of C_4-hydrocarbons

	Boiling point at 0.1 MPa (^{o}C)	V_R*	V_R* in furfural + 4% H_2O
Isobutane	- 11.7	1.209	2.600
Isobutene	- 6.9	1.070	1.666
n-Butene-1	- 6.3	1.040	1.718
1-3 Butadiene	- 4.4	1.000	1.000
n-Butene-2	- 0.5	0.871	2.020
tr-Butene-2	+ 0.9	0.843	1.190
cis-Butene-2	+ 3.7	0.776	1.065

* $p = 0.455$ MPa; $T = 40\ ^{o}C$; $C_{liq} \rightarrow 0$

Figure X.4. gives an indication of the dependency of α_R on the concentration of hydrocarbons in the liquid phase.

A simplified flow diagram of an extractive distillation unit is shown in Figure X.5. The feed is introduced into a distillation column through which the selective solvent flows downward. The preferentially dissolving component leaves the column with the solvent at the bottom and is fed to the second column to separate solvent and solute; the crude extract is distilled further to bring it to specification. The solvent is returned to the extraction column. The top product of this column may be further separated, e.g. to remove traces of solvent. The efficiency of this type of separation depends not only on shifts in α_R but also on reflux ratio and pressure.

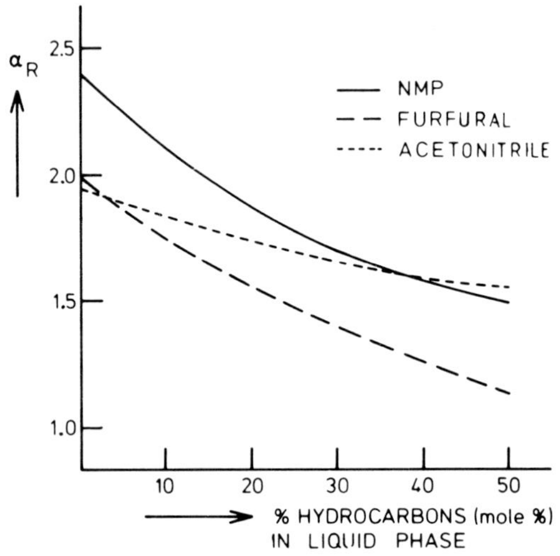

Figure X.4. Relative volatility of n-butene-1 with respect to butadiene as a function of the liquid concentration

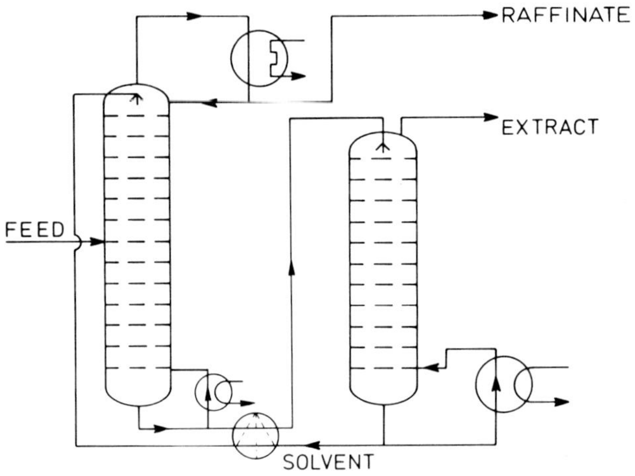

Figure X.5. Principle of extractive distillation

With the low boiling C_4-hydrocarbons, energy and cooling water consumption depend on the operating pressure. Higher pressures, though requiring some compression energy and leading to a somewhat poorer separation, save on operating costs because cooling water of ambient temperature can be used. Otherwise, coolants like propane or freon must be used to condense the top streams. An example of a process working at higher pressure is given in Figure X.6.; this BASF process is suitable for making highly pure butadiene such as is needed for stereospecific polymerizations (see Chapter XII .). In this case N-methylpyrrolidone is used as solvent; other possible extraction agents are listed in Table X.6.

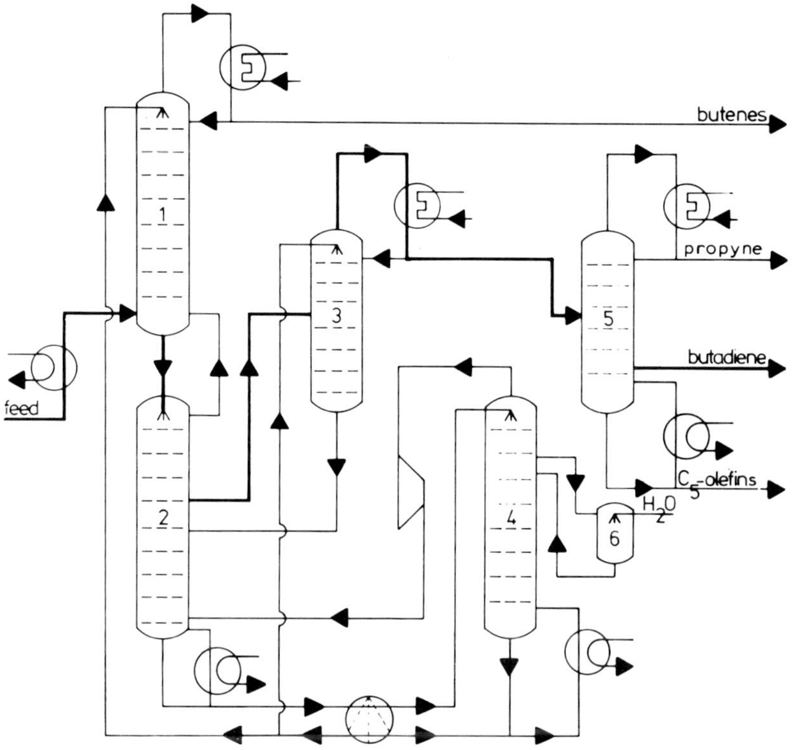

Figure X.6. Butadiene recovery by the NMP-process

1. *absorber*
2. *stripper*
3. *scrubber*
4. *stripper*
5. *purifier*
6. *scrubber*

In the process of Figure X.6. the feed enters the extractor 1; use of reflux leads to a low butadiene concentration in the overhead fraction, the raffinate. In stripper 2 small amounts of butenes are stripped off and returned to the

Table X.6. Selective solvents in butadiene extraction

N-Me pyrrolidone	Dimethyl formamide	Furfural	Dimethyl acetamide	Aceto-nitrile
(B.A.S.F.)	(Jap. Geon.)	(Phillips)	(Union Carbide)	(Shell)
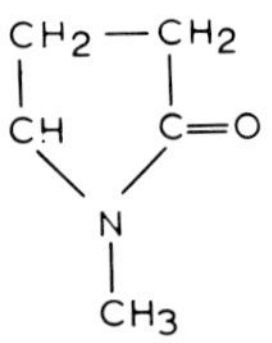	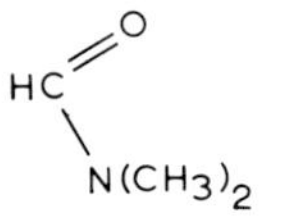	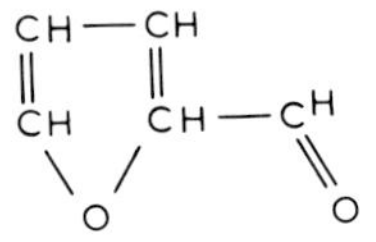	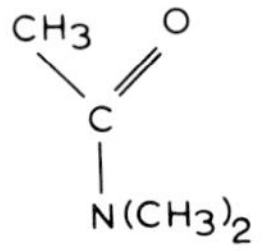	CH_3CN

extractor. From a draw-off in the middle of stripper column 2 a stream is obtained containing butadiene, C_5 hydrocarbons, acetylenes and allenes, which is further treated in scrubber 3 to give crude butadiene as the top product; it is purified further in column 5 and stored under an inert gas blanket after addition of an inhibitor to prevent polymerization. The bottom stream of the scrubber contains much solvent and is returned to stripper 2. The bottoms of stripper 2 are expanded and sent to column 4 where solvent and hydrocarbons are separated. The purified NMP is recycled. The hydrocarbons over the top still contain butadiene and are recompressed and fed back to stripper 2.

The above process has the advantage that reflux flows in the main columns can be obtained by partial condensation of top streams with cooling water. In order for the entire process to operate at 0.1 MPa, reflux streams would have to be condensed with a cold cooling medium requiring much energy. Depending on feed composition and the required purity (related to the intended use) other configurations may be preferred. Integral considerations, including investment, and utility costs (electricity, steam, cooling water) must finally lead to the proper choice.

Alternative routes to butadiene

Although butadiene is recovered in large quantities from the products of naphtha cracking at several locations, the dehydrogenation of butenes to butadiene is also applied, e.g. in the U.S.A. Because the production of isoprene runs along the same lines, and the production of butadiene illustrates several technological principles, the subject is treated here.

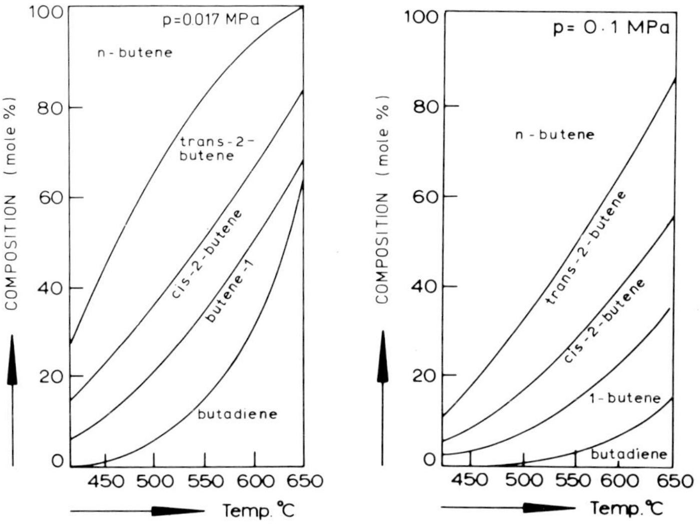

Figure X.7. C_4 dehydrogenation reaction equilibrium composition at two different pressures

Just as is the case for mono-olefins, the direct dehydrogenation equilibrium is unfavorable. Figure X.7. illustrates that only at high temperature and a pressure below atmospheric the degree of conversion is reasonably high. For isoprene the situation is analogous. Using a monoolefin feedstock does not result in significant improvement, since paraffins will also be among the reaction products and, eventually, a mixed feed containing these must be converted anyway. Nevertheless, the direct dehydrogenation of butanes and butenes is practiced industrially for the production of butadiene.

Originally butadiene was synthesized from acetylene in Germany. A first route was based on the aldol condensation of acetaldehyde followed by dehydration, a second method on a Reppe synthesis via production of butynediol from C_2H_2 and formaldehyde, also followed by dehydration. These processes are now too expensive as compared with direct or oxidative dehydrogenation.

The Houdry process makes use of a catalyst consisting of 18-20% Cr_2O_3 on Al_2O_3. The feed, normally a mixture of butanes and butenes, is converted in fixed beds of catalyst at 550-650 ^oC and subatmospheric pressure. At these conditions cracking is accompanied by the formation of coke, resulting in deactivation of the catalyst. The deposited coke is utilized to furnish the necessary heat of reaction:

$$CH_3-CH = CH-CH_3 \rightarrow CH_2 = CH-CH = CH_2 + H_2$$

$$\Delta H^o_s = 90 \text{ kJ/mole}$$

To this end a cyclic process is used: after a cracking period the catalyst is purged with steam to remove hydrocarbons and hydrogen and then regenerated by air. This raises the catalyst temperature to the level where dehydrogenation occurs. Figure X.8. presents a simplified flowchart and Table X.7. gives some process conditions.

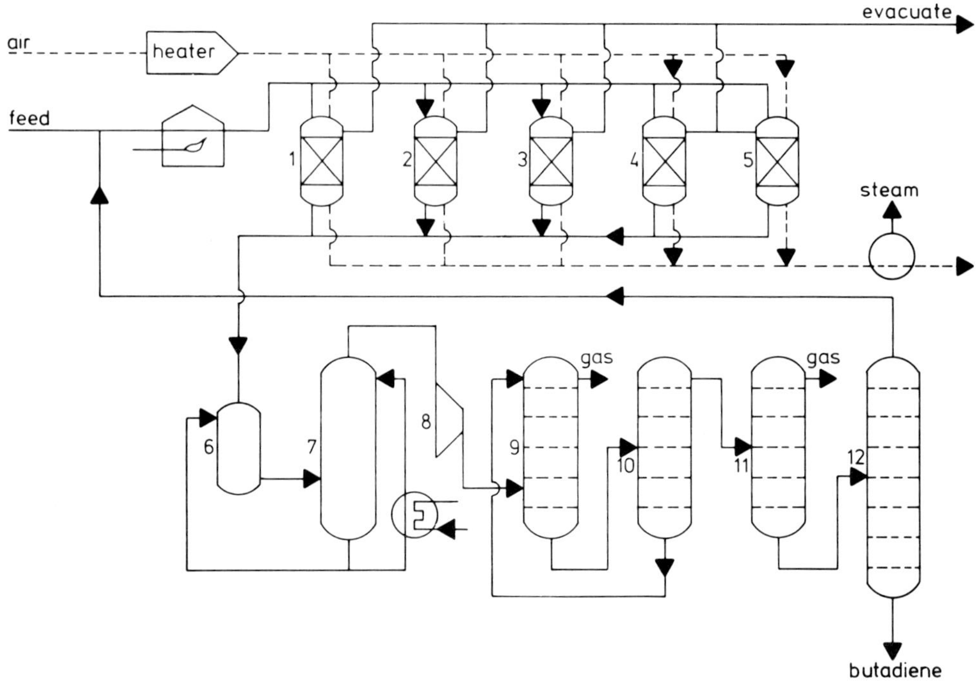

Figure X.8. Production of butadiene by the Houdry process

1-5. reactors

6, 7. quenchers

8. compressor

9. absorber

10. stripper

11. stabilizer

12. extractor

The process has a number of disadvantages:

- the cyclic process requires frequent switching of process streams, hence costly control;

Table X.7. Production of butadiene by the Houdry Process at conditions of maximum butadiene yield (in %w of fresh feed)

Components	Fresh feed	Fuel gas	Products and losses Butadiene concentrate	Losses
Hydrocarbons C_3 and hydrogen		30.3		
Isobutane	5.1			
Isobutene		0.5		
n-Butenes		0.5	1.0	
n-Butane	98.5	1.0		
1-3 Butadiene		0.5	56.7	
C_5^+, CO_x, H_2O		1.5		8.0
Totals	100.0	34.3	57.7	8.0

- the composition of the product gases fluctuates with time;
- the valuable feed is used to generate energy;
- the fluctuating temperatures make high demands on materials of construction and catalyst;
- the coke deposits lower catalytic effectiveness;
- application of low pressures results in bulky apparatus and the danger of leaking air inward;
- sub-atmospheric pressure requires energy.

The Houdry process may be improved upon by using a Fe_2O_3/Al_2O_3 catalyst to which some K_2O is added to reduce the number of overactive cracking sites on the surface. Other improvements are dilution of the hydrocarbon with superheated steam and keeping the degree of conversion per pass low (ξ = 0.20-0.25). In this way the cracking period can be extended to 72 h (fresh catalyst) to 24 h (used catalyst) with a selectivity around 75-80%. Regeneration is still necessary but steam is sufficient and air is superfluous. The frequency of the regeneration increases in time because some coke is deposited which is hard to convert. Once every few months complete regeneration with steam *and* air is necessary. It is interesting to note that steam in the feed has three functions:
- supply of heat, hence an adiabatic reactor can be applied;
- lowering the partial pressure of hydrocarbon;
- keeping the catalyst surface clean and regenerating it.

The oxidative dehydrogenation reaction makes it possible to reach a higher degree of conversion because oxygen shifts the equilibrium to the right hand side:

$$CH_3\text{-}CH = CH\text{-}CH_3 + O_2 \rightleftarrows CH_2 = CH\text{-}CH = CH_2 + H_2O$$

The reaction is now exothermic resulting in a temperature rise in the reactor of about 50 °C with an exit temperature of 550-625 °C. It is necessary to use a highly selective catalyst and to dilute the feed with steam. A suitable catalyst is of the bifunctional type, e.g. Sn-Sb-oxide or Fe-Sb-oxide. In a process developed by Phillips Petroleum Company a selectivity of more than 90% at 70% conversion of the butene feed is claimed. This, of course, facilitates separation of reaction products. It is probable that some unconverted oxygen leaves the reactor and that safety precautions should be taken.

For some time processes were being developed in which an auxiliary agent other than oxygen removes the equilibrium limitation. An example is the use of iodine:

$$2\ C_4H_8 + 2\ I_2 \rightleftarrows 2\ C_4H_6 + 4\ HI$$

$$4\ HI + O_2 \rightleftarrows 2\ I_2 + 2\ H_2O$$

$$\overline{2\ C_4H_8 + O_2 \rightleftarrows 2\ C_4H_6 + 2\ H_2O}$$

The first equilibrium shifts to the right hand side if HI is removed by reaction with another agent or by transferring it to another phase. Another auxiliary material which has been tried is HCl. All these processes (Shell, duPont, Monsanto etc.) have not been viable because of one large common problem, viz: the serious corrosion by halogens and halogen compounds, although these may be overcome by using special construction materials (e.g. titanium) which also resists temperatures of 400 °C and higher. Another problem with the use of iodine is that losses are not permitted because of the costs. From the capital point of view costs, already relatively high, are raised by the use of iodine.

Synthesis of isoprene

Isoprene can be recovered from the C_5-fraction obtained during cracking of naphtha. Nippon Geon Co. employs for this purpose a combination of distillation and extractive distillation. Other components in the C_5-fraction may be dehydrogenated on a $Fe_2O_3/Cr_3O_3/K_2O$ catalyst as in the case of the production of butadiene (Shell). In a number of other processes isoprene is synthesized from smaller molecules.

Goodyear and Scientific Design have developed a process based on propylene. The first step is a dimerization with small amounts of tripropylaluminum as a catalyst:

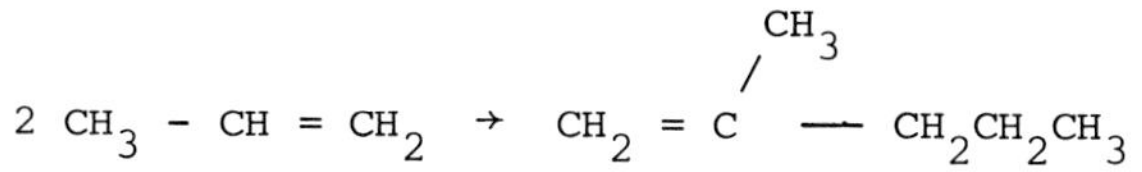

the second one is a nearly quantitative isomerization:

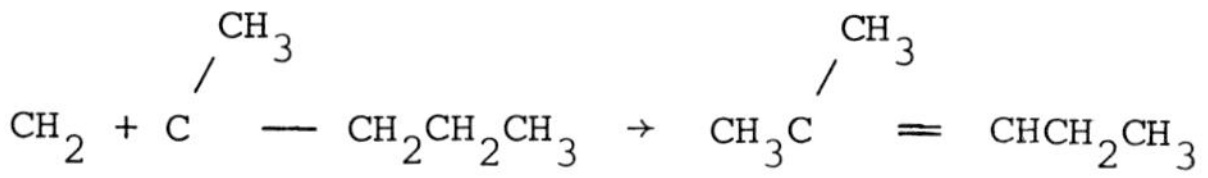

Finally the methylhexene is cracked at 750 °C (!) resulting in isoprene and methane, HBr acting as initiator:

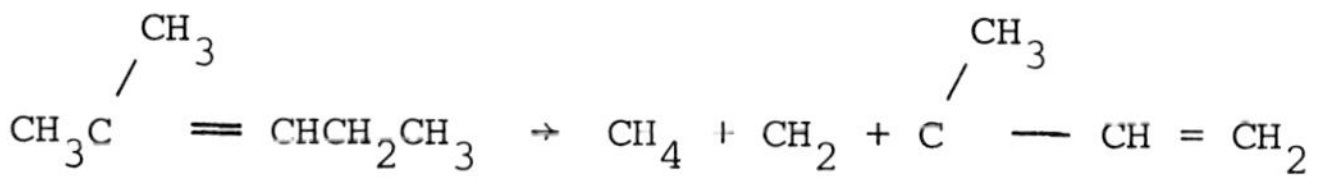

Another example of a technical process (U.S.S.R.) is one based on the Prins reaction between isobutene and formaldehyde:

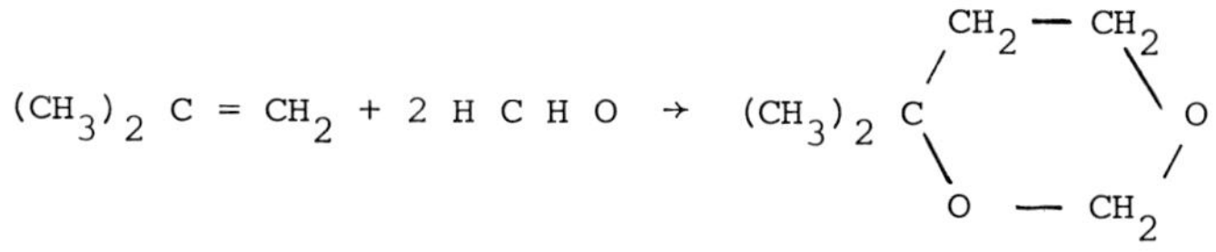

(dimethyldioxane)

This reaction can be carried out in 1.5% sulfuric acid with a yield of 80%. The dioxane is decomposed at 350 °C over a calcium phosphate catalyst:

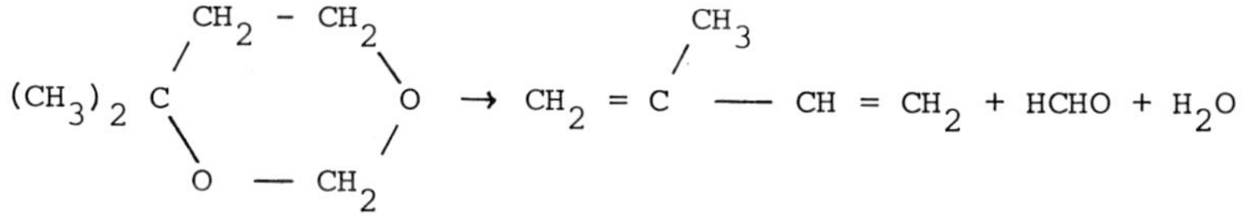

On the basis of acetylene ANIC produces isoprene according to the following reaction with acetone:

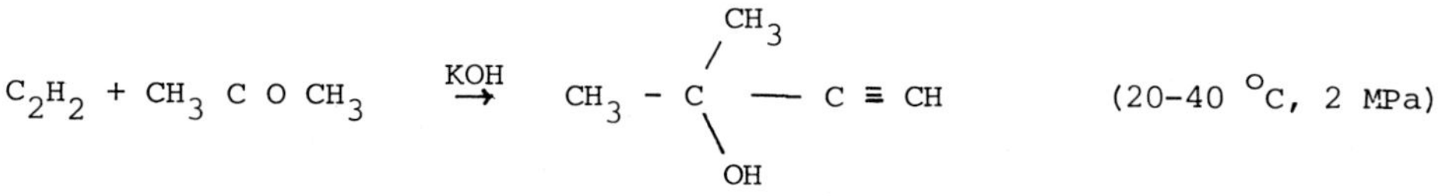

This liquid phase reaction must be carried out with excess acetylene to suppress consecutive reactions. In the second step a selective reduction with hydrogen over Pd results in the hydroxyalkene. Finally this compound is dehydrated at 250-300 °C and atmospheric pressure:

$$CH_3-\overset{\overset{CH_3}{/}}{C}-CH=CH_2 \xrightarrow{Al_2O_3} CH_2=\overset{\overset{CH_3}{/}}{C}-CH=CH_2 + H_2O$$

Distillation is a purification step necessary to upgrade the isoprene to 99.9% to make it suitable for polymerization to cis-polyisoprene, a synthetic rubber with properties which resemble those of natural rubber

Acetylene manufacture

About 1925 an extensive chemical industry, based on coal came into development. In this acetylene played a major role as an intermediate. It was the starting reactant to produce acetaldehyde, acetic acid and acetone. Later it was used for the production of butadiene to manufacture synthetic rubbers, vinyl chloride (one of the first thermoplastics) and chlorinated solvents. This development came to a standstill after the Second World War because of the increasing shift to oil as a chemical feedstock. Ethylene appeared to be cheaper and easier to transport. A number of products classically produced from acetylene could also be made from ethylene by new processes.

Acetylene suffers from a number of disadvantages compared to ethylene:

1. The formation of acetylene from its elements requires much energy. The standard heat of formation is 227 $kJ.mole^{-1}$.
2. Acetylene is stable towards H_2 and C only at temperatures in excess of 1200 ^oC. Below this temperature there is a tendency to decompose to C and H_2. Because of the large heat of reaction this can easily turn into an explosion.
3. As a result of 2, not only very high temperatures are required but it is also necessary to cool the reaction products immediately (quenching) to freeze the equilibrium. This means at the same time dissipation of expensive heat at a high temperature level to a low level.
4. Because of 2, it is impossible to compress acetylene, hence storage in liquid form and transport by pipeline is impossible. Acetylene must be stored as a solution in acetone in cylinders filled with a porous material (e.g. kieselguhr). This makes a large buffer capacity unattractive and necessitates a simultaneous production of acetylene and the chemicals derived from it.
5. For the same reason hot surfaces must be avoided. Separation by distillation is not possible and acetylene must be purified by absorption in a selective solvent.
6. The advantage that acetylene contains a triple bond can only be exploited in a few cases, e.g. when producing certain vinyl compounds.

Some processes for the production of acetylene that have been realized technically are briefly described below:

a. *Production from calcium carbide* is based on:

$3\ C + CaO \rightarrow CaC_2 + CO\ (\Delta H_s^o = 466\ kJ/mole\ CaC_2)$

$CaC_2 + 2\ H_2O \rightarrow C_2H_2 + Ca(OH)_2\ (\Delta H_s^o = -\ 129\ kJ)$

The first reaction is carried out in a furnace, in which an electric arc heats a mixture of coke and lime to about 2000 ^oC. The carbide is drawn off in the molten state. The process has been improved as to the energy required by combining in one reactor coke gasification (with steam and oxygen) with carbide production, without recourse to an electric arc. A by-product is CO, a low-calorie gas of relatively low value.

b. *Methane pyrolysis*: Methane and other light hydrocarbons can be pyrolyzed in an electric arc at a mean temperature of 1700 K producing acetylene and a large amount of soot, hydrogen and ethylene. Heating and cooling is effected in milliseconds. Recovery of C_2H_2 starts with separation of soot by cyclones and flow washing operations with oil and water. One company proposes to use hydrogen at 4000 K to react with hydrocarbons. Such a plasma process can be more selective but requires even more energy.

c. *Recuperative pyrolysis*: Light hydrocarbons, including naphtha, can be pyrolyzed in recuperative furnaces according to the Wulff process. The reactors are used in an alternating mode. At 1300-1500 ^oC a cracking period consumes energy leading to a decrease in temperature (in spite of external heating) because of the endothermic reaction. In a second phase the heavy reaction products, including carbon, are burned with oxygen-rich gases in reverse flow causing the temperature to rise to the original value. Between cracking and regeneration steam is passed through the reactor as a safeguard against explosions. Ethylene and propylene are the main by-products. Lowering the reaction temperature increases the ethylene/acetylene ratio.

d. *Partial oxidation of hydrocarbons* is the modern way of producing acetylene. A number of processes exists, all working above 1200 ^oC. These can be devided into one-step and two-step processes. In the first type oxidation and cracking occur simultaneously, such as in the burner developed by B.A.S.F.

Figure X.9.). A water quench follows the partial oxidation. The mean residence time is a few ms. The acetylene yield is about 15-20%, ethylene (20-30%) is a by-product.

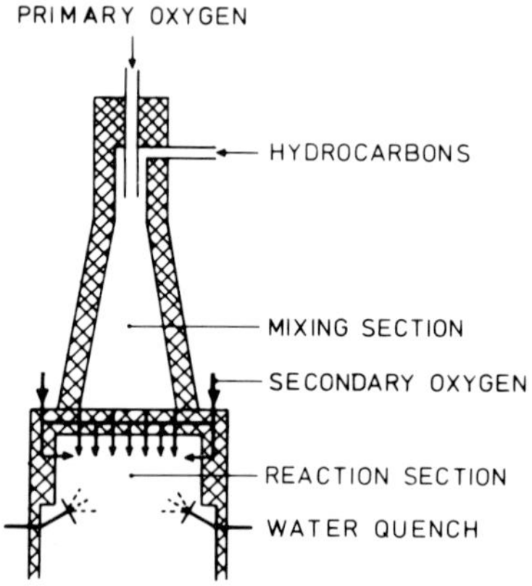

Figure X.9. Cracking of hydrocarbons to acetylene in a BASF-burner

A two-step process bas been realized by Hoechst (HTP = Hochtemperatur-pyrolyse). Figure X.10. presents a simplified block diagram and Figure X.11. is a schematic drawing of the burner. The burner, a reliable piece of equipment, has an exeptionally large heat production (4 $TJ.h^{-1}.m^{-3}$) and is resistant at very high temperatures against the reducing atmosphere. By increasing the reaction temperature and/or lowering the space velocity the ratio acetylene/ethylene can be increased from 0.43 to 2.33.

Acetylene generally is separated by scrubbing with a solvent at low temperature (e.g. acetone, dimethylformamide, ammonia) or by absorption in water at high pressure. Higher acetylenes must be made harmless before valuable olefins can be recovered. In the HTP-process this is done by scrubbing with the feed. Valueless by-products are recycled to the burner.

Another partial oxidation process developed by B.A.S.F. is based on submerged combustion ('Tauchflammenverfahren'). In this case heavy oil can be used as a feed. Oil is partly oxidized in a burner below the liquid level. Soot is a by-product which is recycled to the reactor. The product gas contains only 10% C_2H_2 and is contaminated with CO and CO_2.

In 1978 Union Carbide announced that an advanced flame cracking technology had been developed together with Kureha and Chiyoda from Japan. The burner is similar to the HTP reactor and operates on crude oil or distillate. The residence time is 20 ms and the temperature 2000 °C. The gases and feed droplets are injected into a Venturi tube creating a higher pressure at the point of adiabatic cracking. The main product is ethylene.

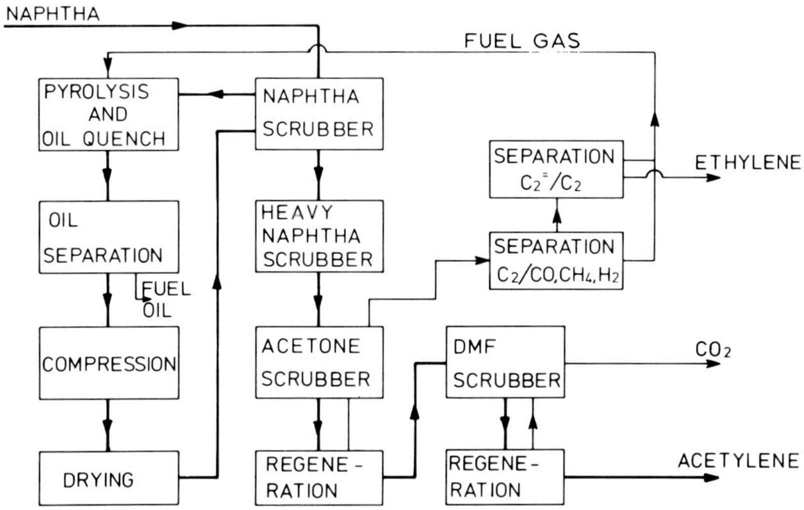

Figure X.10. Simplified block scheme for Hoechst-HTP-process

Applications of acetylene

Worldwide the production of acetylene is decreasing (1 Mt in 1971), in contrast to the steady rise in olefin production (10% in 1970-1975). Because of the large amounts of energy used in the production it is unlikely that, even with the increasing role of coal in the future, acetylene will become a basic chemical. Hydrocarbons and derivatives from coal may eventually be produced on a large scale by synthesis gas. Table X.8. reviews some processes for products originally based on acetylene which are now made from ethylene.

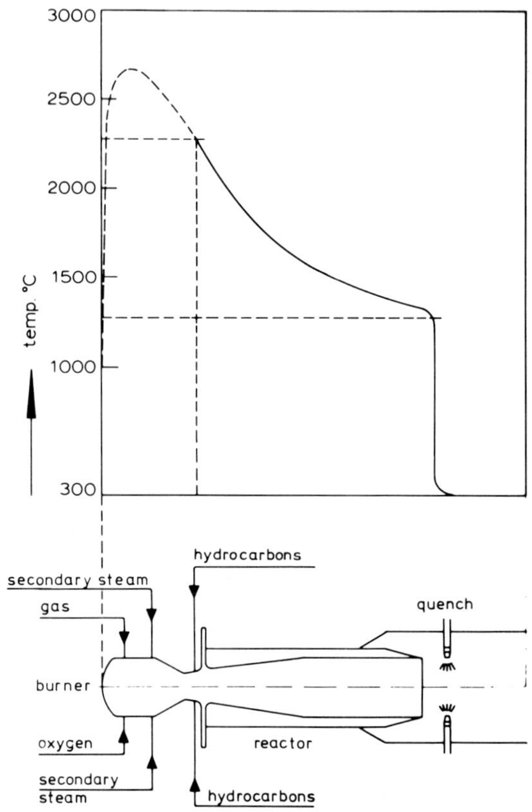

Figure X.11. Burner of the Hoechst-HTP-process

Table X.8. Syntheses based on acetylene vs. olefins

Acetaldehyde	
$C_2H_2 + H_2O \rightarrow CH_3CHO$ 70-100 °C in aq. H_2SO_4 (0.2 MPa, Hg^{++}) Wacker-Hoechst; Chisso	$C_2H_4 + \frac{1}{2}O_2 \rightarrow CH_3CHO$ one-step process (Wacker) 0.3 MPa, 130 °C, O_2, $Cu^{++} + Pd^{++}$ two-step process 105-110 °C, 0.8-1 MPa
Vinyl acetate	
$C_2H_2 + CH_3COOH \rightarrow CH_2{=}CHOOCCH_3$ gas phase, Zn on C, 0.1 MPa, 200 °C	$C_2H_4 + CH_3COOH + \frac{1}{2}O_2 \rightarrow CH_2{=}CHOOCCH_3 + H_2O$ Bayer, gasphase, 0.5-1.0 MPa, 175-200 °C, $\eta(C_2H_4) = 10\%$
Vinyl chloride	
$C_2H_2 + HCl \rightarrow CH_2{=}CHCl$ gas phase, 120-150 °C, 0.1 MPa, Hg^{++} on C	$C_2H_4 + Cl_2 \rightarrow C_2H_4Cl_2$ or 0.3-0.5 MPa, 230 °C, followed by cracking at 550 °C
Acrylic acid	
$C_2H_2 + CO + H_2O \rightarrow CH_2{=}CHCOOH$ 5 MPa, 200 °C, Ni-halide in THF (Rohm and Haas, BASF)	$CH_3CH = CH_2 + 3/2\ O_2 \rightarrow CH_2{=}CHCOOH + H_2O$ 0.1 MPa, 350-400 °C, Bi-Mo-O Analog to production of acrolein
Acrylonitrile	
$C_2H_2 + HCN \rightarrow CH_2{=}CHCN$ liq. phase, 80-90 °C Cu^{++} on C, (Cyanamid, duPont, Monsanto)	$C_3H_6 + NH_3 + 1\frac{1}{2}\ O_2 \rightarrow CH_2{=}CHCN + 3H_2O$ Bi-Mo-O, fluid bed at 0.2 MPa, 400-500 °C (Sohio)
Chloroprene	
$2CH{\equiv}CH \rightarrow CH_2{=}CH\ C{\equiv}CH$ aq. Cu_2Cl_2 in HCl, 80 °C $CH_2{=}CH{-}C{\equiv}CH + HCl \rightarrow$ $CH_2{=}CH{-}\underset{Cl}{C}{=}CH_2$ (duPont, I.G. Farben)	$CH_2{=}CH{-}CH{=}CH_2 + Cl_2 \rightarrow CH_2ClCHClCH{=}CH_2$ + $CH_2ClCH{=}CHCH_2Cl$ 330-420 °C, 0.1 MPa $CH_2ClCH{=}CH{-}CH_2Cl \xrightarrow{Cu} CH_2ClCHClCH{=}CH_2$ $CH_2ClCHClCH{=}CH_2 + NaOH \rightarrow CH_2{=}\underset{Cl}{C}{-}CH\ CH_2$

REFERENCES

General

1. Miller, Editor S.A., 'Ethylene and its industrial derivatives', Ernest Benn Ltd., London 1969.

2. Hancock, Editor E.G., 'Propylene and its industrial derivatives', Ernest Benn Ltd., London 1973.

3. Miller, S.A., 'Acetylene' (Its properties, manufacture and use), Vol. 1-2, Ernest Benn Ltd., London 1965.

4. Asinger, F., 'Mono-olefins' (Chemistry and Technology), transl. by B.J. Hazzard, Pergamon Press Ltd., Oxford 1968.

Special

1. Klein, H., H.M. Weitz, Hydrocarbon Process. 47 (11), 135-138 (1968). Figure X.6. is reproduced from this article with permission.

2. Peters, W.D., R.S. Rogers, Hydrocarbon Process. 47 (11), 131-133 (1968). Figure X.4. is reproduced from this article with permission.

4. Hornaday, G.F., F.M. Ferrell, G.A. Mills, Adv. Petr.Chem.Ref. 4, 451-488 (1961). Figures X.7. and X.8. and Table X.7. are reproduced from this article with permission.

5. Ullmanns Encyklopädie der technischen Chemie, 4th Ed., Vol. 7, 515. Figure X.9. is reproduced from this book with permission.

6. Kamptner, H.K., W.R. Krause, H.P. Schilken, Hydrocarbon Process, 45 (4), 191 (1966). Figure X.11. is reproduced from this article with permission.

7. Winnacker, K., L. Küchler, 'Chemische Technologie', 2nd Ed., Vol. 3, 393. Table X.1. is reproduced from this book with permission.

8. De Blieck, J.L., H.J. Cijfer, R.J. Jungerhans, Erdöl Kohle, 24 452-458 (1971). Table X.3. is reproduced from this article with permission.

9. Ullmanns Encyklopädie der technischen Chemie, 4th Ed., Vol. 8, 173 (1971). Table X.2. is adapted from Tabelle 11 of this part with permission.

XI. SELECTIVE HYDROCARBON OXIDATION PROCESSES

Introduction

Oxidation processes cover a large field ranging from non-selective processes in furnaces and engines to selective processes for the production of industrial chemicals. The first category, where the objective is to raise energy by oxidizing reactants (hydrocarbons, coal, fermentation) completely, is not discussed here; the same applies to the partial oxidation to synthesis gas, treated in Chapter V. This chapter deals with the second category in which oxygen reacts with hydrocarbons or other intermediates in such a way that combustion or complete oxidation is prevented. Examples of the introduction of oxygen into hydrocarbons are the production of ethylene oxide, acrylonitrile and phthalic anhydride. Oxidative dehydrogenation occurs in the production of butadiene and oxidative dealkylation in the conversion of toluene to benzene.

The Gibbs free energy change for total combustion always has a larger negative value than that for partial oxidation. Hence it is necessary to control the oxidation process kinetically by applying the right process conditions and often to employ catalysts to improve selectivity. Kinetic control by mass transfer is often also necessary.

Two main groups of processes exist: oxidation in the liquid phase and in the gas phase. The first group operates at a relatively low temperature (e.g. 120-180 ^{o}C) and often uses a homogeneous catalyst, the second one takes place at higher temperatures (e.g. 250-500 ^{o}C) and is nearly always catalyzed by a heterogeneous contact mass. Table XI.1. lists some important liquid phase processes and Table XI.2. some heterogeneous conversions.

Oxidants

It is usually necessary to choose between air or oxygen as the oxidant. In this respect the following considerations must be taken into account:

Table XI.1. Liquid phase oxidation

Ethylene	→	Acetaldehyde	→	Acetic acid
Propylene	→	Propylene oxide		
Butane	→	Acetic acid		
Cyclohexane	→	Cyclohexanol, Cyclohexanone, Adipic acid		
Toluene	→	Benzoic acid, Benzaldehyde, Phencl		
p-Xylene	→	Terephthalic acid		
Cumene	→	Cumene hydroperoxide	→	Phenol and Acetone

Table XI.2. Gas phase oxidation

Methanol	→	Formaldehyde
Ethylene	→	Ethylene oxide
	→	Vinyl chloride
	→	Acrolein, acrylic acid
Propylene	→	Acetone
	→	Acrylonitrile
Butene	→	Butadiene
	→	Maleic anhydride
Benzene	→	Maleic anhydride
o-Xylene	→	Phthalic anhydride
Naphthalene	→	Phthalic anhydride

1. air is cheaper than oxygen;
2. kinetics of oxidations often depend on the oxygen partial pressure;
3. oxidations are normally strongly exothermic and even more exothermic if consecutive reactions such as combustion occur to a certain degree; temperature control is important, and more difficult with oxygen than with air. Inert gases contribute to the heat capacity and if cooling by evaporation is used, the inert gases increase the cooling capacity. Sometimes, however, inerts are not wanted because the reaction temperature cannot be brought to a sufficiently high level;
4. mixtures of reactants with air or oxygen can be explosive or may even spontaneously detonate at certain circumstances, depending on composition, temperature and pressure;
5. the use of air requires larger reactor volumes and may result in larger recycle streams and processing of more waste gas;

6. the presence of inert gases may influence the separation of reaction products.

A few processes use ozone as an oxidant, which is expensive because of high electricity costs. This is possible only when high-cost compounds like azelaic and pelargonic acid are made. Other oxidants are hydroperoxides, hydrogen peroxide and peracids. It is apparently always necessary to have oxygen in a special form, either adsorbed on a catalyst or in a special configuration, if high selectivities are to be obtained. Other oxidation reactants such as nitric acid and potassium permanganate, used in making adipic acid and saccharin, respectively, are not discussed here.

At present only a few products can be made both by gas-phase and liquid-phase oxidation. In such cases there is a tendency to prefer the gas-phase oxidation because it is inherently safer since residence times are much shorter, causing storage volumes and chemical energy in the reactor to be accordingly smaller. It may well be that processes which have so far been realized only as a liquid-phase oxidation should be converted and developed into gas-phase oxidation processes.

It is of importance commercially that selectivity in oxidation processes is high, preferably close to 100%, because often a multitude of possibilities for thermodynamically much more favorable side reactions exists. If more than one saleable product is formed a classic problem of the chemical industry arises: the markets will normally not absorb the products in the yield ratio(s).

Liquid-phase oxidation - general aspects

The mechanism of liquid-phase reactions is in many cases a chain reaction in which radicals play a role. Four main types of reactions are of importance:

initiation $RH \xrightarrow{(X)} R\cdot + (HX)$

(formation of radicals)

propagation $R\cdot + O_2 \rightarrow ROO\cdot$

$ROO\cdot + RH \rightarrow R\cdot + ROOH$

(number of radicals is constant)

chain branching $ROOH \rightarrow RO\cdot + OH\cdot$

(number of radicals increases)

followed by propagations

$RO\cdot + RH \rightarrow ROH + R\cdot$

$OH\cdot + RH \rightarrow H_2O + R\cdot$

termination $2\,R\,. \rightarrow R\text{-}R$

$2\,ROO\,. \rightarrow ROOR + O_2$

$ROO\,. + R\,. \rightarrow ROOR$

$RO\,. + R\,. \rightarrow ROR$ etc.

(number of radicals decreases)

Initiation can be effected by molecular oxygen (possibly activated) or by initiators, which are relatively unstable compounds (hydroperoxydes, peroxides, peracids, etc.). These form radicals by a thermal or catalytic action. An example of direct catalysis is:

$$RH + Me^{+++} \rightarrow R\,. + H^{+} + Me^{++}$$

Indirectly an unstable compound such as H_2O_2 can be decomposed in the following manner:

$$H_2O_2 + Fe^{++} \rightarrow OH\,. + OH^{-} + Fe^{+++}$$

The initiation is often slow and only after some time (the induction period) the radical concentration does become sufficiently high. Contaminants often react preferentially with radicals, retarding the onset of fast oxidation; these substances are called inhibitors. They must be removed from the feed as completely as possible. Well-known impurities are sulfur compounds, many of which show pronounced negative effects. In products like fatty oils, lubricating oils and polymers, however, these types of compounds are added on purpose to prevent deterioriation by oxidation (antioxidants). The only catalysts used industrially for liquid-phase oxidation processes are of the homogeneous type. Co, Mn, Cr, Fe and Cu ions are often effective and are added as the naphthenates, octoates etc. in non-aqueous media. Their main contribution is a fast decomposition of hydroperoxides:

$$ROOH + Co^{+++} \rightarrow ROO\,. + H^{+} + Co^{++}$$

$$ROOH + Co^{++} \rightarrow RO\,. + Co^{+++} + OH^{-}$$

The amount of catalyst required to obtain rapid reaction is generally low (e.g. 10-100 ppm). This is, however, a disadvantage because the slightest corrosion of metallic surfaces, such as the reactor walls then influences the system - and only seldom in the right way. In serious cases metal ion deactivators must be added, which form stable and catalytically inactive complexes with the unwanted ions. Passivators which adhere to the metal surface may also have a positive influence. Coating the metal with Teflon® or glass is often helpful. A second mechanism uses coordination complexes. An example is the Wacker process for acetaldehyde:

$$C_2H_4 + \tfrac{1}{2}\,O_2 \rightarrow CH_3CHO$$

in which palladium complexes play an important role.

As to the process conditions in homogeneous oxidations, relatively low temperatures are used (70 - 180 °C) resulting in optimal selectivities. If primary products must be made, e.g. hydroperoxides or peracids, much lower temperatures are needed to prevent their decomposition. Yet it may be necessary to keep the conversion per pass low (sometimes only 10%). In the liquid phase heat is very well dissipated, especially if an evaporating solvent (or reactant) is used to carry off reaction heat.

The selectivity of an oxidation process may depend on the thermal stability of the desired product. With hydroperoxides the decomposition rate (i.e. the chain branching step) determines process selectivity to a large extent. The stability of hydroperoxides increases strongly in the sequence:
primary < secondary < tertiary.
Concurrently, the ease of abstraction of a hydrogen atom increases in the same order; hence the tertiary hydroperoxides can be produced in the highest yields.

Table XI.3. lists C-H bond energies for several hydrocarbons. The C-H bond in the CH_3-group of propylene has a much lower energy than that in the CH_2-group. Non-substituted aromatics are stable; the oxidation is much easier if the ring is substituted or dehydrogenated.

With liquid-phase oxidation processes the rate-controlling step is very often the mass transfer of oxygen alone (butane, toluene, cyclohexane etc.) or of oxygen together with the hydrocarbon (ethylene to acetaldehyde or vinyl acetate with palladium chloride as a catalyst).

Table XI.3. C-H bond energies for several hydrocarbons (kJ/mole)

CH_3 - H	435	ϕ - H	435
CH_3CH_2 - H	410	ϕ-CH_2 - H	356
n-$CH_3CH_2CH_2$ - H	410	H - CH_2CHO	368
$(CH_3)_2CH$ - H	396	CH_2=CH - H	435
$(CH_3)_3C$ - H	381	CH_3-CH=CH - H	431
.CH_2 - H	435	CH_2=$CHCH_2$ - H	356

Some liquid-phase oxidation processes

Manufacture of acetaldehyde

Direct oxidation of ethylene is carried out in a water phase with homogeneous catalysts, according to:

a.	$PdCl_4^{2-} + C_2H_4$	$\rightarrow$	$PdCl_3^- \cdot C_2H_4 + Cl^-$	$\Delta H_s^o = -11$ kJ/mole
b.	$PdCl_3^- \cdot C_2H_4 + H_2O$	$\rightarrow$	$HPdCl_2^- + CH_3CHO + HCl$	aldehyde formed
c.	$HPdCl_2^- + 2\ CuCl_2$	$\rightarrow$	$PdCl_4^{2-} + Cu_2Cl_2 + H^+$	
d.	$Cu_2Cl_2 + 2\ HCl + \frac{1}{2}\ O_2$	$\rightarrow$	$2\ CuCl_2 + H_2O$	$\Delta H_s^o = 233$ kJ/mole Cu_2Cl_2 oxidized

Overall $C_2H_4 + \frac{1}{2}\ O_2 \rightarrow CH_3CHO$; $\Delta H_s^o = -244$ kJ/mole CH_3CHO formed

The first reaction step proceeds via a π-complex of C_2H_4:

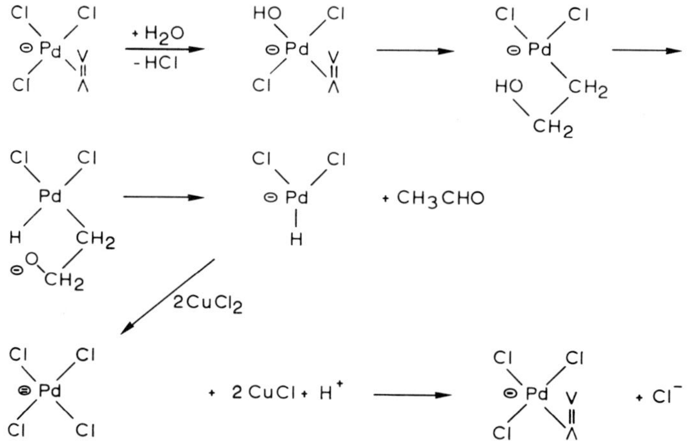

The second step is also complex: in the reaction medium copper (I) is present as CuCl which is oxidized according to:

$2\ CuCl + 2\ HCl + \frac{1}{2}\ O_2 \rightarrow 2\ CuCl_2 + H_2O$

The reaction should be carried out in such a way that no metallic Pd is formed. To obtain fast reaction it is advantageous to operate with excess hydrochloric acid and with a relatively high Cu(II)/Cu(I) ratio. Even then several parallel and consecutive reactions occur:

- addition of hydrochloric acid to ethylene, producing ethyl chloride
- formation of chloroethanol, CH_2ClCH_2OH

- oxidation of HCl to Cl_2 under the catalytic influence of Cu-chloride followed by chlorination of acetaldehyde to predominantly monochloroacetaldehyde (oxychlorination)
- further oxidation of acetaldehyde to acetic acid
- formation of oxalic acid via a combination of chlorination and hydrolysis; this is undesirable because copper oxalate is quite insoluble and may precipitate.

There are two process alternatives for this liquid-phase oxidation, viz. a single-stage and a two-stage process, both of which are applied commercially. The single-stage process (Figure XI.1.) uses a single reactor operating at about 130 ^{o}C and 0.3. MPa. Ethylene and oxygen are introduced in such a way that a large gas/liquid interface results giving high rates of conversion. Excess ethylene must be used for two reasons:

1. the oxygen concentration must be kept below 7% to remain outside the explosion limits;
2. ethylene conversion must be low to keep the ratio oxygen/converted ethylene at about 0.5 because otherwise the conversion of Cu(II) to Cu(I) would be too high or, alternatively, too large a volume of expensive catalyst would have to be used.

Hence only 30-40% of the ethylene is converted and the rest recycled. Because the severe corrosivity of the solution requires that an enameled reactor be used with a high resistance to heat transfer, the considerable heat of reaction is carried off by evaporation of water. Such enamel coatings have to be applied because the basic material of the reactor, steel, corrodes very rapidly on exposure to the reaction mixture, the more so because the reaction temperature must be 130 ^{o}C to obtain a sufficiently high partial pressure of water. A side stream of the catalyst solution is drawn off continuously and is regenerated by air-blowing at a somewhat higher temperature to remove oxalate. The regeneration reactor must be made of expensive titanium alloys because a combination of high resistance to corrosion and good heat transfer is needed.

Pure ethylene and oxygen must be used in this type of reactor for a number of reasons. First, ethane and nitrogen, if present as impurities, would accumulate in the ethylene recycle stream and thus increase the gas load on the reactor. Second, much ethylene would have to be purged or, alternatively, a costly separation of ethylene from the recycle stream would be needed if impure reactants were used. Third, the very well-mixed reactor operates at high oxygen conversions which makes the average rate of reaction per unit volume low, in other words requires a large reactor volume containing an expensive catalyst mixture.

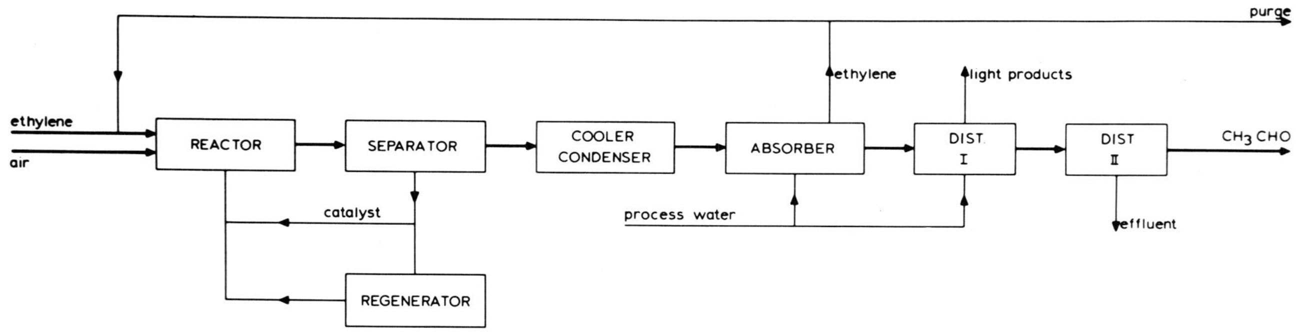

Figure XI.1. The Wacker process for acetaldehyde (one-step process)

The product, acetaldehyde, is carried out of the reactor with the steam; care should be taken to remove entrained catalyst droplets because these would cause corrosion of the equipment downstream of the reactor. The aldehyde is recovered by first cooling the reactor effluent and then washing the product stream with water to adsorb acetaldehyde. Most of the gas is recycled for optimum ethylene utilization but some is purged to prevent the build-up of inert components. The crude acetaldehyde is purified by distillation. It is not economical to recover by-products in the single-stage process.

The two-stage process uses two plug flow reactors (see Figure XI.2.). The first of these is for ethylene absorption; since oxygen is not fed to this reactor reactions a, b and c (see above) proceed and the heat effect is quite small. Ethylene conversion at about 105 ^{o}C and 0.8 to 1 MPa is virtually complete and recycle is not required; therefore, it is possible to use an ethylene/ ethane mixture as feed.

This saves on feed costs because it is not necessary to separate ethylene and ethane by a low-temperature distillation; the separation is now, in fact, accomplished by converting the ethylene to the water-soluble acetaldehyde followed by separation between aldehyde and ethane by a water wash. Since, moreover, reoxidation of the Pd/Cu-chloride solution is performed in a separate second reactor it is not necessary to use oxygen, and air can be fed to the process. A disadvantage is, however, that the two reactors must be made of expensive titanium alloy because in this case evaporative cooling is impossible and heat exchange through the reactor wall is required, preventing the use of enamelled materials of construction. The reactors operate with plug-flow characteristics in order to minimize reactor volume, for cost reasons, and because the Pd/Cu-chloride solution is converted to an appreciable extent, a considerable amount of Cu(I) salts being formed. Therefore, at the outlet of the first reactor the composition of the salt solution is far removed from the composition at which the rate of aldehyde formation is optimal. In fact, the process can better be called a cyclic regenerative one because the 'catalyst' solution undergoes appreciable conversion. The desire to minimize reactor volume is also the reason for operation at pressures higher than those used in the single-stage process, the more so because inerts are present in the hydrocarbon feed and in the oxidant. After the aldehyde reactor the pressure is reduced in a flash tower; the gases containing the desired product are separated by scrubbing and distillation (see Figure XI.2.). Besides acetaldehyde, monochloroacetaldehyde, a valuable by-product, can be obtained. The catalyst solution is returned to the aldehyde

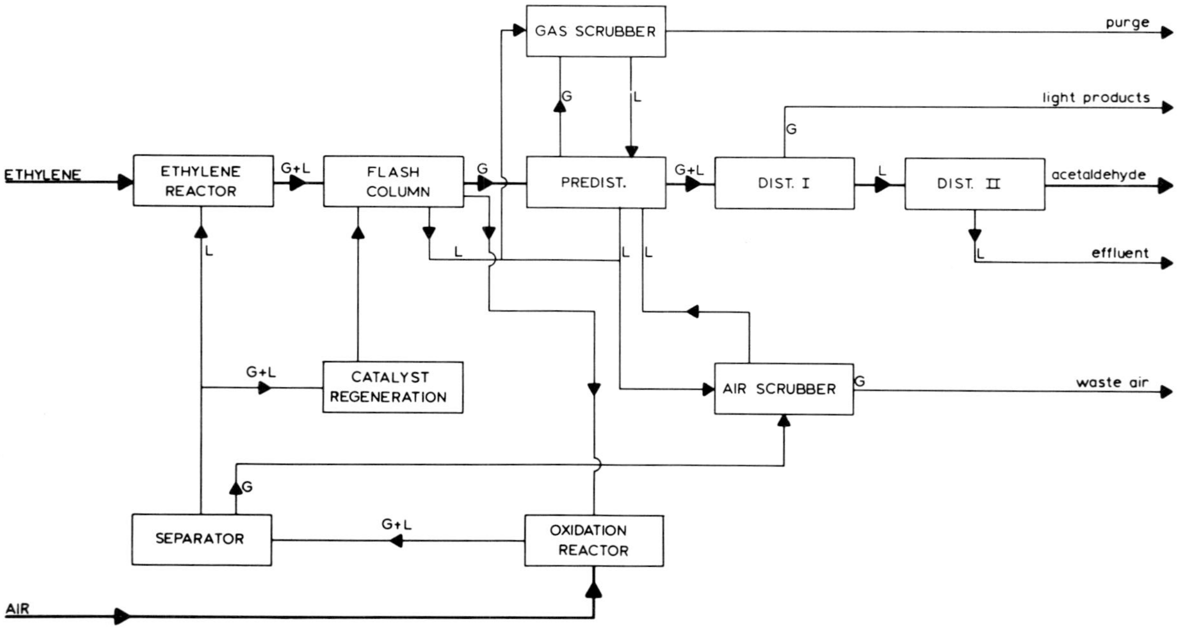

Figure XI.2. The Wacker process for acetaldehyde (two-step process)

reactor via the oxidizer in which Cu(I) is oxidized to Cu(II) with air, the oxygen conversion being almost quantitative. As compared with the single-step process the two-stage process requires higher investments (10-20%) but the extra capital cost can be recouped because of the lower energy consumption and particularly because the feedstock purity can be lower. It depends on the local price structure which of the two forms is the most suitable.

To overcome the main disadvantage of a homogeneous catalyst system, viz. the necessity of separating the catalyst from the reaction products, the attempt has been made to immobilize it on a solid carrier, with the added advantage of increasing the contact area. This idea has been abandoned because of catalyst poisoning by resinous products and the difficulty of reactivating the catalyst.

Acetaldehyde was formerly produced by hydration of acetylene, but this is too expensive nowadays. Moreover, Hg-ions must be used to catalyze the liquid-phase hydration reaction, causing the process effluents to be a serious environmental burden. A potentially better route could be the hydroformylation of methanol at pressures above 30 MPa and about 200 ^{o}C, using cobalt or nickel salts as catalysts. The relevant reaction is:

$$CH_3OH + CO + H_2 \rightarrow CH_3CHO + H_2O$$

The oxidation of ethanol to acetaldehyde is an older and widely used process; it may still be competitive if ethanol can be supplied at a reasonable price. This may be so in developing countries having abundant supplies of feedstocks or waste materials suitable for fermentation processing. The ethanol oxidation process is quite similar to the oxidation of methanol over silver catalyst for the production of formaldehyde which is discussed later in this chapter.

Manufacture of acetic acid

Acetic acid, the most important aliphatic acid in terms of production volume, is used as an intermediate in making e.g. esters, chloroacetic acid and peracetic acid, and also as a solvent for certain oxidations. It can be obtained by liquid-phase oxidation of acetaldehyde, n-butene or naphtha, by gas-phase oxidation of n-butene and hydroformylation of methanol. The processes are briefly discussed below and compared as to their future uses.

Acetic acid from acetaldehyde: The liquid-phase oxidation of acetaldehyde,

$CH_3CHO + \frac{1}{2}\ O_2 \rightarrow CH_3COOH \quad (\Delta H_s^o = -292$ kJ/mole)

can be carried out at 70 °C and about 1 MPa with air and is catalyzed by Mn^{++} ions. The reaction is considered to proceed via the formation of peracetic acid which reacts with acetaldehyde to give acetaldehyde monoperacetate. This monoperester decomposes into two moles of acetic acid. The catalyst ions initiate the chain reaction leading to the formation of peracetic acid from acetaldehyde and also catalyze the decomposition of the monoperacetate. The use of oxygen would give a higher production rate than with air, because the rate of mass transfer is controlling, but air is used for safety reasons. The heat of reaction is carried off by an external recirculation cooler, evaporative cooling being impossible because the initial boiling point of the reaction mixture is higher than the oxidation temperature. The reactor off-gas contains some unconverted acetaldehyde and must be scrubbed with water. Care must be taken to ensure that the liquid product from the reaction section is free of peracetic acid. An easy way to guarantee this is to use two or more reactors in series, increasing the temperature in the cascade up to 70 °C in the last stage to decompose the last traces of peracid. Final purification is either done by azeotropic distillation with ethyl acetate as the entrainer or by direct distillation, depending on the water concentration.

The intermediate peracetic acid can be obtained with a high selectivity at high conversions (both over 90%) when the oxidation of acetaldehyde is performed at 10-30 °C in an ethyl acetate solution using cobalt ions as initiator. The concentration of the peracid in the solution is limited to about 30%w because at higher concentrations the solution can detonate. Another safety precaution is to ensure that the reaction temperature does not go down below 0 °C: at such temperatures acetaldehyde monoperacetate could crystallize, and the crystallites can detonate easily.

The peracetic acid solution obtained must be freed of cobalt ions by e.g. ion exchange, or these ions must be rendered harmless by complexation, if the solution is to be stored or used in subsequent reactions such as epoxidations. Since high conversions to peracetic acid can only be obtained if contact with materials of construction containing Fe, Cr, Ni and/or Mn is avoided, the reactor must be made of high-purity aluminum. Ions of the above metals, if formed by corrosion, would decompose the peracid to give mostly acetic acid.

Acetaldehyde oxidation is a good example of a reaction in which the product composition can be varied by choosing different catalysts and adapting the reaction temperature. A further example of this versatility is that the aldehyde can be oxidized in good yields to acetic anhydride by using a combination of Co and Cu ions as catalysts at about 50 ^{o}C.

Acetic acid from n-butene: In a process developed by Bayer n-butenes are converted to sec-butyl acetate by the addition of acetic acid in the liquid phase followed by oxidation of the butyl acetate formed. The first step, ester formation, proceeds under the catalytic influence of an acid ion exchanger at 100-120 ^{o}C and about 2 MPa, the second step is an air oxidation at 195 ^{o}C and 6 MPa. The total yield approaches 60%.

The direct gas-phase oxidation process of Chemische Werke Hüls is an alternative. It operates at 250-300 ^{o}C and 0.1-0.5 MPa and uses titanium vanadate as the catalyst. The yield is between 35 and 58%; the main by-product is formic acid (2-10%).

Liquid -phase oxidation of n-butane: The Celanese Corp. developed a process for the liquid-phase oxidation at 140-180 ^{o}C and about 5 MPa. Cobalt acetate is used as the catalyst and chromium acetate as co-promoter counteracting the formation of formic acid. A complex mixture of products is obtained as can be understood from the mechanism in which the hydroperoxides formed initially decompose to give an alkoxyradical and a hydroxyl radical. The former radicals are converted to ketone, aldehyde or alcohol:

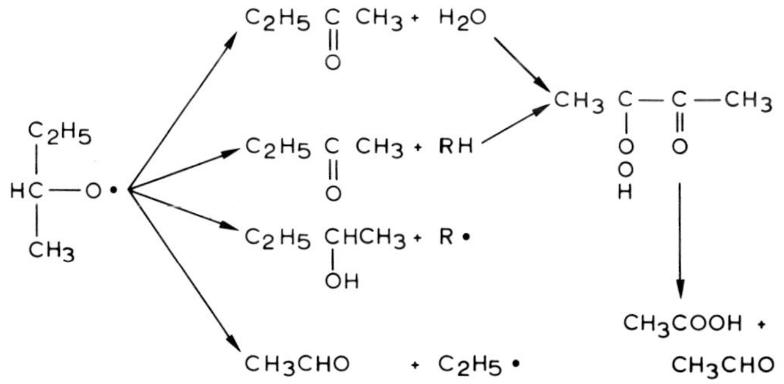

All intermediate products can be further oxidized. Since acetic acid is formed mainly via the oxidation of methyl ethyl ketone and acetaldehyde these by-products must be recirculated if maximum acetic acid yield is to be obtained. Apart from the abovementioned sec-butoxy radical a primary radical is also formed, albeit that the yield is much smaller. The primary radicals give butyric acid, propionic acid and methanol by similarly complex free-radical mechanisms. Furthermore, alcohols react with acids to form esters, extending the spectrum of products.

The complexity of the product mixture necessitates an extensive separation section, as is shown in Figure XI.3. The oxidation products leave reactor 1

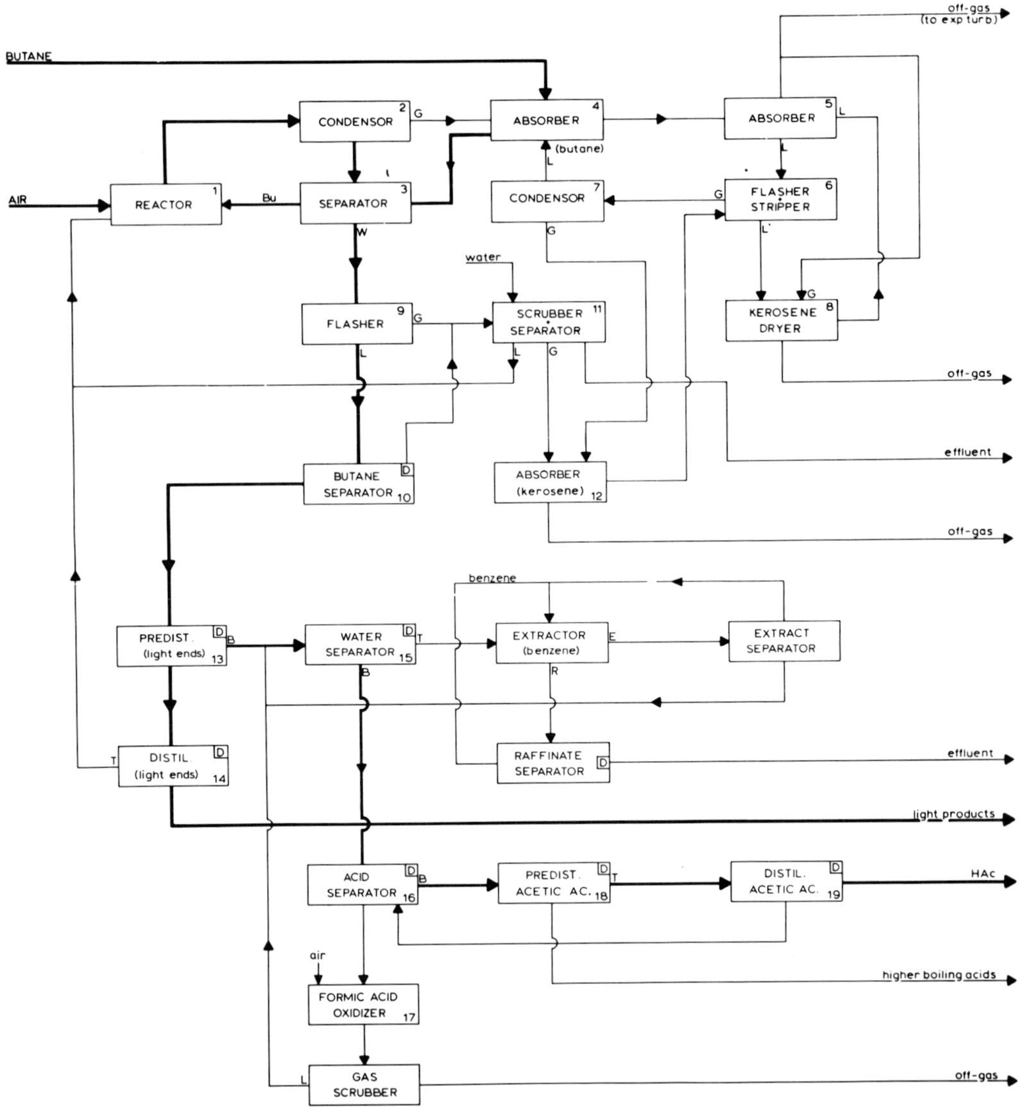

Figure XI.3. Liquid phase butane oxidation
(D = distillation column, T = top product, B = bottom product, G = gas phase, L = liquid phase, E = extract phase, R = raffinate phase)

as vapors together with unconverted butane, whereas the catalyst remains in the reactor and does not present a separation problem. The gases are partly condensed in condenser 2 and separated in separator 3 in a butane-rich phase for recycle to the reactor and an aqueous phase rich in oxidation products. The gases from the condenser should be cleaned because they contain valuable products and butane and cannot be vented because they would give rise to an air pollution problem. The gas is first scrubbed with the butane feed in absorber 4 and then with kerosene (in 5) before venting. Dissolved components are returned to the reactor feed via the flasher/stripper system 6, condenser 7, absorber 4 and separator 3. Part of the purified off-gas is utilized to remove water from the kerosine in dryer 8.

The aqueous phase rich in oxidation products is freed of most of the butane feed and light oxidation products in flasher 9; the gas phase is sent to a purification section (11, 12) to separate it in organics for recycle and purified off-gas. The liquid from the flasher is separated by a series of distillations and an extraction. First, butane is removed in 10 and then a light ends fraction containing acetaldehyde, methyl acetate, ethyl acetate, acetone and methyl ethyl ketone are distilled over in 13. Acetaldehyde is concentrated in column 14 and recycled; the block diagram indicates that the other light oxidation products are drawn off as by-products and sold as solvents after further purification, but it is also possible to recycle them to obtain higher yields of acetic acid. The bottoms of distillation column 13 are dewatered by distillation in 15 and the resulting aqueous top product containing some acids is purified by extraction with benzene to give an effluent and an extract which is recycled to the dewatering column 15. The bottoms of this column contain C_1-C_4 acids and water. Quantitative separation between formic and acetic acid being difficult because of the formation of a ternary azeotrope with water, the top product of distillation tower 16 is fed to catalytic reactor 17 in which formic acid is oxidized to CO_2 and water over a copper-on-silica catalyst. If the formic acid conversion is not quantitative there is little loss of acetic acid but the partial conversion increases the load on columns 15 and 16 because the partially purified effluent from the formic acid decomposer must be returned to the dewatering column. Finally, pure acetic acid is produced in a yield of more than 50% (basis: carbon in the butane feed). If the light oxidation product stream mentioned before is not recycled but further separated into individual components for sale as solvents, the carbon-based yield of useful products is about 75%.

Acetic acid by hydroformylation of methanol may become a competitive process for the above oxidation methods. It was developed by Reppe at BASF, operates at

210 - 250 °C and 50 - 70 MPa and is based on the reaction:

$$CH_3OH + CO \rightarrow CH_3COOH \qquad (\Delta H^o_s = -130 \text{ kJ/mole})$$

The reaction is catalyzed by cobalt iodide and gives a selectivity of 90 - 95% at methanol conversions approaching 90%. A newer development is by Monsanto, who found that the pressure can be reduced to about 1.5 MPa at roughly the same temperature if rhodium iodide is used as the catalyst. Very high selectivity is possible (up to 99%) resulting in much less complex product recovery in addition to savings in energy.

Future developments

In general, the cost of raw materials and energy are the most important items in the manufacture of acetic acid. Therefore, processes with relatively cheap feed-stocks giving high yields are expected to have the best prospects. In this connection, the Monsanto methanol carbonylation process is expected to have some advantage over the other processes, but the difference probably is too small to make new plants based on this process competitive over existing plants based on e.g. butane oxidation. As for acetaldehyde oxidation, whether based on acetaldehyde made from ethanol or on ethylene, it does not seem likely that this route will grow in importance. This is particularly so if oil prices rise more rapidly than the price of coal. However, small plants based on ethanol made by fermentation may still be competitive, particularly in developing countries where relatively abundant supplies of fermentation feedstocks are available.

An uncertain factor is that it may also become feasible to produce acetic acid directly by bio-processing. If this route is to become important, the problems of recovering acetic acid from the dilute reactor product, subsequent purification and in particular the disposal of waste liquor from fermentation plants must be solved.

Manufacture of cyclohexanol and adipic acid

Cyclohexanol and adipic acid are important in the production of nylons. To obtain cyclohexanol, cyclohexane is oxidized by air in the liquid phase at 150 - 160 °C (hence a pressure of 0.8 MPa) with 60 ppm cobalt catalyst. A large number of products results and to keep the selectivity as high as possible it is necessary to limit the degree of conversion to as low as 5 - 10%. Cyclohexanol and cyclohexanone are the main products with a total selectivity of 80% (ratio cyclohexanol : cyclohexanone = 65 : 35). By-products are adipic acid, lower mono- and dicarbonic acids, caprolactone, esters and acetals. The products from the reactor are separated in an aqueous layer containing acids and an organic phase. The latter is treated with alkali to remove acid impurities. The purified stream is then further purified in two distillations: first cyclohexane is recovered and finally the cyclohexanone-cyclohexanol mixture.

The process uses much energy because more of the unconverted cyclohexane must be evaporated than is necessary for the removal of heat. Therefore, attempts have been made to increase the conversion in the oxidation reaction without decrease in selectivity. This has been tried by adding boric acid which forms a complex with cyclohexyl hydroperoxide, resulting in the formation of the boric acid ester of cyclohexanol. The ratio of cyclohexanol : cyclohexanone is raised to 9 : 1 and the conversion level can be increased to 10 - 15%. The boric acid ester is hydrolyzed. The 'catalyst' has now been converted to orthoboric acid which is crystallized and reconverted to metaboric acid by thermal dehydration. As such it is added to the reactor. Unexpectedly the process uses even more energy, probably because of the thermal dehydration.

Adipic acid is made from cyclohexanol by oxidation with nitric acid. It is, however, also possible to use air as an oxidant if barium and manganese acetate are used as catalysts and the reactants are free of water. The reaction time then is extremely long (about 6 hrs) and the yield is 80% compared with 95% in the nitric acid route.

Manufacture of benzoic acid and phenol

Toluene can be effectively oxidized with air to benzoic acid ($S > 90\%$) in a bubble column. About 40 ppm of cobalt octoate is sufficient at 140 °C and 0.4 - 0.5 MPa to reach a degree of conversion of about 50%. The large heat of reaction is carried off by evaporation of toluene. The vapours are condensed, after which layers of toluene and water separate. Toluene is recycled. The gases should not contain more than 0.4% of oxygen to avoid the formation of explosive mixtures with toluene vapor. The off-gas is treated with activated carbon to remove toluene completely.

The benzoic acid is crystallized from toluene or toluene is evaporated; of course, the first method gives a purer product. The benzoic acid may be oxidized to phenol according to a process developed by Dow.

Molten benzoic acid is oxidized at about 230 °C with air in the presence of copper benzoate and magnesium oxide. At the same time steam is injected, which volatilizes the phenol. The phenol yield is 80% and quite a large amount of tar is formed. The product is finally distilled to separate tar and catalyst is recovered by burning the tar. The mechanism of this oxidation is very complex and does not only include Cu(II) and Cu(I) benzoate but also the formation of salicyclic acid-benzoate which splits off carbon dioxide and reacts further to give phenol.

Manufacture of phenol from cumene

Cumene (isopropyl benzene) is easily oxidized without catalysts. By such an autoxidation cumene hydroperoxide is formed. The hydroperoxide is decomposed in a second step according to:

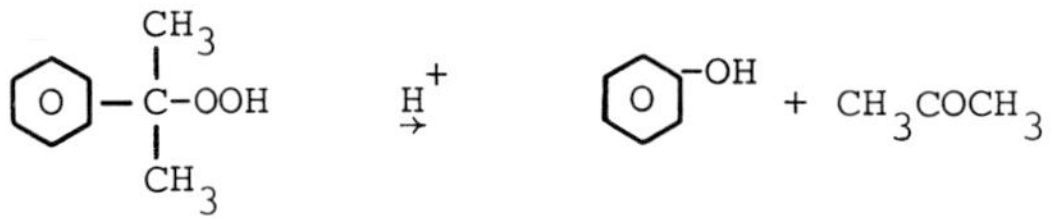

The hydroperoxide is relatively stable and a real intermediate. Nevertheless care should be taken that it is not decomposed via radicals which initiate further hydroperoxide decomposition. Such chain branching is avoided by good heat control and keeping the hydroperoxide concentration below 25%.

Addition of some caustic soda reduces the induction time due to traces of sulfur compounds in the feed and small amounts of phenol formed as a by-product. Metals like Ni, Cr, Fe, Co, Mn and Cu must not be present, because these catalyze the breakdown of the hydroperoxide. Sometimes a reactor cascade is used with decreasing temperature levels to increase the selectivity by avoiding chain branching due to thermal hydroperoxide decomposition. Cumene hydroperoxide is decomposed in the next step by the catalytic action of sulfuric acid. This reaction is very fast and strongly exothermic. This part of the process operates at 65 ^{o}C in acetone, i.e. the second product of the reaction, as solvent. Evaporation of acetone controls the reaction temperature. If this is not done effectively the reaction will shift into an explosive region. Also, the concentration of peroxide must be kept low; at such conditions, viz. low concentration of reactant and very high degree of conversion, the continuous tank reactor is most suitable. A flow scheme is given in Figure XI.4.

One of the by-products is α-methylstyrene, resulting from dehydration of dimethyl phenyl carbinol. It may also be hydrogenated and recycled. The process feed, cumene, is produced by alkylation of benzene with propylene on a solid catalyst (H_3PO_4 on SiO_2 as a carrier, 65 wt% P_2O_5) at 250 ^{o}C and 2 - 2.5 MPa:

$$C_6H_6 + CH_3CH{=}CH_2 \rightleftarrows C_6H_6(CH_3)_2 \qquad (\Delta H^o_s = -106 \text{ kJ/mole})$$

Possible parallel and consecutive reactions include the formation of diisopropylbenzenes and oligomerization of propylene. Hence the process is operated with

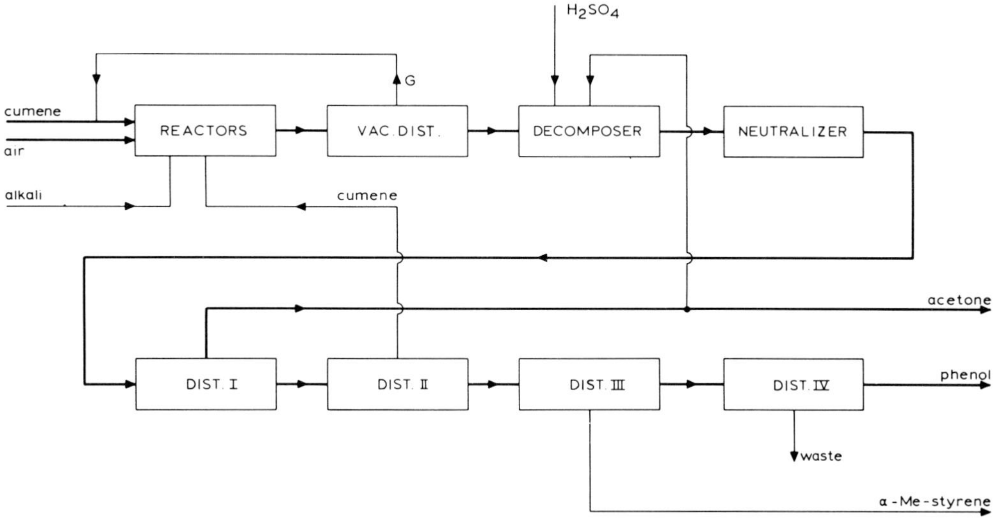

Figure XI.4. Cumene oxidation to phenol and acetone

excess benzene and at a pressure high enough to shift the equilibrium to the cumene side. The reactor consists of several adiabatic catalyst beds with cooling between stages by injection of cold feed.

The H_3PO_4/SiO_2 catalyst has a lifetime of one year. The water concentration in the system must be carefully controlled: insufficient water increases acid strength (formation of coke) and too much water decreases the rate too much. The yield of cumene is 75%.

Manufacture of terephthalic acid

p-Xylene is not easily oxidized in the liquid phase by air, but at temperatures of 175 - 230 ^{o}C and corresponding pressures (2.5 - 3.0 MPa) cobalt and manganese ions catalyze the reaction if bromine derivatives are present as promoters. These assist the initiation by the reaction: $HBr \xrightarrow{X} HX + Br$. Remarkably, the reaction is best performed in acetic acid medium. The process makes a purer product than the other commercial process in which 30 - 40 wt% HNO_3 is used to oxidize p-xylene.

Gas-phase oxidation

General aspects

Oxidation of organic compounds in the gas phase with excess air or oxygen leads to the combustion products CO, CO_2 and water, unless selective catalysts are applied. Even if this is done, only compounds having a π-electron system or unshared electrons can be oxidized selectively. Examples of the first class of compounds are olefins and aromatics, of the second methanol, ethanol and ammonia. Alkanes do not contain such structures and catalysts for their selective oxidation have so far not been found. Among the olefins, ethylene occupies a special position in that it is the only one that can be oxidized to the corresponding epoxide in high yields. The reason is that very reactive allylic hydrogen atoms are absent in ethylene. Silver is a unique and extremely stable catalyst for this oxidation. There is, however, another possibility for ethylene oxidation, viz. its conversion to acetaldehyde. Although the mechanism of this oxidation is closely similar to the homogeneous oxidation to acetaldehyde as practiced in the Wacker process, heterogeneous catalysts containing palladium have also been found.

Propylene can also be oxidized to several important industrial chemicals if the right catalyst is selected. It can give propylene oxide over silver catalysts, but the yield is too low to make this process economically feasible. Much more interesting is the type of oxidation in which propylene is oxidized via an allylic intermediate to give acrolein, acrylic acid and acrylonitrile. Other significant industrial chemicals are produced in this manner, combinations of metal oxides, e.g. Bi_2O_3/MoO_3, often being applied as catalysts. Careful selection of the ingredients and optimal methods of catalyst production are applied to obtain the best possible selectivity to a given product.

Such oxide catalysts are also effective in converting n-butenes and iso-butene. Isobutene is similar to propylene in that it can be oxidized to methacrolein. n-Butenes offer more possibilities: firstly, the relative stability of butadiene caused by resonance makes it possible to dehydrogenate n-butenes oxidatively. Several of the metal oxide combinations known from propylene oxidations can be used. It is also possible to increase the extent of oxidation by choosing the proper catalyst compositions: butenes and butadiene may then be oxidized selectively to maleic anhydride and other compounds. Vanadates are very suitable for this purpose. These catalysts are also effective in the oxidation of aromatics.

Either the ring or the side chains are attacked. The oxidations of naphthalene or o-xylene to phthalic anhydride are examples used industrially.

The partial oxidation of the above hydrocarbons proceeds step-wise according to a kinetic model resembling a rake:

$$\begin{array}{lccccccc} \text{GAS} & A & & B & & C & & D \\ & \downarrow\uparrow & & \downarrow\uparrow & & \downarrow\uparrow & & \downarrow\uparrow \\ \text{SURFACE} & A & \rightarrow & B & \rightarrow & C & \rightarrow & D \end{array}$$

The reaction of adsorbed A to adsorbed B is the first step in the sequence. B can either desorb or react further to give C. The final steps of the rake are oxidation to combustion products and desorption of carbon oxides and water. The selectivity of each step is obviously determined by the rates of reaction and desorption. If some oxidation steps are very rapid, the rake model reverts to a combination of consecutive and parallel reactions:

$$\begin{array}{llll} & & B \rightarrow & \\ & \nearrow & & \\ A & & & \text{etc.} \\ & \searrow & & \\ & & C \rightarrow & \end{array}$$

The redox mechanism originally proposed by Mars and Van Krevelen has been shown to occur with a large group of metal oxide catalysts. The oxidation of hydrocarbons, methanol and SO_2 is brought about by oxygen supplied by the catalyst, very often by oxygen from the crystal lattice down to several layers below the surface. A vacancy results which is reoxidized by oxygen from the gas phase:

$$A + MO_n \rightarrow AO + MO_{n-1}$$

$$MO_{n-1} + \tfrac{1}{2}\, O_2 \rightarrow MO_n$$

The replenishment of the vacancy can be direct, from the gas phase, or indirect, from the bulk of the catalyst.

The chemical composition as well as the texture of oxidation catalysts should be selected with care. In general, catalysts with a relatively low surface area, low porosity and high thermal stability are preferred for the following reasons:

- oxidation reactions to the desired product are often quite fast, which would result in limitation of the overall process rate by pore diffusion if porous catalysts are applied. In oxidations proceeding according to the rake model outlined previously, the result would be decreased selectivity;

- if the number of active sites per unit volume of catalyst is larger, the rate of oxidation would be high and lead to the evolution of much heat of reaction. This not only decreases the selectivity but it may also cause sintering and deactivation.

An example of a catalyst with a low surface is the Ag catalyst applied in the oxidations of methanol and ethanol to the corresponding aldehyde. Both Ag gauze and silver particles having a low degree of porosity are suitable. An added advantage is the high thermal conductivity of silver which prevents the occurrence of local hot spots in the catalyst.

Oxidation of methanol to formaldehyde

Two processes are used in industry with different catalysts and process conditions. Adiabatic reactors containing silver gauze or shallow beds of silver grains are applied at 600 - 700 ^{O}C using a feed containing a relatively high methanol/oxygen ratio for reasons of safety. Iron molybdates are active at much lower temperatures (about 350 ^{O}C) and are used with low methanol/oxygen ratios. Intermediate reactant ratios in the feed cannot be applied unless inerts are added to prevent the occurrence of explosive gas mixtures (see Figure XI.5.).

The older silver-based process is still used in many plants. Yields of over 90% are reported; combustion to carbon dioxide and water is the main side reaction. The significant amounts of hydrogen found in the product gas have long been attributed to the occurrence of dehydrogenation followed and accelerated by combustion of part of the hydrogen formed. Recently, however, several investigators came to the conclusion that part of the methanol reacts directly with oxygen on

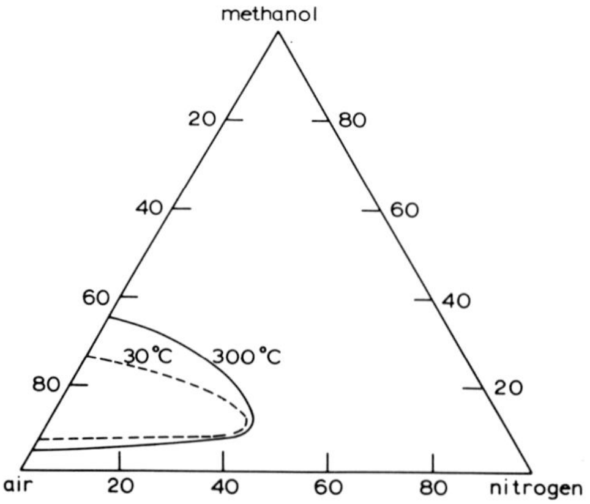

Figure XI.5. Explosion limits for the system methanol-air-nitrogen

the catalyst to give formaldehyde, another part oxidizes completely, the dehydrogenation being a parallel reaction:

$$CH_3OH + \tfrac{1}{2}\, O_2 \rightarrow HCHO + H_2O \qquad (\Delta H^o_s = -159\ \text{kJ/mole})$$

$$CH_3OH + \frac{3}{2}\, O_2 \rightarrow CO_2 + 2\ H_2O \qquad (\Delta H^o_s = -731\ \text{kJ/mole})$$

$$CH_3OH \rightleftarrows HCHO + H_2 \qquad (\Delta H^o_s = +\ 84\ \text{kJ/mole})$$

At industrial conditions the temperature of the reactor is controlled by the amount of reactant burned. A very high degree of conversion (near 100%) can be reached if the methanol is diluted with water which also acts as phlegmatizer against explosions. If water is not used the degree of conversion is much lower (77 - 87%) and formaldehyde must be separated from methanol. Separation between formaldehyde and the permanent gases in the product stream is accomplished by absorption with water. Care must be taken that explosive mixtures are not formed anywhere. This also applies during start-up, misoperation or technical troubles. The gas composition should never be within the explosive region of Figure XI.6., which applies to mixtures of methanol, air and nitrogen. Of course, the presence of hydrogen and formaldehyde should also be taken into account. The newer process using iron molybdate (Fe_2O_3/MoO_3) catalysts demands excess air and temperatures around 400 ^oC to give conversions of about 95% in reactors of acceptable dimensions. The selectivity is between 90 and 95% over fresh catalysts but diminishes a little with catalyst age due to losses of the somewhat volatile molybdenum component. A multitubular reactor is used.

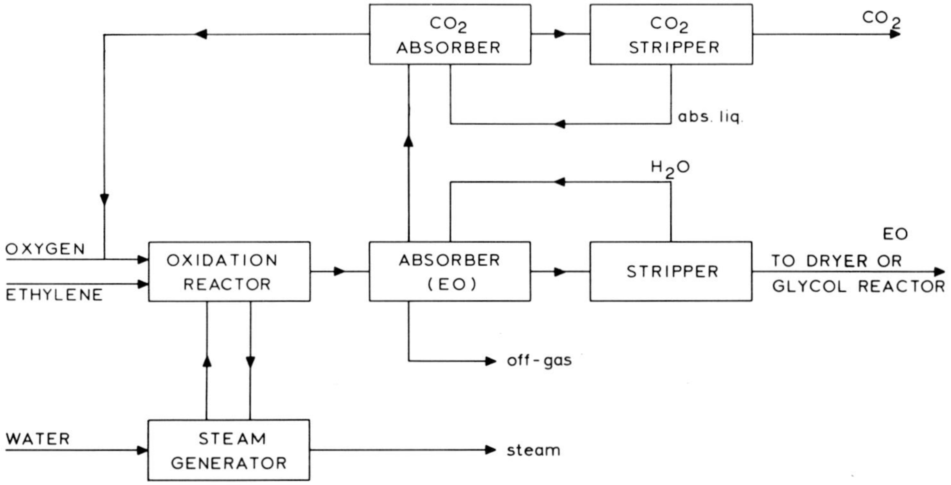

Figure XI.6. Process for ethylene oxide

Crude methanol containing dimethyl ether may be used as feed because the ether is also oxidized to formaldehyde in high yields according to

$$CH_3OCH_3 + O_2 \rightarrow 2\ HCHO + H_2O \qquad (\Delta H^o_s = -285\ kJ/mole\ ether)$$

Only when the product is to be used for making poly(acetals) such as 'Delrin' this is not acceptable. Of the other impurities in crude methanol higher alcohols present in low concentrations are converted to the corresponding aldehydes.

Formaldehyde is mainly used in making thermosetting resins by reaction with urea, phenol or melamine, and for the production of polymers like poly(formaldehyde) and polytrioxane.

The process of making acetaldehyde from ethanol is very similar to the conversion of methanol into formaldehyde. Silver catalysts are commonly applied.

Partial oxidation of ethylene to ethylene oxide

The only major intermediate produced from ethylene by catalytic gas-phase oxidation is ethylene oxide. Silver catalysts promoted with alkali and earth alkali metals are selective as well as active for oxidizing ethylene to ethylene oxide, presumably because of their property to activate oxygen as O_2^-. The main by-products, carbon dioxide and water, are formed by parallel combustion of ethylene and complete oxidation of ethylene oxide:

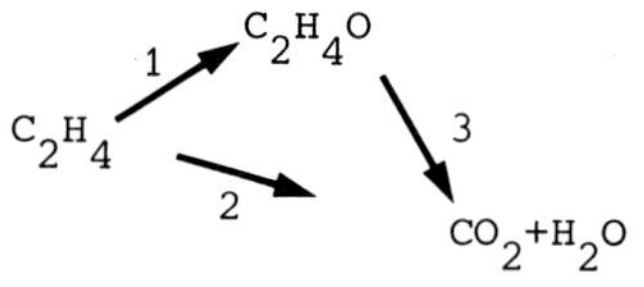

1: ΔH^o_s = - 120 kJ/mole

2: ΔH^o_s = -1324 kJ/mole

3: ΔH^o_s = -1204 kJ/mole

Yields of 65 - 70% are reported when working with commercial catalysts and air at 230 - 270 ^oC and 1.0 - 2.0 MPa total pressure. With silver catalysts containing heavy alkali metals and using oxygen a maximum yield of 75% is reached. Such high selectivities can only be obtained if small amounts (a few ppm) of moderator, usually chloroalkanes, are added to the feed to poison catalytic sites active for total combustion by deposition of chlorine, and when the conversion per pass is kept below 10%. If oxygen is used as an oxidant care has to be taken that the concentration stays below the explosion limit, hence part of the CO_2 formed by total combustion must be recycled. There is a tendency to prefer oxygen because the added cost of an oxygen plant is more than compensated by the lower cost of separating the inerts from the reactor effluent. In either case exact control of the reactor temperature is mandatory because the selectivity decreases if the temperature becomes too high. Excellent cooling is all the more necessary because the heat of total combustion is much larger than the heat of epoxidation. If total oxidation increases, 'runaway' conditions may occur, i.e. conditions where the cooling capacity of the reactor is inadequate causing the temperature to get out of hand. Hot spot formation is avoided by using a multitubular reactor in which the tubes containing the catalyst are externally cooled with a rapidly circulating cooling medium such as a molten salt mixture. Fluid bed reactors have so far not been a success, not only because they show some degree of mixing leading to lower selectivity, but also because silver catalysts are deactivated by attrition and catalyst losses by formation of fines occur. Still, it may be worthwhile to develop stronger catalysts and improved designs of fluid-bed reactors having a high length/diameter ratio which causes approach to plug flow characteristics in the bubble phase. Apart from excellent temperature control fluid-bed reactors offer a safety advantage in that incipient explosions are effectively smothered in the fluid bed.

Figure XI.6. shows a simplified flowchart of an ethylene oxide plant. The product is recovered by absorption in water, stripping from the absorbent and upgrading by rectification. The crude epoxide may either be fed directly into a glycol reactor or be purified further by distillation. Since the atmospheric boiling point of ethylene oxide is 10.7 ^{o}C, the purification section is operated under pressure to allow the use of water as a coolant.

The older ethylene oxide process via ethylene chlorohydrin (described later in this chapter) is no longer used. Old plants have mostly been converted to the production of propylene oxide because this intermediate cannot be obtained by catalytic gas-phase epoxidation in acceptable yields. The main applications of ethylene oxide are in producing anti-freeze, polyesters, ethanolamines and non-ionic detergents (poly(ethylene oxide)).

Partial oxidation of propylene is at present the most extensively studied catalytic gas-phase oxidation. Although the selective conversion of propylene to acrolein over cuprous oxide has been known for a fairly long time, the discovery of bismuth molybdates as highly active and selective catalysts for oxidation to acrolein, and also for ammoxidation to acrylonitrile, has opened a new era in oxidation catalysis.

There are several other propylene oxidation processes besides the above allylic oxidations. Table XI.4. summarizes the most important of these processes and products which are discussed below. Partial oxidations giving compounds with less than three carbon atoms as the main product cannot be regarded as selective and are not considered here.

The widespread commercial interest in the main group of processes listed in the table, allylic oxidations, has led to the development of an extraordinarily large number of metal oxide combinations as catalysts. These systems are also a fruitful field for basic catalytic studies which are contributing greatly to our knowledge of complex multicomponent catalysts. Typical binary oxide combinations forming the basis of most complex oxidation catalysts are: Bi-Mo-O, U-Sb-O, Sn-Sb-O and Fe-Sb-O. Characteristic features of allylic oxidations are the initiation by abstraction of an allylic hydrogen atom, the formation of a symmetric allylic intermediate and the role of the catalyst as donor of oxygen to the substrate and acceptor of oxygen from the gas phase according to a redox mechanism. Oxidative dimerization of propylene has recently attracted attention, both from a fundamental point of view and as a means for synthesizing aromatics from low-molecular weight olefins. The reaction is essentially a combination of two allyl radicals, the oxidation being limited to abstraction of a single hydrogen atom. Typically, the catalysts applied here do not contain MoO_3 or a similar component promoting the selective incorporation of oxygen.

Table XI.4. Selective propylene oxidation processes and products

Nature of reaction	Main products
oxidation of the allylic CH_3 group	acrolein, acrylic acid,
oxyhydration	acetone
oxidative dimerization	hexadiene, benzene
ammoxidation	acrylonitrile

The oxyhydration of propylene to acetone occurs at rather lower temperatures than allylic oxidation and requires the presence of excess steam. The reaction is initiated by the addition of a proton from the catalyst surface, acetone formation involving an oxygen atom derived from water.

Oxidation to acrolein: The conversion of propylene to acrolein is in many ways representative of gas-phase oxidation of olefins containing an allylic methyl group:

$CH_3-CH=CH_2 + O_2 \rightarrow OHC-CH=CH_2 + H_2O \qquad (\Delta H_s^o = -369 \text{ kJ/mole})$

The oxidation is carried out between 350 and 450 ^oC and at atmospheric pressure. Cu_2O was applied as a catalyst in the first generation of processes, but the discovery at Sohio that bismuth molybdates are much more efficient proved to be a major breakthrough in applied catalysis as well as process technology. If propylene is fed to the reactor in a mixture with air and steam in a ratio of 1 : 10 : 2 - 5, selectivities of up to 90% can be reached at conversions of 80% or even higher. The high air content of the feed gas is needed to keep the oxidation state of the catalyst at or above a certain level. If this is not done, the selectivity decreases because of catalyst reduction. Addition of steam improves the selectivity and also keeps the mixture of reactants outside the explosive region. Efficient temperature control is needed in the reactor as well as in downstream equipment. Because the temperature is rather high and in view of the reactivity of intermediates and products, any post-catalytic volume at high temperature would act as a homogeneous reactor, with the consequence that combustion occurs. The reactor effluent should be cooled very rapidly, if necessary by quenching with water, and the volume of dead space behind the reactor must be minimal and reduced further by applying e.g. inert fillings.

Acrolein is used for the production of allyl alcohol, and for the synthesis of methionin, an essential amino acid added to cattle feed.

Acrylic acid

Acrylic acid, an important monomer for the manufacture of various polymers, is obtained by oxidation of propylene over complex oxidic catalysts according to what is essentially a two-step reaction:

$CH_2=CHCH_3 + O_2 \rightarrow CH_2=CHCHO$ and $CH_2=CHCHO + O_2 \rightarrow CH_2=CHCOOH$

It is possible to produce the acid in a single reactor, but improved yields can be attained by using two reactors in series containing essentially the same catalyst but operating at different conditions. The temperature in the first reactor (about 350 ^oC) is 70 - 80 ^oC higher than in the second reactor. Since the conversion per pass can be high (> 95%) without the risk of low selectivity, the product is condensed and recovered from the aqueous solution by solvent extraction only.

Heat removal and the necessity of avoiding the formation of explosive gas mixtures determine the design of the reactor to a large extent. It is necessary to add steam to the feed for reasons of safety.

Dimerization and aromatization

Several single oxides and binary oxide mixtures have the capacity to oxidize propylene to dimerization products. The primary product, 1.5-hexadiene, may give benzene by dehydrocyclization. The initial step in the reaction sequence is the same as in the allylic oxidation to acrolein, viz. abstraction of hydrogen. The symmetrical allyl intermediate reacts with another propylene molecule to give 1.5-hexadiene. The oxygen concentration is low enough to avoid the formation of acrolein.

$$CH_2 = CHCH_3 \longrightarrow \overset{\cdots\cdots\cdots}{CH_2CHCH_2} \xrightarrow{+\text{ propylene}} CH_2 = CHCH_2CH = CH_3$$

$$\xrightarrow{+\text{ oxygen}} CH_2 = CHCHO$$

Acetone by oxyhydration of propylene

Among the process routes to acetone the oxyhydration of propylene is a recent one. Several binary oxides, in particular SnO_2-MoO_3, catalyze the formation of acetone from propylene and air in the presence of steam. Low temperatures, i.e. about 150 ^{o}C, must be used. It is too early yet to decide whether this process can compete with by-product acetone from the manufacture of phenol via the cumene route, by oxidative dehydrogenation of isopropyl alcohol over Ag or Cu catalysts or by liquid-phase oxidation of propylene to acetone, analogous to the Wacker process for acetaldehyde.

Acrylonitrile

The most important development in the field of propylene oxidation has been the ammoxidation to acrylonitrile. It is the most attractive commercial process for making the monomer from which the wool-like poly(acrylonitrile) fibers and ABS resins are derived. The reaction conditions are similar to those applied in the oxidation to acrolein: the same catalysts and similar conditions of temperature and space velocity are used. At temperatures of 400 - 500 ^{o}C yields of up to 80% can be obtained from feeds containing the requisite amount of ammonia. The superior selectivity of ammoxidation as compared with oxidation is mainly caused by the fact that acrylonitrile is oxidized much less readily than acrolein. The main by-products are acetonitrile (⩽ 2%) and HCN (⩽ 7%). The reactions are:

$$CH_3CH{=}CH_2 + NH_3 + 1\tfrac{1}{2}\ O_2 \rightarrow CH_2{=}CHCN + 3\ H_2O \qquad (\Delta H^o_s = -515\ kJ/mole)$$

$$2\ CH_3CH{=}CH_2 + 3\ NH_3 + 3\ O_2 \rightarrow 3\ CH_3CN + 6\ H_2O \qquad (\Delta H^o_s = -360\ kJ/mole\ CH_3CN)$$

$$CH_3CH{=}CH_2 + 3\ NH_3 + 3\ O_2 \rightarrow 3\ HCN + 6\ H_2O \qquad (\Delta H^o_s = -314\ kJ/mole\ HCN)$$

Figure XI.7. shows a simplified flow diagram. The reactor is a fluid bed with vertical cooling tubes because of its advantages in controlling the temperature of the very exothermic process. The residence time is a few seconds. The gases are cooled and then scrubbed with sulfuric acid to neutralize NH_3 and to prevent untimely polymerization. Absorption in water then gives an aqueous solution from which HCN and acetonitrile are distilled off as top products. Further rectification results in pure acrylonitrile and HCN. See Figure XI.7.

Partial oxidation of higher olefins becomes increasingly difficult to accomplish with any great selectivity as their molecular weight increases, due to the many more ways of incorporating oxygen in the molecule, the increased likelihood of C-C bond cleavage and easier isomerization and dehydrogenation the higher the carbon number. Moreover, some catalysts have a certain capacity for inducing dimerization and aromatization.

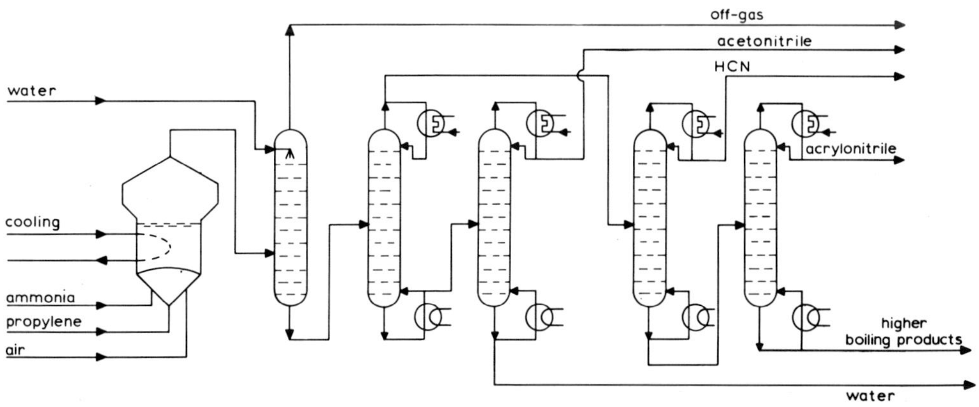

Fig. XI.7. Sohio process for acrylonitrile

The oxidation of isobutene to methacrolein can best be carried out at the conditions which are also optimal for the production of acrolein from propylene. A similar conclusion applies to methacrylic acid and methyl methacrylate. Isopentene can be dehydrogenated oxidatively to produce isoprene, an important monomer for high-quality synthetic rubber, but this is not (yet) done industrially.

The straight-chain butenes offer more possibilities than isobutene. With mild oxidation catalysts the reaction can be directed towards dehydrogenation, e.g. for 1-butene:

$$CH_2{=}CHCH_2CH_3 + \tfrac{1}{2}\,O_2 \rightarrow CH_2{=}CHCH{=}CH_2 + H_2O$$

The stabilization by the conjugated system of butadiene makes this reaction energetically favorable. However, it is also clear that the activated butene intermediate has a tendency to isomerize, with the consequence that oxidation products are formed to some extent.

With the allylic intermediate introduction of oxygen is also possible. One of the first products to be formed is crotonic aldehyde, but it is not very stable at the conditions of oxidation and reacts further to crotonic acid and the relatively more stable furan. Under more severe oxidation conditions maleic acid and its anhydride are formed.

Figure XI.8. summarizes the selective oxidation reactions of butenes. Besides these selective oxidations side reactions occur, i.e. breakdown of aldehydes

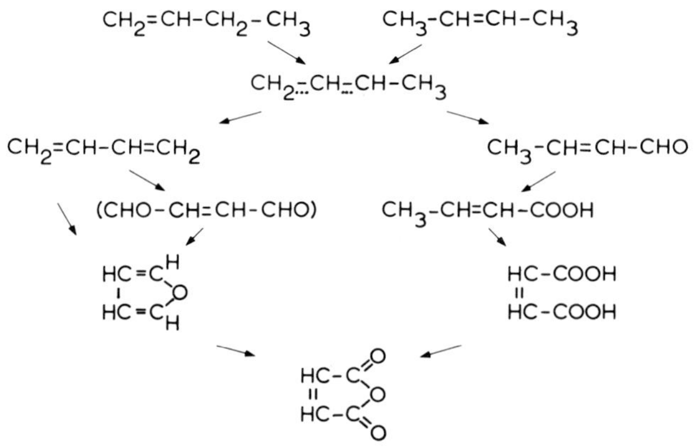

Figure XI.8. Selective oxidation reactions of butenes

and acids to compounds with less than 4 carbon atoms, and complete combustion. The relative importance of these side reactions, which in fact occur with each component of the diagram, depends on the particular catalyst and reaction conditions applied.

The main industrial applications of selective reactions of butenes with oxygen or air are the production of butadiene and oxidation to maleic anhydride. The very selective oxidative butene dehydrogenation has received considerable attention in the literature and is also being applied industrially. Two types of catalysts are effective: (a) mild-acting catalysts, e.g. the same catalysts that oxidize propylene to acrolein and isobutene to methacrolein, and (b) iron oxide catalysts and iron-containing spinels, which accelerate selective dehydrogenation and complete combustion but not the selective introduction of oxygen into the butene molecule. Bismuth molybdates are very active for oxidative dehydrogenation of n-butenes at 430 - 500 ^{o}C.

More extensive oxidation to products like furan and, in particular, maleic anhydride requires stronger oxidation catalysts than those mentioned under (a), above. At the same time, these catalysts should retain the capacity to transfer oxygen selectively to the organic molecule. This does not occur with the (b)-type catalysts. Moreover, the surface should probably be more acidic if an acid like maleic acid is to be produced. The most interesting catalysts for these 'deeper' oxidations are base on V_2O_5 or MoO_3 or contain molybdates.

For details of a process for the production of butadiene by oxidative dehydrogenation (Phillips Petroleum) the reader is referred to the literature. It should be emphasized here that butadiene made by processes of this type cannot compete very effectively with by-product butadiene recovered from the C_4-fraction obtained in thermal cracking of naphtha. Therefore, oxidative dehydrogenation is most likely to be introduced in areas where other means of making ethylene (e.g. from natural gas) and propylene (from catalytic cracking operations) are available.

As compared with direct butene dehydrogenation, oxidative dehydrogenation has the advantage that high conversions and high selectivities can be reached. Moreover, a distillate fraction containing other hydrocarbons may be used as feed. A composition from a cracker is : 7 - 12%w butane, 1 - 3%w isobutane, 45 - 47%w isobutene, 26 - 28%w 1-butene and 14 - 16%w 2-butene. Substantial savings in costs of feed preparation and product separation are obtained. The butadiene is recovered by a combination of normal and extractive distillation.

Partial oxidation of aromatics

Because of the abundance of aromatic hydrocarbons originating from coal, oxidation of these products was already developed at an early stage of the heavy organic chemical industry. An important step was the discovery that the V_2O_5 catalyst used in the sulfuric acid process is a selective catalyst in the oxidation of naphthalene to phthalic anhydride. A similar process is the production of maleic anhydride from benzene.

In general, two classes of aromatic oxidations may be discerned:
- chain oxidation of alkyl aromatics, which is a mild oxidation yielding aldehydes and acids. The mechanism is similar to the oxidation of propylene and butenes and the process is carried out over similar catalysts;
- acid anhydrides result from more severe oxidation over V_2O_5-based catalysts. These compounds are produced by oxidative attack on two ortho methyl groups or on the aromatic nucleus.

Ammoxidation can be applied to methyl aromatics (e.g. toluene and xylenes) but at the moment aromatic nitriles are only of minor importance.

Oxidation of o-xylene and naphthalene to phthalic anhydride: Phthalic anhydride can be produced by oxidation of naphthalene by a strongly exothermic reaction:

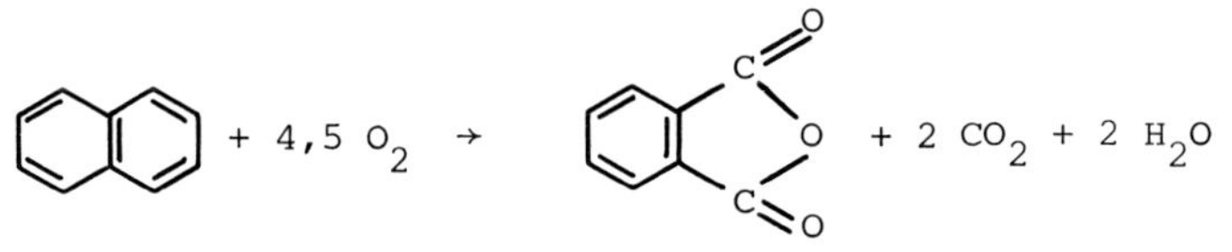

$(\Delta H_s^o = -1860$ kJ/mole)

o-Xylene has become a competitive feedstock. Although naphthalene may be produced by dehydrogenation and dealkylation of kerosene fractions, it is still mainly a coal chemical. o-Xylene can be separated from catalytic reformates obtained from petroleum naphtha (see chapter XIII).

Benzene oxidation to maleic anhydride: Catalysts based on V_2O_5 (e.g. a combination of V_2O_5 and K_2SO_4) make it possible to produce maleic anhydride from benzene with a selectivity up to 80% if a short residence time (about 0.1 sec) is used. The best temperature region is 400 - 450 ^oC. The process is closely analogous to the conversion of o-xylene and naphthalene to phthalic anhydride.

The hydrocarbon feed is evaporated by the second reactant, air, which is preheated. The mixture is kept outside the explosion limits by using excess air. At 350 - 400 ^oC the reactants are brought into contact with a V_2O_5 based catalyst, using a short residence time. Because of the formation of byproducts (carbon oxides, naphtoquinone) the selectivity is 70 - 80% depending on the feed type.

The formation of combustion products is even more exothermic than the selective oxidation ($\Delta H_{c,s}$ = -5100 kJ/mole naphthalene), hence temperature control of the reactor is a very important aspect. Either a tubular reactor with a large number (e.g. 5000) of small-bore tubes is necessary, or a fluid bed must be used. A cooling circuit with molten salts (potassium nitrate, sodium nitrate) carries the heat away and generates steam.

The anhydride is recovered from the gas stream by desublimation. The products are first cooled to about 200 ^{o}C by heat exchange with the feed; in a second step condensers separate the anhydride. The condensers are heated from time to time to melt down the product (switch-condensers). Figure XI.9. gives an impression of the process.

Ammoxidation of aromatics: Ammoxidation of toluene and xylenes lead to nitriles, analogous to the production of acrylonitrile from propylene. In particular, the manufacture of p-benzodinitrile might become interesting as an intermediate for terephthalic acid. The latter is nowadays produced only by liquid-phase oxidation of p-xylene, which is quite tedious.

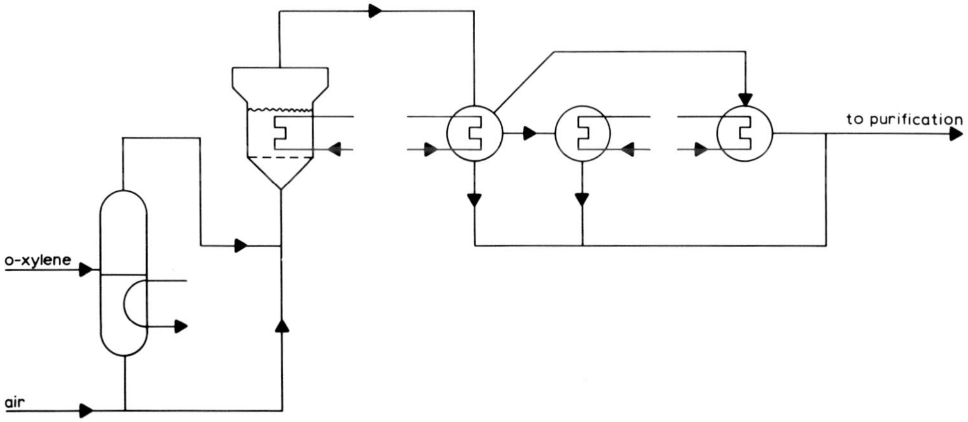

Figure XI.9. Phthalic anhydride process

Oxychlorination processes

Chlorobenzene

A special form of a selective oxidation process is an oxychlorination, of which the oldest example is the production of chlorobenzene from benzene according to:

$$C_6H_6 + HCl + \frac{1}{2} O_2 \rightarrow C_6H_5Cl + H_2O \qquad \Delta H_s^o = -222 \text{ kJ/mole}$$

In this reaction the chlorination of benzene is preceded by the oxidation of hydrochloric acid in situ, a modern form of the old Deacon process:

$$2\ HCl + \frac{1}{2} O_2 \rightleftarrows Cl_2 + H_2O \qquad \Delta H_s^o = -\ 51 \text{ kJ/mole } Cl_2$$

This process, developed in the early 19th century but improved about 100 years later by Shell, used $CuCl_2$ as a catalyst. This improvements existed of (1) finding a good carrier for the $CuCl_2$ catalyst, SiO_2 which stood up to attrition, (2) lowering the vapor pressure of $CuCl_2$ by using a complex salt of $CuCl_2$ with KCl, (3) activating the catalyst by admixing rare earth metals resulting in a lower reaction temperature and hence a better equilibrium conversion, and (4) the use of a fluid bed reactor.

Copper chloride catalysts on Al_2O_3 (often together with Fe of Co chlorides) are used in the oxychlorination of benzene. Care must be taken that the temperature of the reactor is well controlled and that the degree of conversion is kept low because otherwise a high percentage of higher chlorinated compounds and even combustion products occur. The temperature is kept at about 240 ^oC and the degree of conversion at 10 - 20%. The surplus of benzene is separated from other products by cooling and absorption in water on HCl. The condensed products are separated (after washing) by rectification. At temperatures below the dew point of the reaction products corrosion problems are so large that ceramic construction materials must be used. Care must be taken that the off-gases (N_2 and CO_2) are cleaned by absorption to make sure that no chlorinated compounds are emitted. The process has been of importance in the production of phenol.

Vinyl chloride

In principle all hydrocarbons can be oxychlorinated but technically ethylene is the most important. In the classic process for production of vinyl chloride from acetylene no co-products were formed:

$$HC{\equiv}CH + HCl \rightarrow H_2C = CH - Cl \qquad \Delta H_s^o = -99 \text{ kJ/mole}$$

The change to ethylene as a basic component, however, introduced the problem of all chlorinations, hydrochloric acid:

$$H_2C{=}CH_2 + Cl_2 \rightarrow ClH_2C{-}CH_2Cl$$

$$ClH_2C{-}CH_2Cl \rightarrow H_2C{=}CHCl + HCl \qquad \Delta H_s^o = 70 \text{ kJ/mole}$$

One of the possible solutions is to replace only part of the acetylene by ethylene. In that case the following combination can be made (I.G. Farben):

$$C_2H_4 + Cl_2 \rightarrow C_2H_4Cl_2$$

$$C_2H_4Cl_2 \rightarrow C_2H_3Cl + HCl$$

$$C_2H_2 + HCl \rightarrow C_2H_3Cl$$

$$\overline{C_2H_4 + C_2H_2 + Cl_2 \rightarrow 2\ C_2H_3Cl}$$

Figure XI.10. presents a block diagram.

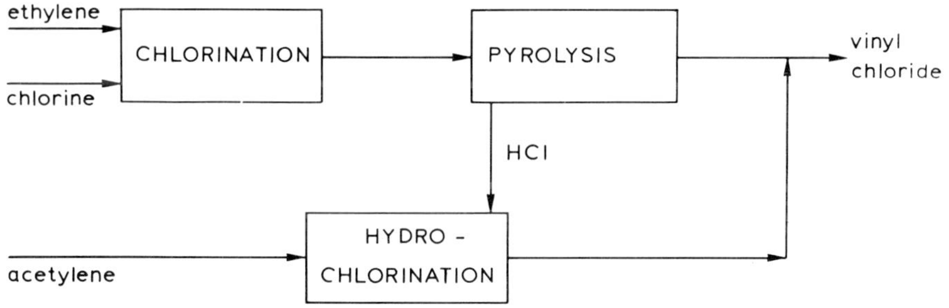

Figure XI.10. Production of vinyl chloride without hydrochloric acid

Processes developed by Kureha and others in which a mixture of impure ethylene and acetylene is produced from crude oil or naphtha by thermal cracking give another possibility. See Figure XI.11.

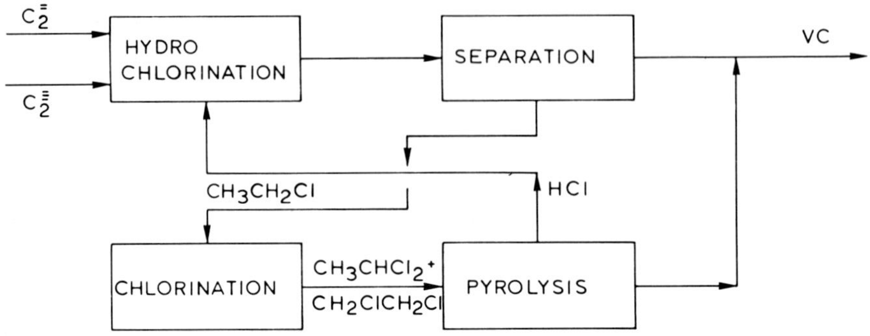

Figure XI.11. The use of impure $C_2^{=}$ *in a vinyl chloride process*

Because the price difference between C_2H_4 and C_2H_2 is large a more attractive method is given in Figure XI.12. In this way ethylene is converted by chlorination as well as oxychlorination to dichloroethane followed by combination of both product streams and subsequent cracking:

$$C_2H_4 + Cl_2 \rightarrow C_2H_4Cl_2$$

$$C_2H_4 + 2\ HCl + \tfrac{1}{2}\ O_2 \rightarrow C_2H_4Cl_2 + H_2O$$

$$2\ C_2H_4Cl_2 \rightarrow 2\ C_2H_3Cl + 2\ HCl$$

$$2\ C_2H_4 + Cl_2 + \tfrac{1}{2}\ O_2 \rightarrow 2\ C_2H_3Cl + H_2O$$

Other ways to combine HCl-producing plants with Cl_2-consuming plants are based on inorganic conversions of HCl, e.g. the Shell Deacon process. It is also possible to oxidize HCl by nitrogen oxides (corrosion!) and to electrolyze HCl (Hoechst and de Nora).

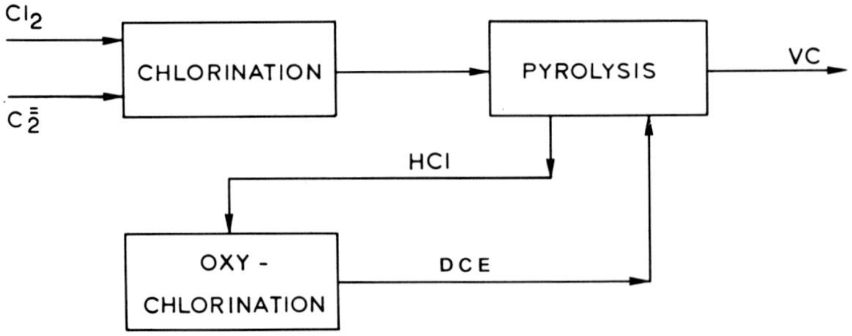

Figure XI.12. Vinyl chloride production by oxychlorination

The oxychlorination is catalyzed by Deacon-type combinations of $CuCl_2$ with alkali metal chlorides supported by Al_2O_3 or SiO_2. Optimal temperatures are in the range 250 - 350 °C, preferably 250 °C because consecutive reactions have to be prevented as far as possible and the volatility of the catalyst is still acceptable.

The feed contains ethylene, air and hydrogen chloride in near stoichiometric ratio, although some small excess of one of the reactants may be used to optimize selectivity. Air is preferred to oxygen because temperature control is easier. Close temperature control and good heat transfer are also the reason why fluid beds are often used (Goodrich). The gaseous feed is heated by the heat of reaction before entering the reactor. A fixed-bed reactor is used by Stauffer.

The space velocity is an important process variable and depends on the ratio height/diameter of the fluid bed. It is normally between 30 and 300 Nm^3 gas.min^{-1} kg^{-1} cat. The yield is 95% in dichloroethane, 1,1,2-trichloroethane being a by-product.

To pyrolyze dichloroethane, catalytic and non-catalytic processes are applied. In the non-catalyzed process the reactor tubes are heated to 450 - 650 °C, taking extreme care not to overheat the walls because otherwise a carbonaceous product is deposited. The residence time is approximately 15 sec. Pressures of 1.0 - 4.0 MPa are used mainly because of easy separation of the products. At 2.5 MPa the product stream may even be condensed at 0 °C. Figure XI.13. gives a flowchart of the vinyl chloride production. In this process part of the heat of reaction is taken up by a cooling medium (e.g. 'molten salt') from a coil in the fluid bed, raising steam in a heat exchanger. The reaction products are first scrubbed with water and then with alkali. The bottom product is separated into an aqueous phase and crude dichloroethane. After drying the product by azeotropic distillation two further distillations are required to separate light and heavy ends. The DCE is then converted to VC in a cracking furnace at an overall conversion of about 50%. After quenching the hydrochloric acid is obtained overhead as an anhydrous product in a distillation column. A final distillation separates vinyl chloride and DCE.

It seems probable that a catalyst can bring down the cracking temperature, reducing the danger of carbon deposition. SBA claims a conversion per pass of 60% at 400 °C and a selectivity of 99%. However, in packed tubes the realization

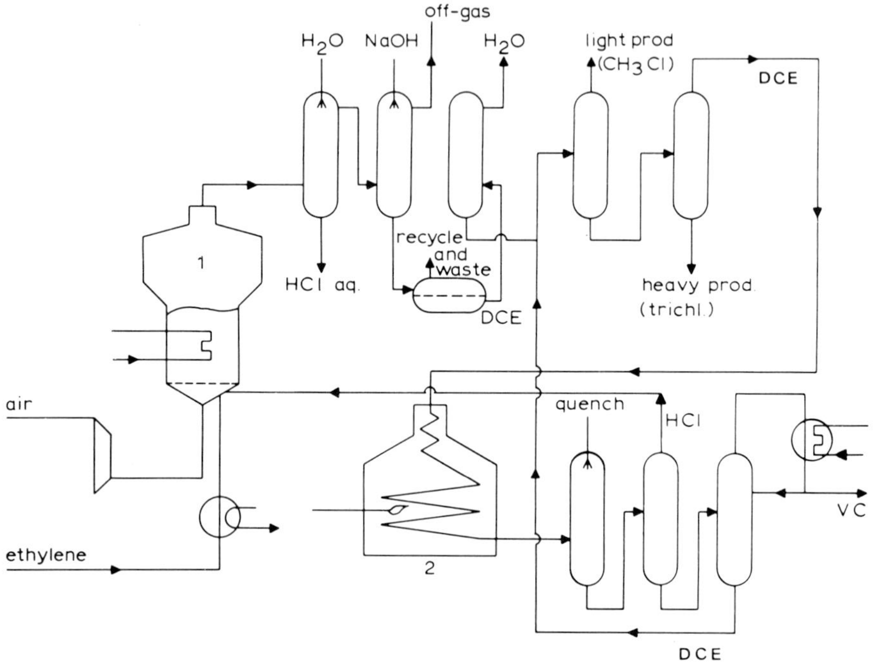

Figure XI.13. Oxychlorination of ethylene and production of vinyl chloride
1. fluid bed
2. pyrolysis furnace

of optimal temperature profiles is more difficult.

Oxidants other than air or oxygen

The previous discussion on oxidation processes in which air or oxygen is applied shows that the selectivity is lower than is economically viable in a number of cases, particularly when more complex oxygen compounds must be produced. In such cases such oxidants as hydroperoxides, peracids and hydrogen peroxide can be used to effect the oxidation in the liquid phase under rather mild conditions of temperature and pressure. The oxidations with peracids and hydroperoxides are designed as *two-step processes*: in the first step, the oxidizing agent is produced, usually from air or oxygen and another organic compound. In the second, the oxidant reacts with the substance to be oxidized. Another characteristic of such processes is that a second product is obtained: an alcohol in the case of a hydroperoxide and a carboxylic acid if a percarboxylic acid is used. This may lead to problems of disposal: the by-product must be sold or reduced and recycled. Another characteristic of these two-step procedures is that investment costs are inevitably high. Thus, single-step oxidations with air or oxygen are usually preferred, unless the selectivity is so low that two-step processes perform better. An example of the latter situation is found with propylene oxide; in this case a two-step process using a hydroperoxide as the oxidant is superior to catalytic oxidation.

In principle, it is also possible to co-oxidize two components, one of which forms a hydroperoxide or a peracid, which then reacts with the other feed component to give the desired product. This type of oxidation has so far not been a success because it is difficult to find conditions where both reactions proceed with acceptable selectivity at levels of conversion suitable for a commercial process. In the two-step type of process, each individual step can be optimized separately. Therefore, the discussion below deals with two-step processes only.

Epoxidation

With hydroperoxides: Halcon discovered the possibility of producing epoxides via a two-step process involving a hydroperoxide and an olefinic compound:

$$\mathrm{RH + O_2 \rightarrow ROOH}$$

$$\mathrm{ROOH + R'CH{=}CH_2 \rightarrow \underset{\diagdown\, O\, \diagup}{R'CH{-}CH_2} + ROH}$$

The first step, an autoxidation, is carried out in the liquid phase using air as oxidant. The commercial process based on this reaction, which was developed jointly by Halcon and Atlantic Richfield, employs tert-butyl hydroperoxide or ethyl benzyl hydroperoxide. An attractive feature of the use of hydroperoxides is that the by-product can be dehydrated and then hydrogenated. The resulting hydrocarbon can be recycled to the autoxidation. An alternative is to dehydrate the methyl carbinol formed from ethyl benzene hydroperoxide and to sell the by-product styrene, if there is a market for it. However, the first step of the process is not fully selective: for instance, acetone and tert-butyl alcohol are formed as by-products. The autoxidation, which is carried out under pressure (3.0 - 4.0 MPa; 120 - 140 ^{o}C) to ensure that the vapor phase remains outside the explosion limits, should be carried out to relatively low conversions up to 25 - 30% to reach adequate selectivity. An initiator is added to eliminate the induction period. For the same reason, the hydrocarbon feed should be free of impurities like traces of sulfur compounds. These would reduce the hydroperoxide yield by the formation of strong acids promoting acid-catalyzed hydroperoxide decomposition, in particular with aromatic hydroperoxides.

The crude product can either be used directly in the second step, the epoxidation, or after removal of part of the unconverted hydrocarbon feed. The epoxidation is catalyzed by heavy metal ions or complexes (e.g. Mo) at about the same temperature and pressure as the first step.

The products are separated in several distillations. The exact form of the recovery section depends on the compound to be oxidized, the solvent and the hydroperoxide applied. The presence of small amounts of by-products formed in the autoxidation should also be taken into account. The hydroperoxide and any other peroxy-compounds should be decomposed entirely to avoid their uncontrolled decomposition during distillation.

The use of a heterogeneous catalyst has recently been claimed. High rates of epoxidation can be reached and the catalyst is easily separated from the products. Of course, adequate means of contacting catalyst and liquid phase must be provided, which is somewhat more difficult than with homogeneous catalysts. Examples of applications are:
Propylene oxide is produced industrially by epoxidation with tert-butyl hydroperoxide and, in other plants, ethyl benzene hydroperoxide. New developments in the field of propylene oxide are the use of peracids, probably peracetic acid (Daicel, Solvay and others) to convert propylene. Another development is by Olin Corp. in which isopropanol is oxidized to yield a complex stream containing hydrogen peroxide which reacts with propylene. Still another one is the liquid phase reaction between propylene and oxygen in acetic acid with TeO_2 as a catalyst (150 ^{o}C). The reaction products consist of about 50% propylene glycol monoacetates, the rest being glycol and diacetate. This stream is cracked catalytically to propylene oxide and acetic acid.
Epichlorohydrin can be obtained by epoxidizing allyl chloride with a hydroperoxide. Since the presence of the chlorine atom reduces the rate of epoxidation relative to propylene, fairly long residence times or rather severe conditions must be applied. This is a disadvantage because hydrolysis of epichlorohydrin is enhanced. For the same reason, the recovery of the product must be carried out at moderate temperatures, in other words by distillation under reduced pressure. This process has not yet been built on an industrial scale.

Epoxides via the chlorohydrin route: Strictly speaking the chlorohydrin synthesis of epoxides is not an oxidation but it consists of the addition of HOCl to an olefinic double bond followed by dehydrochlorination:

$$Cl_2 + H_2O \rightleftarrows HOCl + HCl$$

$$RCH{=}CH_2 + HOCl \rightarrow \underset{\displaystyle Cl}{RCH}{-}CH_2OH \text{ and } \underset{\displaystyle OH}{RCH}CH_2Cl$$

$$\underset{\displaystyle Cl}{RCH}{-}CH_2OH \text{ and } \underset{\displaystyle OH}{RCH}CH_2Cl + Ca(OH)_2 \rightarrow \underset{\displaystyle \diagdown O \diagup}{RCHCH_2} + CaCl_2 + 2H_2O$$

The process was originally developed to produce ethylene oxide. It was abandoned when the much more economical direct oxidation processes over Ag catalysts came into general use. The chlorohydrin process has the following disadvantages:

- chlorine is devalued to useless calcium chloride; this means that the energy requirement of the chlorohydrin route is very much higher than for direct oxidation because one mole Cl_2 is needed per mole epoxide produced;
- lime must be used as a reactant in dehydrochlorinating the ethylene chlorohydrin;
- the process effluents contain not only $CaCl_2$ but also some chlorinated hydrocarbons.

However, this type of process is still in use to make propylene oxide because a direct oxidation route having high selectivity has not been found to date. In fact many plants formerly used for ethylene oxide have been converted to making propylene oxide. Propylene is treated with an aqueous chlorine solution in a packed column at about 40 $^{\circ}C$. The reaction products are converted without prior isolation into propylene oxide by adding slaked lime. The residence time of the propylene oxide formed should be short to prevent it being hydrolyzed. Here, too, chlorinated hydrocarbons are formed as by-products. In spite of this, the overall selectivity of converting propylene into its epoxide is about 90%. This is one of the reasons why this process is still applied.

A certain capacity for chlorohydrination is also necessary for making *epichlorohydrin* from allyl chloride. This compound is less reactive than olefins because of the presence of the chlorine atom. In this case allyl chloride is reacted with chlorine and water in a very dilute system at about 20 $^{\circ}C$. A well-cooled recycle reactor is required. The reason is that the formation of a separate non-aqueous phase from the allyl chloride must be avoided because otherwise over-chlorination produces highly substituted water-soluble compounds containing Cl as well as OH substituents. These by-products are very difficult to remove from the aqueous effluent and present an environmental hazard.

The crude chlorohydrins formed are dehydrochlorinated with lime in such a manner that the epichlorohydrin is distilled off as soon as possible, in the form of an azeotrope with water. The pressure should be reduced to keep the temperature below 50 °C to avoid hydrolysis of the product. The same applies to the final rectification. The undesirable effluent produced in this process makes it particularly desirable to find better ways of producing this important intermediate for resins and detergents.

Epoxidation with percarboxylic acids: Percarboxylic acids are more reactive in the epoxidation of olefinic compounds than hydroperoxides. Peracetic acid is preferred because of the ready availability of reactants from which it can be obtained, viz.:

- acetic acid; peracetic acid is obtained by reaction with H_2O_2, strong mineral acids acting as catalysts. This method is not very attractive commercially because of the high price of H_2O_2 and in view of the difficulty of removing water and strong acid from the product. Water interferes in the epoxidation step.
- acetaldehyde can be oxidized to peracetic acid in a solution of e.g. ethyl acetate, using air as the oxidant and Co^{+++} as catalyst. This reaction, which was discussed earlier in the present chapter, proceeds smoothly at 10-30 °C and 0.5 - 2.5 MPa total pressure.

The two steps of this route to epoxides are:

$$CH_3CHO + O_2 \xrightarrow{Co^{+++}} CH_3COOOH$$

$$CH_3COOOH + RCH=CH_2 \rightarrow \underset{\diagdown O \diagup}{RCH-CH_2} + CH_3COOH$$

Disadvantages of processes based on these reactions are that a catalyst, required in the first step, is not needed and may even be undesirable in the second step. Cobalt ions catalyze the decomposition of peracetic acid into acetic acid and oxygen at the temperatures (60 - 100 °C) usually needed for the epoxidation of olefinic compounds which are difficult to oxidize. An example is allyl chloride; the radicals formed in the decomposition of the peracid can induce polymerization of allyl chloride. Thus, it is necessary to remove the Co ions, e.g. by ion exchange, or to render them harmless by adding a metal deactivator which must be stable towards peracetic acid. Another drawback is that a market must be found for the by-product acetic acid, which is not possible at every location.

Still, this route is used in the USA for the epoxidation of allyl alcohol obtained from propylene oxide by isomerization; hydrolysis eventually leads to glycerin. A surprisingly large amount of acetic acid is formed as the by-product, viz. 2½ ton CH_3COOH/ton glycerin. Another possible application is in the production of epichlorohydrin from allyl chloride; this uncatalyzed epoxidation proceeds with selectivities of well over 80% when peracetic acid is used. Since, moreover, selectivities of up to 95% can be attained in making peracetic acid this process offers advantages over the current chlorohydrin route, provided that the acetic acid can be disposed of profitably. Very strict safety measures should be taken to prevent explosive decomposition of peracetic acid.

Future trends

General predictions about future developments in the field of oxidation processes are difficult to make, mainly because what is valid for one product is not necessarily true for another. Besides, these developments depend to a large extent on progress made in developing new catalysts and the discovery of novel reactions. Some general statements can, however, be made.

Gas-phase oxidations with air or oxygen will continue to be preferred to other processes if acceptable selectivities can be reached because

- it is not necessary to prepare reagents like hydroperoxides or peracids in a separate process step;
- the disposal of co-products is rarely necessary;
- separation between catalyst and products is easy;
- heat removal is relatively simple, particularly when fluidized beds can be applied.

This leads to the expectation that efforts to find more selective catalysts and optimal conditions for e.g. the production of propylene oxide are likely to continue; the alternative processes discussed in this chapter are less economic than direct gas-phase oxidations of even moderate selectivity.

Liquid-phase oxidations with air or oxygen are second best. Heat removal is easier than with gas phase oxidation. Moreover, reactant inventory in plants based on liquid-phase oxidation is usually greater, with the consequence that additional safety measures should be taken. If only small amounts of homogeneous catalysts are applied, separation from the product stream is usually not a problem. Nevertheless, there is a growing tendency to try and replace homogeneous catalysts by supported liquid phase catalysts for liquid-phase oxidation. Apart from their better separation and lower consumption, improved selectivity is the objective of development work in this field.

Co-oxidation in the liquid phase is rarely possible with adequate selectivity. An exception is the process for caprolactone which depends on the co-oxidation of cyclohexanone with a gas consisting of air and acetaldehyde. This process uses Mn, Co or other metal ion catalysts at about 30 ^{o}C; the peracetic acid formed from the aldehyde reacts *in situ* with cyclohexanone to the lactone. An advantage of this method is that the amount of peracetic acid present in the plant at any time is small, at the expense of a moderate selectivity. One obtains two moles of the by-product acetic acid per mole of caprolactone. The direct oxidation processes discussed above are, however, not very suitable for making more complex oxygen compounds in high yields. This applies in particular to relatively reactive substances; when such products must be produced two-step processes have more chance of performing well. This is illustrated by the difficulty of selectively oxidizing hydrocarbons with five or more carbon atoms.

The above picture becomes even more complicated when other means of synthesizing oxygen compounds are considered, as was done when discussing the manufacture of acetic acid. Examples of such reactions are *hydroformylation* and *carbonylations*; in both cases CO and another reagent are used to make an oxygen compound from a reactant containing an olefinic bond. Examples of these reactions are:

$$RCH=CH_2 + CO + H_2 \rightarrow RCH_2CH_2CHO \text{ and } \underset{\displaystyle CHO}{\underset{|}{RCH}}CH_3 \qquad \text{(Hydroformylation)}$$

$$RCH=CH_2 + CO + H_2O \rightarrow \overset{\displaystyle COOH}{\overset{|}{RCH}} - CH_3$$

$$RCH=CH_2 + CO + R'OH \rightarrow \overset{\displaystyle COOR'}{\overset{|}{RCH}}-CH_3 \qquad \text{(Carbonylation)}$$

Hydroformylation is applied in industry on a large scale. The main product made by this reaction is n-butyraldehyde, which is used to prepare higher alcohols from which plasticizers for e.g. PVC are prepared. In the first generation of processes, equimolar mixtures of CO and hydrogen were reacted with an olefin, mostly propylene, at pressures of about 20 MPa and 150 ^{o}C using $HCo(CO)_4$ as the catalyst.

Mixtures of α-olefins produced by thermal cracking of straight-chain paraffins are also used as feedstocks. The products, named oxo-alcohols after the original name of the process, can be used in making biodegradable detergents, i.e. detergent materials that can be broken down by microorganisms because they consist predominantly of straight-chain components.

The disadvantage of the earlier types of hydroformylation was that a sizeable amount of branched-chain product was also formed. Moreover, the high pressure was economically unattractive. Improvements have been attained by using Rh as the metal component of the catalyst; this leads to a reduction in the catalyst requirement by a factor of more than 100, the possibility of operating at much lower pressure and to much lower iso/normal ratios in the product. A further improvement consists of using supported Rh-complexes with phosphorus containing ligands. In this form, the hydroformylations can be carried out at pressures as low as 1.0 - 3.0 MPa. Further applications to other unsaturated compounds are likely to be found.

Carbonylation is also used industrially to make branched-chain carboxylic acids, useful ingredients of alkyd resins and lacquers. A commercial process employs H_3PO_4/BF_3 catalysts at 70 ^{o}C and 7 MPa pressure.

In the more distant future, *biochemical oxidations* may become more important, both in the form of reactions brought about by microorganisms and of processes catalyzed by supported or unsupported enzymes. Difficulties to be overcome include adequate temperature control, which is not easy in view of the limited temperature range for these reactions, and isolation of the products from the relatively dilute solutions produced. Processes of this type will probably be limited to products of high added value.

REFERENCES

General

1. Hucknall, D.J., 'Selective Oxidation of Hydrocarbons', Academic Press, New-York 1974.

2. Bamford, Editor C.H., C.F.H. Tipper, 'Comprehensive Chemical Kinetics', Vol. 20, 123-262 (1978), Amsterdam.

Special

Ullmanns Encyklopädie der technischen Chemie, 3rd Ed., Vol. 7, 659 (1953). Figure XI.6. is reproduced from this book with permission.

XII. MONOMERS AND POLYMERS

Introduction and definitions

In the preceding chapters hardly any attention has been paid to the interrelation between process type and product quality. An area in which product properties are extremely important is the production of plastics. Because these are often used as construction materials, physical properties are at least as important as chemical properties. In typical intermediates treated in the foregoing chapters the situation is quite different.

Plastics consist completely or partly of polymers. These polymers were originally made by modification of natural products (e.g. rayon from cellulose) but nowadays synthetic materials have taken over the lead. Polymers are macromolecules with a structure based on the repetitive sequences of one (or a few) monomers. The macromolecules in one plastic differ in molecular weight, hence a polymer must be chemically characterized by a mean molecular weight as well as a molecular weight distribution. Both are very important in product properties.

Macromolecular products occur also in nature e.g. wood, cotton, silk, rubber. In the 19th century these products had already been modified to improve their properties, as in the case of the vulcanization of rubber with sulfur. Extreme vulcanization of rubber leads to ebonite. Nitration of cellulose gives xylonite. Baekeland made a fully synthetic product in 1909 by treating phenol with formaldehyde. The resulting insoluble products did not melt and could be used as a construction material. In 1885 A. von Baeyer remarked that such a polycondensation resulted 'in a resin only'. Because of the pioneering work of Staudinger it became clear in 1922 that such resins consist of very large molecules which are bound to each other by chemical or physical forces. From that moment onward a polymer chemistry and technology was rapidly developed. One of the highlights has been the invention of Ziegler and Natta, who discovered around 1953 that ethylene and propylene can be polymerized at low pressures with metal-organic catalysts.

The properties of polymers depend on the structure of the macromolecules and their mutual interaction. In that connection chainlike molecules with relatively sparse side-chains differ widely from three-dimensional networks. The first group can be transformed by heating (thermoplastics), the second remains rigid (thermosets). Sometimes chains are crosslinked, making possible a type of network having rubber properties. In the application of plastics the most important properties are those of the solid phase, hardness, brittleness, tensile strength and flexibility, all as a function of time and temperature. For processing the behavior of the molten phase is important, characterized by viscosity, melt elasticity, chemical stability etc.

In chain polymers the degree of order is important. Irregularly built chains do not crystallize and polymers containing these can be a glass, a rubber or a liquid, depending on the molecular structure, on the temperature and on the presence of plasticizers or solvents. The glass state is found at the lowest temperature. In this situation the movement of molecular chains is strongly restricted. The free volume in the material is too low to make it possible for chain segments of the size of one or more monomer units to move. Hence externally applied energy cannot be taken up and the product is rigid or even brittle. Above the transition point called the glass point, movements are possible and a rubber or a viscous liquid results. Even well flowing polymer melts, however, exhibit an elasticity. Such non-Newtonian behavior is called viscoelasticity. In a rubber elasticity dominates while viscous properties predominate in the liquid state. Temperature and time have a great effect on the possibility of transition between these states. Vulcanization of a rubber prevents viscous flow.

The glass transition temperature depends on the macromolecular structure i.e. rigidity of the chains (e.g. polystyrene is more rigid than polypropylene) and the interaction between chains (e.g. polyacrylonitrile is more rigid than polypropylene). The transition temperature is also influenced by the presence of low-molecular weight products. These can be monomers, oligomers or plasticizers. As examples of the latter, dioctyl phthalate and tricresyl phosphate are often used.

If the chain molecules have a regular character such a polymer is able to crystallize. Complete crystallinity does not generally occur because crystallization commences in different places. One molecule may be part of more than one crystallite. Other factors influencing the order are chain branches, chain ends or foreign structural elements. Non-crystalline material is always found between crystallites and it is of great importance whether this material is in the glassy or rubbery state. In the glassy state the product has a much lower impact strength. This can be improved by copolymerization, or by blending with a small amount of rubber.

The properties of the liquid phase are decisive in the processing of polymers to finished products. Molecular weight and molecular weight distribution as well as chain structure determine the melt behavior and the chemical stability. In order to improve processing, lubricants, stabilizers, antioxidants etc. are added.

Continuous processing is done by e.g. extrusion and calendering, discontinuous processing by injection and vacuum molding. A special process is spinning to make thin filaments. Linear macromolecules with a very low number of branches, a high molecular weight and a high crystallinity give a high tensile strength due to molecular orientations, which are further increased by subsequent stretching. In this way fibers are produced which can be used in textiles.

Chemically there are three mechanisms of reaction to build macromolecules: polymerization, polycondensation and polyaddition. Some monomers e.g. ethylene, propylene and butadiene can be polymerized in more than one way.

Polymerization

Polymerization is a linking of monomers without splitting off small molecules as a coproduct (e.g. water) and without migration of atoms or groups of atoms. It is also called addition polymerization. If two different monomers are combined, the process is called copolymerization; if three, it is termed terpolymerization. The reaction may proceed by a radical, ionic or coordination complex mechanism.

In a radical addition polymerization a chain reaction occurs in which the following reaction steps can be distinguished:

initiation: peroxides, persulfates and azo compounds decompose (thermally or catalytically) with formation of radicals, viz.

```
ØC-O-O-CØ → 2 ØC-O.
 ‖      ‖       ‖           (Ø = phenyl)
 O      O       O
```

propagation: (radical R.)

```
R. + C = C → R - C - C.      etc.
     ‖   ‖       ‖   ‖
     XY  PQ      XY  PQ
```

chain transfer:

```
     monomer         monomer.
R! + polymer → R'H + polymer.
     solvent         solvent.
```

termination:

R!' + R!'' → R'' - R''' or

R!' + R!'' → olefine + saturated molecule

Examples of a radical addition polymerization are found in Table XII.1.

By copolymerization two monomers can be built into the polymer randomly, viz.:

AAABABBAAAAB.....

If longer sequences of equal monomers occur (by using special reaction conditions) block polymers are formed:

A----AB----BA----A----

By grafting (special conditions again) graft-polymers are produced with this structure:

```
A---┬---┬---A
    |   |
    B   B
    .   .
    .   .
    .   .
    B   B
```

With a monomer PQ head-tail isomers can arise:

Table XII.1. Polymers made by radical addition

Polyvinyl-polymers based on $CH_2=CHX$ as a starting material

X	polymer	symbol
H	polyethylene	PE
Cl	polyvinylchloride	PVC
OH	polyvinylalcohol	PVA
CN	polyacrylonitrile	PAN
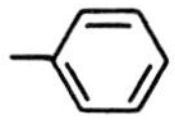	polystyrene	PS
$-O-C(=O)-CH_3$	polyvinylacetate	PVA
$-C(=O)-O-CH_3$	polymethylacrylate	PMA

Polyvinyl-polymers based on $CX_2=CY_2$

X	Y	polymer	
F	F	polytetrafluoroethylene	PTFE
Cl	H	polyvinylidene chloride	PVDC

Polydiene-polymers based on $CH_2=C(X)-CH=CH_2$

X	polymer
H	polybutadiene
Cl	polychloroprene

X[PQQP]Y X[QPPQ]Y X[PQPQ]Y

tail-tail head-head head-tail

Apart from chains containing C-atoms it is also possible to introduce other types of atom (e.g. polyethylene oxide, polyformaldehyde).

In an ionic mechanism two possibilities exist, i.e. cationic (via a carbonium ion) and anionic (via a carbanion). An example of the first group is the production of polyisobutene:

$$(CH_3)_3C^+ + CH_2 = C(CH_3)_2 \rightarrow PIB$$

Initiation takes place with Lewis or Brønstedt acids (e.g. BF_3 and H_2SO_4). The reaction velocity is very high. Usually termination occurs by transfer to some reactant in the system. When highly purified monomers are used termination may not proceed; if chain transfer to other positions in the monomer does not occur (e.g. styrene) the product is essentially monodisperse with regard to molecular weight, still contains active end groups, and is therefore called a 'living polymer'. Termination may then be brought about by adding an agent such as water. More useful is the addition of a comonomer. Block copolymers are produced in this way industrially. An example of the second group is the formation of polyisoprene:

$$R^- + CH_2{=}CH{-}\underset{\displaystyle CH_3}{\underset{|}{C}}{=}CH_2 \rightarrow PIR$$

Initiation succeeds with lithium alkyls.

Polymerization with coordination complexes uses activation of metals in complexes. For example it is possible to activate compounds with Ti(IV) aluminum alkyls, in which Al tends to form complexes with chlorine bridges as well as Ti. During activation the coordination number (4) is changed to (6) with simultaneous weakening of the Ti-Cl-bond:

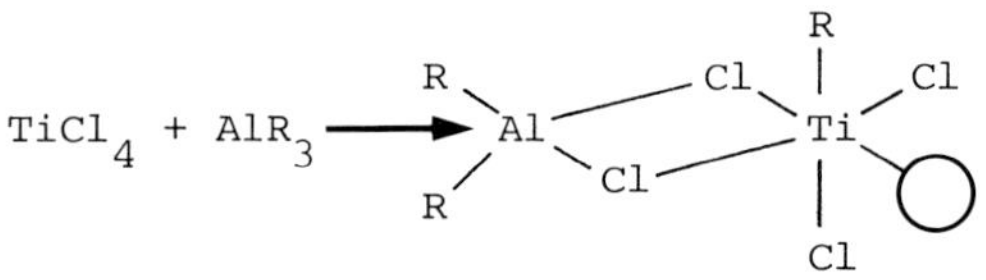

○ is a vacancy which can be filled by a monomer (e.g. ethylene). Then migration of a ligand takes place:

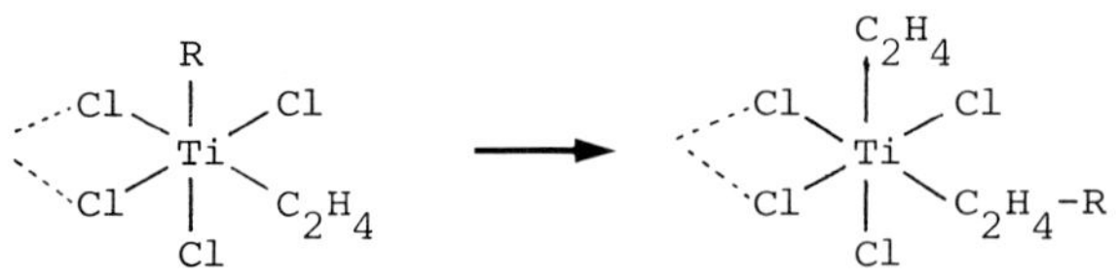

The process can be continued for a long time, resulting in high molecular weights. Such polymerizations occur homogeneously in solution or heterogeneously on the surface of solid metal catalysts (e.g. $TiCl_3$, VCl_3 etc.). Merit should be accorded to Ziegler, who first studied the catalyst-polymer coordination complexes, and also to Natta, who showed that use of the complex catalysts allows the production of stereospecific macromolecules. For example, propylene is always connected to the catalyst lattice in the same manner. In the final polymer all CH_3-groups point in the same direction (isotactic structure). Soluble catalysts result in syndiotactic polymers. A non-ordered product is atactic.

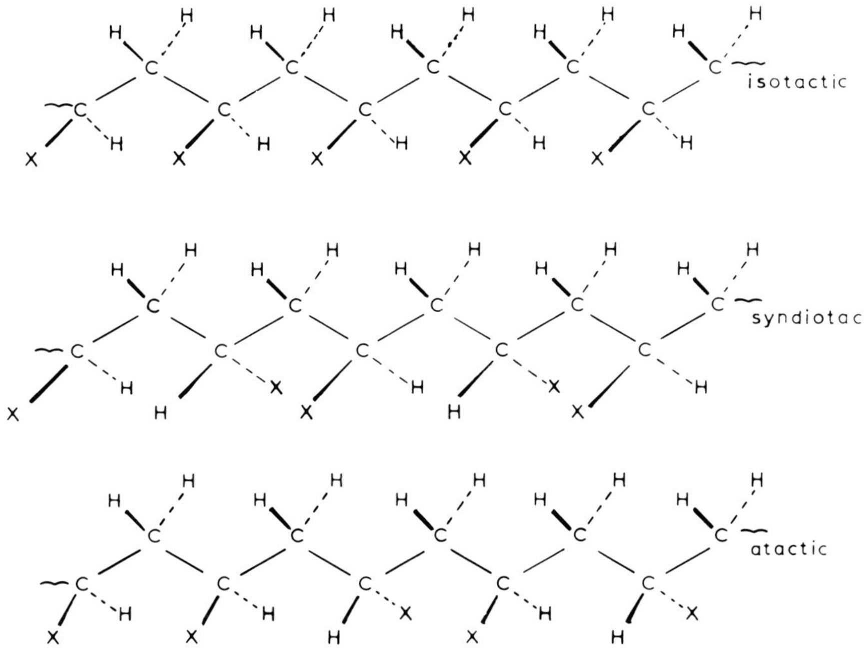

The degree of order is important in building up crystal lattices. Dienes can demonstrate cis-trans isomerism, viz.

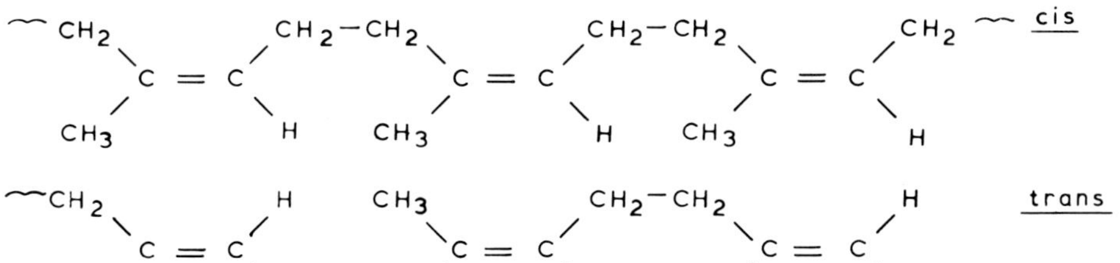

Coordination complexes come into play also in the medium pressure polymerization of ethylene (Phillips). With CrO_3 on a SiO_2-Al_2O_3 carrier, polymerization to highly ordered macromolecules is possible.

Polycondensation

Polycondensation leads always to the sequence ABABAB..... because bifunctional monomers react with simultaneous splitting off of small molecules. In this way polyesters may be synthesized from a bifunctional alcohol and a bifunctional acid. With a polyfunctional alcohol a 3-dimensional network results.

Silicones are polysiloxanes made by polycondensation according to

$$n\ \mathrm{HO-\underset{CH_3}{\overset{CH_3}{\underset{|}{\overset{|}{Si}}}}-OH} \rightarrow \mathrm{HO}\left[\mathrm{-\underset{CH_3}{\overset{CH_3}{\underset{|}{\overset{|}{Si}}}}-O}\right]_n\mathrm{-H} + (n-1)\mathrm{H_2O}$$

Nylons (polyamides) can also be produced by a polycondensation, viz.

$$n\ \mathrm{HOOC(CH_2)_4\,COOH} + n\ \mathrm{NH_2(CH_2)_6\,NH_2} \rightarrow -\left[\mathrm{NH(CH_2)_6\,NH\underset{O}{\underset{\|}{C}}\ (CH_2)_4\ \underset{O}{\underset{\|}{C}}}\right]_n- + \mathrm{H_2O}$$

Starting with a lactam, however, an addition polymerization is the mechanism

$$\overline{\mathrm{HN(CH_2)_5C}} = \mathrm{O} \rightarrow -\left[\mathrm{NH - (CH_2)_5\ \underset{O}{\underset{\|}{C}}}\right]_n- \qquad \text{(nylon - 6)}$$

Polyaddition

Polyaddition is a reaction in which monomers are combined in such a manner that atoms are shifted from one monomer to another. An example is the synthesis of polyurethanes:

step 1 : $\underset{\text{diamine}}{\mathrm{H_2N - R - NH_2}} + \mathrm{2COCl_2} \rightarrow \mathrm{O = C = N - R - N = C = O}$

+ 4HCl

step 2 : polyaddition, viz.

$$\mathrm{HO\text{-}R'\text{-}OH + OCNRNCO + HO\text{-}R'\text{-}OH \rightarrow HO-(R'O\underset{O}{\underset{\|}{C}}\overset{H}{\overset{|}{N}}R\overset{H}{\overset{|}{N}}\underset{O}{\underset{\|}{C}}O)R'OH}$$

(linear polyurethane)

Important monomers

Quite a number of monomers and their production have been treated in foregoing chapters: ethylene, propylene, butadiene, isoprene, vinyl chloride, acrylonitrile, styrene and vinylacetate. Other basic materials are adipic acid, terephthalic acid, glycol and others. Only the production of caprolactam and diamines will be treated here.

The essential feature of a lactam is a ring structure containing nitrogen in the ring. There are two technical possibilities to synthesize such a ring.

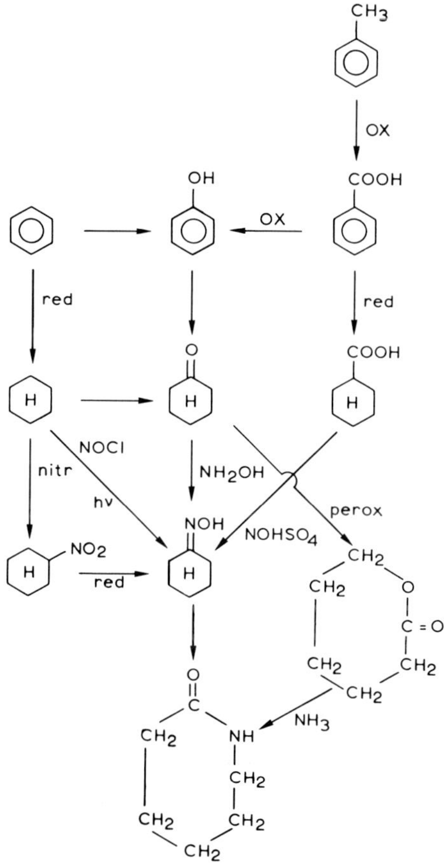

Figure XII.1. Routes to caprolactam

One, not yet in commercial practice, uses caprolactone, synthesized from cyclohexanone and a per-acid is converted to a lactam under high ammonia pressure. The second is based on a Beckmann rearrangement of cyclohexanone oxime:

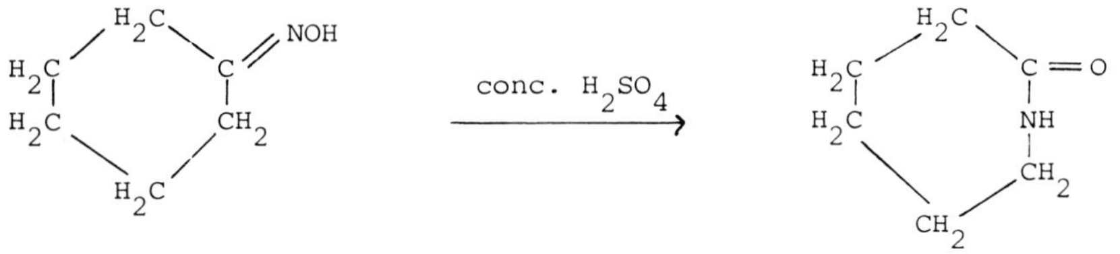

The necessary oxime can be made by a number of routes, as shown in Figure XII.1. One route starts with phenol which is hydrogenated to cyclohexanone. Another route consists of the oxidation or the nitration of cyclohexane. In a third one, hydrogenated benzoic acid is converted by nitrosyl sulfuric acid to the oxime. Figure XII.2. gives an idea of the lactam synthesis starting from phenol.

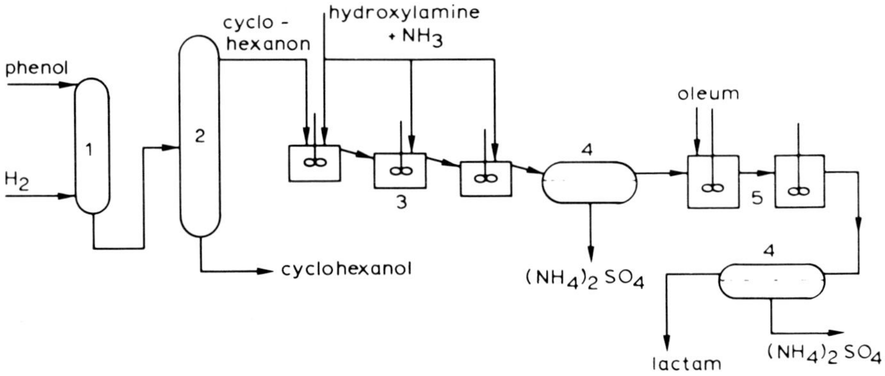

Figure XII.2. Synthesis of caprolactam

1. hydrogenation to cyclohexanone

2. distillation

3. oximation

4. separators

5. Beckmann conversion

Diamines are made by reacting adipic acid or higher dibasic acids with ammonia over a dehydrating catalyst to give the dinitrile, which is hydrogenated to the corresponding diamine. An interesting route also is the electrochemical conversion of acrylonitrile to adiponitrile:

$$CH_2{=}CH\text{-}CN + H_2O + e \rightarrow .\ CH_2CH_2\text{-}CN + OH^-$$

$$.\ CH_2CH_2CN + e \rightarrow :\ CH_2CH_2CN$$

$$:\ CH_2CH_2CN + \underset{(+)}{CH_2} - \underset{(-)}{\overline{C}H} - CN \rightarrow CN(CH_2)_3\ \overline{C}H\ CN \rightarrow$$

$$\rightarrow CN(CH_2)_4CN + OH^-$$

Polymerization processes

Six types of polymerization process can be distinguished: bulk-, solution-, precipitation-, emulsion-, suspension-, and gas-phase polymerization.

Bulk polymerization

With this type of process no diluting agent is used. The reaction is started by an initiator. If the polymer solidifies the reactor often has the form of the end-product (sheet or rod). The heat evolution must be controlled very effectively, since otherwise bubbles and cavities develop because of the poor heat conduction. One application of the technique is found in the production of polymethylmethacrylate. If the polymer is soluble in the monomer a continuous process can be realized (e.g. high-pressure polyethylene). Often the heat effects may be larger than in other processes.

Solution polymerization

The polymerization can be carried out in many cases in such a manner that polymer and monomer both are soluble in a medium. The heat of reaction can be removed easily by evaporation and the low viscosity of the solvent (e.g. naphtha). The solvent is condensed and recycled. The solvent may have an effect on the polymerization because of radical chain transfer. In general high molecular weights cannot be produced in this way. The process is a natural choice for lacquers and glues. Homogeneous Ziegler-Natta catalysts require this type of process.

Precipitation polymerization

This method is a variant on the foregoing one, with the essential difference that only the monomer is soluble, not the polymer. The advantage is clearly that the polymer can easily be separated from the mixture. A difficulty may be agglomeration of the polymer. With Ziegler-Natta catalysts the system becomes heterogeneous at high molecular weights.

Emulsion polymerization

A medium in which most monomers do not dissolve is water. To keep the monomer in a dispersed state emulsifiers are used. Initiator and catalyst (H_2O_2 or persulfate and Fe^{3+}) are dissolved in the water. Often moderators to influence the molecular weight are added, making the system as a whole rather complicated. The amount of emulsifier (a few %) must be sufficient to stabilize the monomer as droplets as well as to build a number of micelles in which the monomer is solubilized. Radicals react with monomer, obtaining a degree of surface activity in the process. These enter the micelles and react further. The number of micelles is so much larger than the number of droplets that these are the most important. The micelles grow more and more into microglobules (latex). The droplets take care of the transport to the micelles via diffusion through water. See Figure XII.3.

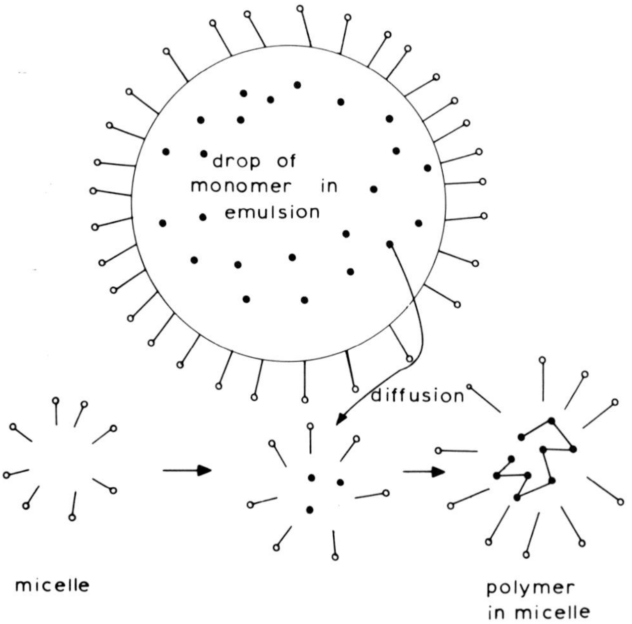

Figure XII.3. Emulsion polymerization

To produce a polymer with a narrow molecular weight distribution a cascade of reactors is often used. In the final processing step the latex is coagulated or the water is evaporated.

Suspension polymerization

In this process water again is used as a dispersing medium but the reaction takes place *within* the droplets of monomer. Effective stirring and possibly a protective colloid are necessary to counteract agglomeration. The initiator is added to the monomer. Solid polymer granules (pearls) are washed and dried after separation.

Production of polymers

Polyethylene: four processes exist for the production of polyethylene and are distinguished by the applied pressure: 150, 5 and 0.1 MPa, and the phase in which the polymerization takes place. The high pressure process is the oldest and dates from 1939 (ICI). At a temperature of 200 °C traces of oxygen trigger a radical polymerization. The amount of oxygen is extremely important in order to maintain a proper stationary state. Too little oxygen extinguishes the reaction, too much leads to an explosion. Figure XII.4. gives a schematic impression. At these pressures the system is supercritical and the polymer dissolves in the monomer. After the ICI development of stirred autoclaves BASF introduced a long tubular reactor to control the process better. In all cases the degree of conversion is kept low and ethylene is recycled. Because of chain transfer the polymer molecules are highly branched. The medium pressure process

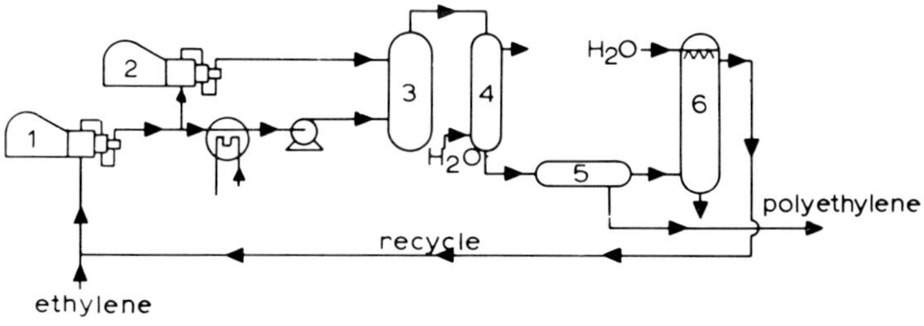

Figure XII.4. High pressure polymerization of ethylene

1, 2. compressors

3, 4. autoclaves

5. separator

6. scrubber

(Phillips) uses a fixed bed of solid catalyst consisting of reduced chromium oxide. A fluid bed or suspension may also be used. The polymerization is carried out at about 165 °C in naphtha as a solvent. The crystallinity of the polymer is very high (little branching) and hence the density is higher than in the high pressure process.

The low pressure process uses Ziegler-Natta catalysts at a temperature of about 100 °C. Aliphatic hydrocarbons are used as solvents but the polymer separates if molecular weights are high. The polymerization is terminated by treatment with methanol. It is important to wash the product effectively because residues of catalyst give rise to coloration and instability of the product. The product has a high crystallinity because of a low degree of branching (nearly as high as for the medium pressure process). This has an effect on mechanical properties and stability. Tertiary hydrogen atoms are abstracted more readily than primary or secondary hydrogen atoms and influence the sensitivity to oxygen and UV-light.

At the time of the development of the medium and low pressure process it was expected that the days of the ICI process were numbered because of the costs of high pressure equipment and use of energy. This has not occurred due to extra costs of catalyst, washing, purification and polymerization. Even more important is the fact that the mechanical properties differ in such a way that the products have different applications.

A fourth process uses a fluid bed to polymerize ethylene in the gas phase at 0.7-2 MPa and about 100 °C. Ethylene is circulating and removes the heat of reaction, besides supplying monomer. The polymer is removed from the reactor as a powder. The modified Ziegler-catalyst is so active that it is unnecessary to separate the catalyst from the polymer.

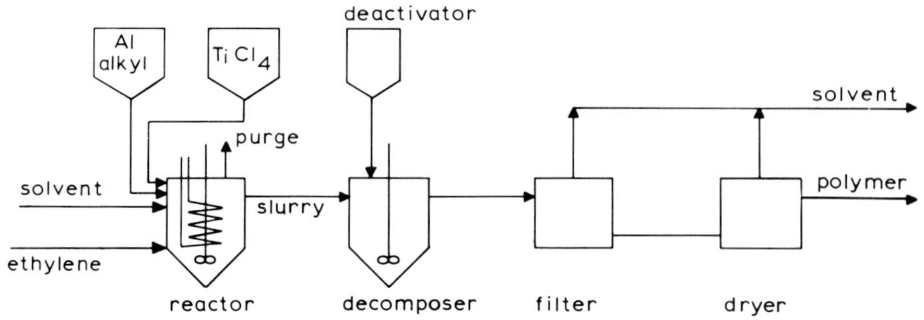

Figure XII.5. Polymerization of olefins at low pressure

Polypropylene and ethylene-propylene copolymers: these polymers are also made in processes analogous to the low pressure polyethylene process with Ziegler-Natta catalysts. Polypropylene has a higher melting point, a higher elasticity modulus and a lower density than polyethylene.

Ethylene may be copolymerized with vinyl acetate, acrylates and others. A very interesting copolymer is one with propylene. The ratio of ethylene to propylene strongly influences the properties of the final product. At a ratio $C_2 : C_3 = 2 : 1$ a rubberlike product results. Vulcanization is a special problem for this product, which has been solved in the following way: a first solution was to create active sites with a peroxy compound to make crosslinking possible. A better solution is to use a third monomer which still contains C = C double bonds after polymerization. Isoprene and butadiene are too active but dicyclopentadiene or 1,4-hexadiene suit the purpose.

Polyvinylchloride: PVC can be produced by bulk-, emulsion- and suspension polymerization. Because the monomer has a boiling point of -14 $^{\circ}C$ at 0.1 MPa these processes always work under pressure (0.5 - 1 MPa). The temperature is about 50 $^{\circ}C$. Care must be taken to prevent oxygen from being built into the polymer because this lowers the stability. Suspension polymerization gives a purer polymer, resulting in a clearer product and better electrical insulation properties. It must be carried out, however, by a discontinuous method.

Although the glass transition temperature is not very high (about 90 $^{\circ}C$) the area of application is large. Often plasticizers are added (20 - 60%) to improve flexibility and impact strength. Esters of phthalic acid, sebacic acid etc. are used. It is also possible then to lower the processing temperature, whereby splitting off of hydrochloric acid can be avoided. Degradation of the finished product by oxygen and UV light is counteracted by stabilizers.

A number of copolymers are produced, e.g. with vinylidene chloride, propylene, acrylonitrile and vinyl acetate. The melt flow properties of the latter are such that this is the material of which gramophone records are made.

General purpose rubbers

Natural rubber: two types of polyisoprene occur in nature. The cis-isomer is natural rubber, and the trans-isomer gutta-percha. Only the first one has elastic properties because only in the cis-conformation do the chains have the necessary freedom. The dense packing of chains in gutta-percha causes it to crystallize more readily so that it behaves like a thermoplast. The production of synthetic rubbers is much larger than of natural rubber.

Styrene-butadiene rubbers are synthesized by emulsion polymerization of styrene and butadiene (and have been since 1937, when I.G. Farben introduced Buna S). Hydroperoxides or persulfate are initiators which control the product properties together with reaction temperature, and the process parameters of temperature and concentration distribution over a cascade of mixers. These effects influence the emulsification and chain reactions. The large variation in possibilities makes the product very suitable for use in tires, conveyor belts, cables, shoes etc. Many fillers are used of which carbon black and oil are the most common. SBR can be blended with other elastomers, e.g. with polybutadiene. Other general-purpose rubbers include *polybutadiene* and *polyisoprene* produced by stereospecific polymerization.

Specialty rubbers

Butyl rubbers have a remarkably low permeability for gases (tire inner tubes). Friedel-Crafts catalysts, such as BF_3 polymerize isobutene at low temperatures to a rubberlike polymer. This material does not possess, however, residual double bonds. Therefore 0.5 - 3 mole% isoprene is copolymerized at -100 ^{o}C and normal vulcanization with sulfur can be used, albeit somewhat more slowly. Very high molecular weights make processing difficult, hence moderators such as diisobutene are added. These promote chain transfer. Because of the relatively low number of double bonds, the polymer is more stable against O_2 and O_3 than most of the other rubbers.

Nitrile rubbers are made by copolymerization of butadiene and acrylonitrile. An increasing percentage of the polar -C≡N group diminishes the solubility in oil and gasoline making it attractive in seals, tubes, gaskets etc. Other specialty rubbers are poly(2-chlorobutadiene) ('Neoprene'®) which does not corrode easily and an elastic poly(glycoldiisocyanate) (one of the polyurethanes) which can be spun ('Lycra'®). Isoprene can be polymerized stereospecifically. If the product contains not more than 92% cis-isomer the rubber has an excellent flow behavior at higher temperatures, which facilitates injection molding. It is possible to increase the cis-percentage to 96%, coming very close to natural rubber (97 - 98%).

Polyamides

In search of synthetic macromolecules which could equal wool and silk N.H. Carothers of E.I. du Pont de Nemours in 1930 found products which were characterized by a -NHCO- bridge between monomer units, the same as is found in the natural materials. He started with ε-aminocaproic acid:

$H_2N(CH_2)_4\ CO_2H$

By condensation a polymer was obtained which could be spun, although a new spinning method had to be developed (melt spinning). Technically it is more interesting to react hexamethylenediamine and adipic acid:

$$NH_2(CH_2)_6\ NH_2 + HOOC(CH_2)_4\ COOH \rightarrow NH_2(CH_2)_6\ NHCO(CH_2)_4\ CONH.....+ x\ H_2O$$

In 1939 Dupont were the first to present 'nylon'.

At I.G. Farben in Germany a parallel development took place. In 1939 an analogous polymer was synthesized from ε-caprolactam:

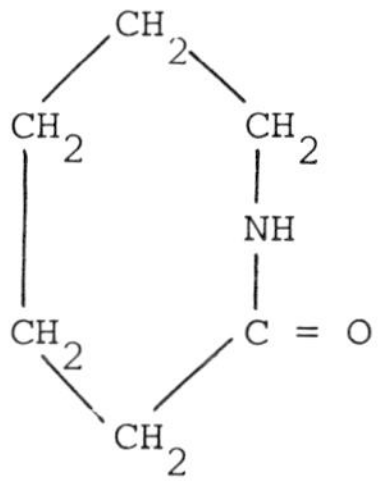

$$\rightarrow\ NH_2(CH_2)_5\ CONH\ (CH_2)_5\ CONH.....$$

The product ('Nylon 6') has similar properties to the 'Nylon 66' although not exactly alike. For example the melting point of the first is 215 ^{o}C in comparison with 251 ^{o}C for the latter. The field of application is the same (textile fibers as well as industrial fibers). For a long time though the main product in the U.S.A. has been 'Nylon 66' and in Europe 'Nylon 6'. Nowadays more 'nylons' have been developed. They are indicated by a code related to the number of C-atoms in the monomers: e.g. 'Nylon 6, 10' from hexamethylene diamine and sebacic acid. The main properties are: high tensile strength, high melting point, relatively durable as a construction material, low density and reasonably good elastic properties. Woven products exhibit a large resistance against wear, tear and creasing. If the properties of a number of nylons are compared it can be concluded that the number of CH_2 groups has a strong influence (Table XII.2.).

Table XII.2. Properties of polyamides

Nylon	Density $g \cdot ml^{-1}$	Melting pt. ^oC
5	1.13	240 - 250
6	1.14	235
7	1.10	223
8	1.08	233
9	1.09	209
12	1.01	179
6,6	1.13	265
6,10	1.08	223

An increasing number of CH_2 groups gives a tendency to shift into the direction of a polyolefin. The melting points are shown in Figure XII.6. in which different diamines have been coupled to adipic acid (solid line) or sebacic acid (broken line).

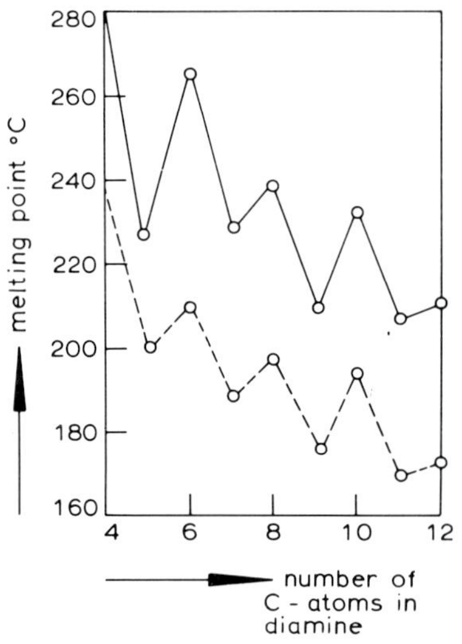

Figure XII.6. Dependence melting point of nylons on type of acid and base

Chains with an odd number of CH_2 groups show a lower melting point than those with an even number. This can be understood by paying attention to the crystal structure (Figure XII.7.).

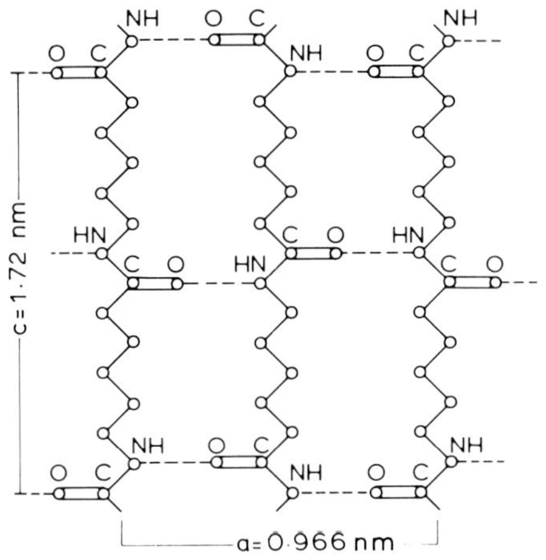

Figure XII.7. Crystal structure of nylon

The chains (which are already rather stable) are connected by hydrogen bridges. The strength is accordingly determined by the number of these bridges. Especially orientation in one direction will have a strong influence. This can be promoted by using regular chains (e.g. avoid the use of more than one dibasic acid) and keeping the number of side chains low. Another positive effect is to keep the structural units rather short. The number of hydrogen bridges in Nylon 6 is half the number in Nylon 66 with the consequence that the melting point is much lower. Because of the strong bridges the molecular weight can be relatively low. For all nylons these are in the region 15000-25000, compared with 50000-100000 in a polyolefin fiber and about 300000 in a natural cellulose fiber. The length of the nylon molecule must be > 1000 Å.

The effect of orientation is evident from Table XII.3. in which the tensile strength is given before and after stretching.

Table XII.3. The influence of stretching on tensile strenght (MPa)

	nylon 6,6	nylon 6
before stretching	75	55
after stretching	350	300
ultimate strain before	85%	325%
ultimate strain after	25%	30%

Nylon 66 is produced by polycondensation from dicarbonic acids and diamines. The reaction goes to completion if the by-product water is removed (usually by vacuum). Care has to be taken, however, as the water can result from an intermolecular reaction as well as an intramolecular condensation. The risk of the latter is only small if the chains are long (e.g. Nylon 11). The production of Nylon 66 is carried out via an intermediate ('Nylon salt'). This is a neutral salt from acid and diamine i.e. hexamethylene diammonium adipate. The acid and diamine are dissolved in methanol from which the nylon salt precipitates. The advantage is great purity and neutrality. The condensation is carried out discontinuously at 220 - 280 ^{o}C in an autoclave. Water is blown off at constant pressure.

Caprolactam is polymerized continuously with an initiation by the ring opening by water:

$$\underbrace{C(=O)-(CH_2)_5-NH}_{\text{ring}} + H_2O \rightleftarrows HOOC(CH_2)_5NH_2$$

The ease with which the ring opens depends on the number of methylene groups. Dodecalactam is converted at 260 ^{o}C to a degree of only 50% in a reaction time of 30 hrs, which is sufficient for caprolactam. The polymerization is carried out in a vertical tubular reactor.

Polyesters

Coupling of multivalent acids and alcohols leads to polyesters. Divalent components result in thermoplastics which can be spun. A higher valency produces a network and the resulting polymer is a thermosets. Because the strength of the latter type is not sufficient for use as a construction material these polyesters are combined with glass fibers.

Although polyesters based on divalent acids and alcohols do not have the strong hydrogen bonds of polyamides, by proper choice of the components polymers can be produced having sufficient strength. The crystallinity is also high and the material can be spun into fibers or processed into biaxially oriented films. Polyesters based on terephthalic acid and ethylene glycol have a large share of the market.

Two methods are used in the polymerization: a discontinuous one based on the conversion of dimethyl terephthalate (DMT) with ethylene glycol into a glycol terephthalate followed by catalytic polycondensation, and a continuous process, also based on DMT. The reason for using DMT as a basis is the difficult oxidation of the second methyl group in p-xylene. This can be improved, however, by oxidizing the methyl ester of p-toluic acid in a second stage.

REFERENCES

General

1 Jenkins, A.D., 'Polymer Science', Vol. 1-2, North Holland Publ. Co., Amsterdam 1972.

2. Billmeyer, F.W., 'Textbook of Polymer Science', J. Wiley and Sons, New-York 1962.

3. Mark, Editors H.F., S.M. Atlas, E. Cernia, 'Man-made Fibers', Vol. 1-3, Interscience Publ., New-York 1967.

4. Boor Jr., J., 'Ziegler - Natta Catalysts and Polymerization', Academic Press, New-York 1979.

5. Kaufman, Editors H.S., J.J. Falcetta, 'Introduction to Polymer Science and Technology', J. Wiley and Sons, New-York 1977.

6. Mark, Editors H.F., N.G. Gaylord, N.M. Bikales, 'Encyclopedia of Polymer Science and Technology', Interscience Publ., New-York 1964 -

7. Morton, M., 'Rubber Technology', Van Nostrand Reinhold Co., New York 1973.

Special

1. Hopff, H., A. Müller, F. Wenger, 'Die Polyamide', Springer Verlag, Berlin 1954.
Figure XII.6. is reproduced from this book with permission.

2. Bunn, C.W., E.V. Garner, Proc.Roy.Soc. (London), A 189, 39 (1947).
Figure XII.7. is reproduced from this article with permission.

XIII. MANUFACTURE OF GASOLINE COMPONENTS

Introduction

The composition of many petroleum products such as gasolines and lubricants is related in a very complex manner to product performance and type of application. Moreover, it is usually impossible to determine product composition in terms of individual chemical compounds. Although a simpler type of analysis is possible in which the composition is determined in terms of groups of components (e.g. aromatics, olefins, saturated hydrocarbons), the results of such analyses seldom reflect product behavior. It is, therefore, necessary to devise methods of testing products in which the performance characteristics are assessed as accurately as possible. This is a difficult problem for every complex product because performance tests should not only predict the behavior in a certain application, but also take as little time and effort as possible. Otherwise, new products cannot be developed in acceptable periods of time and quality control during processing is impossible.

This makes it often necessary to accelerate tests. A good example is the determination of oxidation stability of insulating oils. Modern oils are of such a high quality that their useful life in transformers often exceeds ten years. Yet it is necessary to test them in but a few days time; this is accomplished by accelerating the test by using higher temperatures, oxidation catalysts, higher oxidant concentration and/or more intense mixing. The danger then is that the results are no longer representative of the performance of the product in practise. For instance, the distribution of products obtained in an oxidation often depends on the ratio between the rates of chemical reaction and oxygen transfer from gas to liquid phase. Another problem is that the ratios of the rates of various steps of a complex set of oxidation reactions varies with temperature.

A further complicating factor is that the type of equipment in which the product is applied may vary widely. An example is that motor oils for use in internal combustion engines must perform well in many makes of engine. Moreover, conditions of use, e.g. the type of driving and the area of use, may differ vastly. Thus, in view of the extreme complexity of the field of application, motor oils are usually tested in a few standard laboratory tests in the earliest phase of their development. After a first selection of adequate products has been made, more extensive testing is required, e.g. in laboratory test engines, to limit the number of candidate products still further. The final phase, the only one in which the true performance in the field can be predicted with any great reliability, consists of road tests of relatively long duration. Even then, if the product holds up well, it is usually necessary to perform a market trial in a restricted area before full-scale marketing is attempted.

There are also, of course, simpler products, such as fuel oils. These are supplied on the basis of relatively short specification in which the most important criteria are viscosity and sulfur content. As a general rule, the price of a fuel is higher the lower its viscosity and sulfur content. Other significant properties are carbon residue remaining after a standardized pyrolysis test and pour point, i.e. the temperature at which the oil just ceases to flow because of a very high viscosity or the formation of a network of wax crystals. This is not only a complex criterion, because it depends on the thermal history of the fuel, but also because its relation with pumpability at low temperatures is not a straightforward one.

Another basic feature of the petroleum industry is that a relatively large number of products is made from a single feedstock, crude oil. The type of crude oil influences the distribution of straight-run products, i.e. products obtained by distillative separations only. Since the market usually demands a different ratio and because the quality is rarely adequate, conversion and refining processes must be applied to change product ratios and improve quality. The yield of gasoline should normally be raised and that of heavy fuel lowered. An additional complication is that the demand ratios vary in time and with the location. An example is shown in Table XIII.1.: gasoline demand in the USA is much higher and that of fuel oil lower than in other parts of the world. This may change to some extent as the pressure on car manufacturers to produce more economic engines and smaller models increases. Other factors contributing to future changes are tax structure, cost of other fuels like natural gas or coal, possibilities of converting e.g. some power stations to coal, the development of electricity generation in nuclear power stations, customer behavior, and the like. Generally speaking, petroleum refiners should therefore aim at a certain operational flexibility of their refineries. Since this causes capital investment to rise, careful forecasting of future market developments is a must.

Table XIII.1. Demand structure for oil products (years: 1975-1976)

	Approximate demand (in vol%) in:	
	World, including Communist Countries and USA	USA
Gasoline	25	45
Aviation fuels	5	5-10
Middle distillates *	20-25	20-25
Fuel oil	25-30	10
Miscellaneous	15-20	15-20

* gas oil and kerosene

This chapter discusses the manufacture of gasoline components in some detail. After a short review of the relation between performance properties and composition, a number of processes for the manufacture of gasoline components is described. The chapter ends with a brief survey of new developments and their significance for the oil industry, both as regards the type of processes needed and the effects on refinery structure as a whole.

Gasoline quality requirements

Anti-knock quality

Gasolines should have good anti-knock characteristics to perform well in modern high-compression ratio engines. Knock is caused by too rapid combustion of the fuel. After compression and ignition by the spark, a flame front is formed which travels through the unburned gasoline/air mixture in the combustion chamber. It is preceded by a pressure wave and a heat wave, which cause the unreacted mixture ahead of the flame front to ignite if the anti-knock characteristics of the fuel are inadequate. Another possibility is that the homogeneous oxidation reactions ahead of the flame front are accelerated, thereby increasing the speed of the flame front. The pressure wave caused by this phenomenon and the knock, a metallic sound produced under these circumstances, have the same frequency. Knock leads to power loss and, in serious cases, to engine failure.

There is yet another phenomenon with consequences similar to those of knock, viz. pre-ignition. It is defined as the early formation of one or more flame fronts in the compressed fuel/air mixture, hot deposits and engine parts acting as ignition sources. The result is power loss because ignition starts before the piston has completely compressed the mixture. Unlike knock, pre-ignition is not directly related to gasoline quality.

Many engine conditions influence knock. Apart from gasoline composition, engine load and speed, temperature and relative humidity of the inlet air, the construction of the engine is an important factor. In fact, knock became a major problem when engine manufacturers raised the compression ratio to increase efficiency, in other words to decrease fuel consumption. To develop the improved fuels needed to combat knock, a test method was needed to determine anti-knock quality. In view of the complexity of the phenomenon the only reliable method is to use a rigidly standardized engine test. It was developed in the USA jointly by the industries concerned, who formed the Cooperative Fuel Research Committee. The resulting single-cylinder engine, the CFR engine, is now used throughout the world.

Anti-knock quality is expressed as octane numbers. The number 100 was assigned somewhat arbitrarily, to a hydrocarbon having excellent anti-knock characteristics (2,2,4-trimethylpentane, isooctane) and another showing poor performance (n-heptane) was given the octane number 0. The gasoline to be tested is compared with a mixture of these two hydrocarbons; if it shows the same anti-knock properties as a blend of N vol.isooctane and (100-N) vol. n-heptane, it is assigned the octane number N. Octane numbers above 100 and below 0 can be determined by adding known amounts of n-heptane or isooctane to the product to be tested and then determining the octane number.

Two different octane numbers are used most frequently:
- the Research Octane Number (RON, designated as F-1), determined by ASTM method D 908 under relatively mild engine conditions;
- the Motor Octane Number (MON, also F-2) found by applying ASTM method D 357. The conditions of determination are more severe than for the RON in respect of engine load and speed, air temperature and relative humidity.

The MON is a better yardstick of performance under high-speed driving conditions. The difference (RON-MON) is called the sensitivity of the fuel. It is usually positive and indicates how sensitive the fuel is to increases of severity in engine conditions. Table XIII.2. shows how dependent the octane number is on hydrocarbon structure. The following general rules apply:
- the octane number of alkanes and alkenes is higher as the molecular weight

Table XIII.2. Octane numbers of hydrocarbons

Hydrocarbon	ON(F1-0)*	Sensitivity**	Hydrocarbon	ON(F1-0)*	Sensitivity**
n-alkanes			Cyclanes		
n-butane	94	+5	cyclopentane	102½	+17½
n-pentane	62	0	cyclohexane	83	+ 5
n-hexane	25	-1	methylcyclopentane	91½	+11½
n-heptane	0	0	methylcyclohexane	67	+ 6
Isoalkanes			Alkenes		
2-methylpentane	73½	0	n-hex-1-ene	76½	+13
3-methylpentane	74½	0	n-hex-2-ene	92½	+12
2,2-dimethylbutane	92	-1½	2-methylpent-1-ene	94	+13
2,3-dimethylbutane	103½	+9	2-methylpent-2-ene	98	+15
			4-methylpent-2-ene	99½	+15
			2,3-dimethylbut-2-ene	97½	+17
Aromatics					
benzene	106	+3½			
toluene	120	+16½			
ethylbenzene	107	+ 9			
m-xylene	117½	+ 2½			
p-xylene	116½	+ 6½			
n-propylbenzene	111	+12			
isopropylbenzene	113	+14			

* ON(F1-0) = research octane number unleaded

** sensitivity = research octane number minus motor octane number (both unleaded)

*** the trans isomer

decreases and the degree of branching increases;

- alkenes usually have higher octane numbers than the corresponding alkanes,
- aromatics and, to a lesser extent, cycloalkanes are desirable components of a gasoline;
- high sensitivity is shown by most aromatics and olefins.

The overall reason behind the above rules is that those components have high octane numbers from which a hydrogen atom is abstracted relatively easily to give a rather stable radical. With such behavior oxidation starts at a relatively low temperature; further reaction is then rather slow because of the higher stability of the intermediates. For example, high-molecular weight n-alkanes do not contain tertiary hydrogen which is abstracted most readily. If a radical is formed, at a somewhat higher temperature, it tends to be so unstable that very rapid decomposition leading to strong chain branching sets in. This accelerates the reaction to such an extent that knock may occur.

It is possible to raise the octane number of gasolines by additives; tetra-alkyl lead compounds (alkyl groups: methyl and ethyl) are used for this purpose. Although many other compounds, e.g. containing manganese, show similar effects, lead compounds are the most economic. The improvement of the octane number is most striking at low octane level. Addition of lead alkyls enables the refiner to use a cheaper, lower-quality base gasoline. It is necessary to add a scavenger to convert the lead into relatively volatile salts; alkyl chlorides and bromides have this effect. This prevents excessive build-up of lead-containing deposits in the engine. An unwanted side effect is, however, emission of lead salts. Table XIII.3. shows how the changes in operating conditions affect octane requirements, i.e. the variation in research octane numbers to be applied to ensure knock-free running.

Table XIII.3. Effects of changes in engine conditions on octane requirement

Variable	Change*	RON required*
Relative humidity	+	-
Engine speed	+	-
Air temperature	+	+
Spark advance**	+	+
Cooling temperature	+	+
Altitude	+	-
Engine fouling	+	+

* + denotes an increase and - a decrease

** change by a few degrees

Volatility characteristics

The boiling range of a gasoline is important in several respects:
- the initial boiling point and the temperature at which 10%v has distilled

over should be sufficiently low to obtain a good startability at low temperatures;

- the temperature at which 50%v has distilled over is related to the period required for engine warm-up. This can also be expressed in the form of Figure XIII.1.a. where the minimum percentage distilled at 70 °C needed for adequate warm-up rates is plotted as a function of the lowest ambient temperature. It follows that a gasoline should contain more volatile components in winter. Sufficiently rapid warm-up is needed because corrosive wear and rust formation occur to a marked extent as long as the operating temperature is low. This is caused by the formation of an acidic and highly corrosive aqueous phase in the engine at low temperatures;

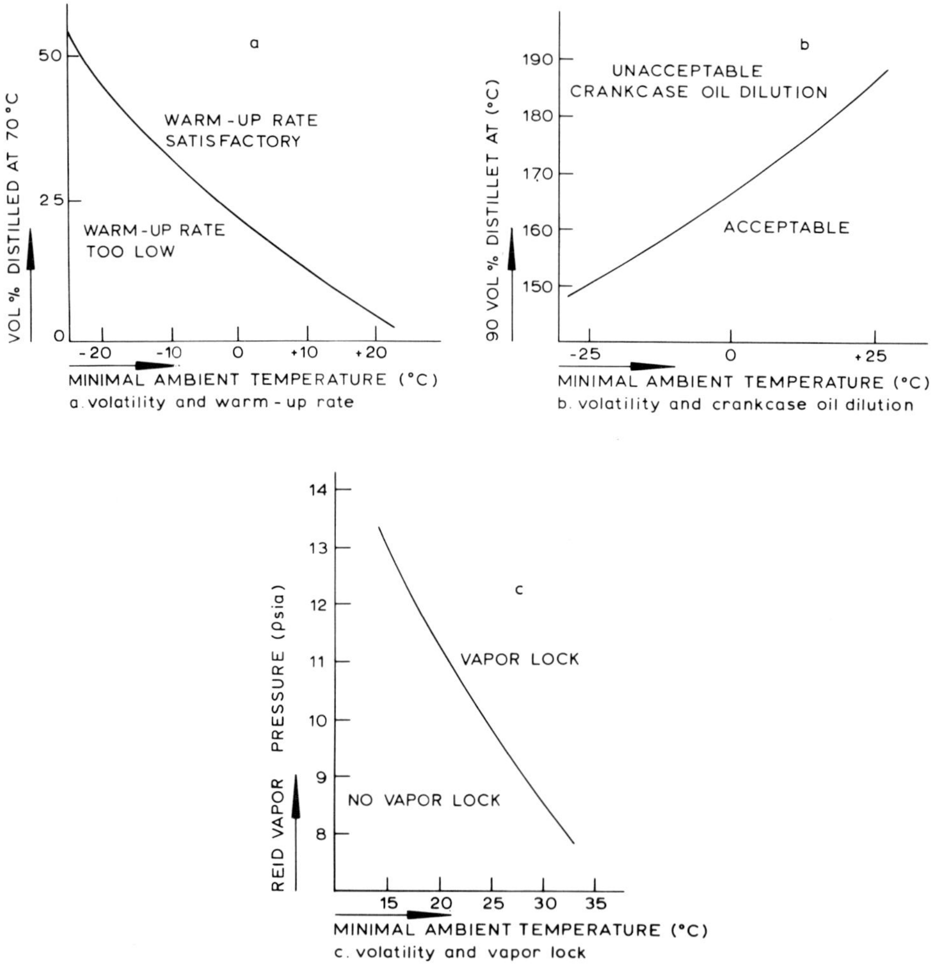

Figure XIII.1. Effects of volatility on gasoline performance characteristics

- the product should not contain substantial amounts of high-boiling components because otherwise engine fouling and excessive crankcase oil dilution by fuel components will occur. The maximum 90%v distillation temperature needed to avoid this problem is shown in Figure XIII.1.b. as a function of ambient temperature;
- the tendency to vapor lock, engine stalling due to the formation of vapor bubbles in the tubing upstream of the carburetor, is related to the vapor pressure. Figure XIII.1.c. shows the vapor pressures above which vapor lock occurs.

The above volatility characteristics are determined by a rigidly standardized ASTM method.

Another desirable property of gasolines is that the octane quality of low-boiling as well as high-boiling fractions of the gasoline are not too different. If, for instance, the low-boiling fractions have a low RON, those cylinders receiving more volatile fuel components may suffer from knock, although the gasoline as a whole meets the specification. Such maldistribution frequently occurs in multi-cylinder engines when the temperature varies appreciably over the inlet manifold.

Other product properties

There are many other requirements with which a gasoline has to comply. Usually, additives are present in the product to achieve this. Examples are:
- the tendency to preignition as a result of deposit formation in the combustion chamber or on the piston crown increases with engine age. It can be combated to some extent by the addition of e.g. phosphorus compounds which modify the deposits and also reduce the amount of deposit formed;
- anti-oxidants and metal deactivators are added to reduce gasoline oxidation and oxidation catalysis by metals, respectively. Oxidation may lead to the formation of gums and deposits in the inlet system. Both interfere with engine operation;
- corrosion inhibitors are added to minimize corrosion in the tank and the fuel system. Tubing is made of copper, which is a strong oxidation catalyst if dissolved in the fuel as ions as a result of corrosion;
- anti-icing additives serve to prevent the formation of ice in the carburetor caused upon cooling by gasoline evaporation. Some of these lower the melting point (isopropanol) and others prevent the adhesion of tiny crystals which would otherwise grow to sizes large enough to cause blockage;

- detergents and other types of additives are added to reduce the formation of deposits in the carburetor, on the inlet valves, etc.

Processes for gasoline components

Among the processes for making gasoline components cracking processes and reforming processes are the most important. In cracking processes, the hydrocarbons in the feedstock, usually a heavier distillate or sometimes a long residue, are broken down into fragments, most of which have atmospheric boiling points in the gasoline range. Reforming does not cause much difference in boiling range of molecular weight, but changes the naphtha feed into hydrocarbons having much better anti-knock properties. Apart from these two main categories of processes, there are others like alkylation and polymerization, by which high-quality gasoline components can be made from low-molecular weight hydrocarbons. Oil refineries usually comprise still other processes intended to change the yield ratios of the three major product classes: gasoline, middle distillates and fuel oil. Some of the conversion processes are listed in Table XIII.4., together with processes for making gasoline components. The latter will now be discussed first; later on, the other refining tools are touched upon very briefly in connection with the overall structure of oil refineries.

Thermal cracking

Thermal cracking of residual oils was introduced around 1920 when the straight-run product used up to that time became inadequate quality- and quantity-wise. The process yields gasoline, light gaseous components, a residue and coke. However, in view of the absence of isomerization in thermal cracking (see Chapter VII.) the product consisted of lightly branched or straight-chain paraffins and olefins, components having rather low octane numbers. Moreover, the yield was rather low and the formation of large amount of almost worthless coke troublesome from an operational point of view. Therefore, the process was abandoned when better methods like catalytic cracking and reforming became available in the late 1930s or early 40s. The same applies to thermal reforming, a mild thermal treatment aiming at some improvement of octane quality

Catalytic cracking

General: When heavy distillates are cracked in the presence of acidic catalysts products are obtained having much higher octane numbers than thermally cracked

Table XIII.4. Conversion processes

Process	Feedstock(s)	Main product(s)	Remarks
Cracking processes			
Thermal cracking	vacuum residues	gasoline components	obsolete, see text
Thermal cracking	s.r.naphtha or gas oil	C_2-C_4 unsaturates	see Chapter VII
Visbreaking	vacuum residues	fuel oil of lower viscosity; middle distillates	fairly mild conditions for product stability
Delayed coking	vacuum residues	fuel oil of lower viscosity; middle distillates, coke	more severe conditions than visbreaking; see text
Flexicoking	vacuum residues	middle distillates fuel gas	gasification combined with thermal cracking; see text
Catalytic cracking	heavy distillates	gasoline components	see text
Hydrocracking	distillates	hydrocracked naphtha and kerosenes	catalytic process at high H_2 pressures; see text
Hydrocracking	distillates and deasphalted oils	low-sulfur fuels and distillates; lubricating oil base stocks *	as above
Reforming processes			
Thermal reforming	naphtha	gasoline components	virtually obsolete
Catalytic reforming	naphtha	gasoline components	see text
Synthesis processes			
Polymerization	light olefins (C_2-C_4)	C_6-C_{12} branched olefins; for gasoline	formation of oligomers; see text
Alkylation	light olefin + light isoalkane	strongly branched alkanes; for gasoline	see text
Other processes			
Isomerization	light paraffins	light isoparaffins	usually hydroisomerization. See text

* often named hydrotreatment

Table XIII.5. Catalytic cracking reactions

1. Carbonium ion formation (initiation):

- addition of H^+ to e.g. an alkene:

$$RCH_2CH_2CH_2CH{=}CHCH_2CH_2CH_2R' + H^+ \rightarrow RCH_2CH_2CH_2\overset{+}{C}HCH_2CH_2CH_2CH_2R'$$

- hydride ion abstraction of a saturated structure:

$$RCH_2CH_2CH_2CH_2CH_2CH_2CH_2CH_2R' \rightarrow RCH_2CH_2CH_2\overset{+}{C}HCH_2CH_2CH_2CH_2R' + H^-$$

2. Subsequent reactions (propagation):

- β-scission:

$$RCH_2CH_2CH_2\overset{+}{C}HCH_2CH_2CH_2CH_2R' \rightarrow RCH{=}CH_2 + CH_3\overset{+}{C}HCH_2CH_2CH_2CH_2R'_2 \rightarrow R_1^{=} + CH_3\overset{+}{C}HCH_2R_2$$

or $RCH_2CH_2CH_2CH_2\overset{+}{C}HCH_3 + CH_2{=}CHR'$

- isomerizations:

$$R\,CH_2\overset{+}{C}HCH_2CH_2R' \rightarrow R\,CH_2CH_2\overset{+}{C}HCH_2R'$$

$$RCH_2\overset{+}{C}HCH_2CH_2R' \rightarrow R\,CH_2\overset{\displaystyle CH_3}{\underset{\displaystyle CH_3}{C}}CH_2R'$$

- hydrogen transfer:

$$R\,CH_2\overset{+}{C}HCH_2CH_2R' + R''H \rightarrow R\,CH_2CH_2CH_2CH_2R' + R''^+$$

- carbon formation by exhaustive dehydrogenation

3. Termination:

carbonium ions $\rightarrow$ hydrocarbons + H^+

gasolines. This is due to much higher contents of aromatics, branched paraffins and similar olefins. Moreover, the amount of coke formed is much smaller and the conditions of processing are milder. Therefore, catalytic cracking replaced thermal cracking very rapidly from about 1940 onwards. The process is the largest single means of producing gasoline components: in the USA alone, more than one million tons of feedstock are processed by it every day. Apart from the resulting higher ON of the gasoline pool, the gasoline yield is increased at the expense of the fuel oil yield.

Mechanism: Acidic catalysts are able to accelerate cracking reactions by a factor of 100-1000 as compared with thermal cracking. Carbonium ions are formed on the surface in the initiation phase of the complex reaction mechanism (see Table XIII.5.). The ions are formed from olefins or aromatics by addition of a proton supplied by a Brønsted site on the catalyst surface, or by hydride ion (H^-) abstraction from a saturated structure by the action of a Lewis site (see under 'Catalysts'). Traces of olefins may be formed by incipient thermal cracking at the reaction temperatures 475-515 ^{o}C. The reactions shown in the table are very similar to those of thermal cracking, except for the fact that in this case carbonium ions act as intermediates. Moreover, not only is the overall reaction very much faster, but some steps of the reaction sequence occur much more readily. In particular, isomerizations of primary carbonium ions formed by β-scission (Table XIII.5.) are much more rapid than the corresponding thermal reaction with radicals. This has important consequences for the structure of the product: whereas in thermal cracking ethylene or, more generally, α-olefins are major product components, these compounds do not occur so abundantly in catalytically cracked gasolines. Instead, catalytic cracking yields large amounts of propylene and butenes because the isomerization of primary carbonium ions to secondary ions is so much faster than the β-scission. This clearly follows from the comparison between the composition of thermally and catalytically cracked products formed from n-hexadecane shown in Figure XIII.2. Another conclusion drawn from this figure is that the degree of branching in the product is much greater in the catalytically cracked material. This is quite beneficial for the octane number of the product, as follows from the discussion in the preceding section. The formation of branched structures arises from the greater stability of tertiary carbonium ions as compared with secondary and primary ions, in combination with the high rates of chain isomerization.

Acidic catalysts give rise to coke formation with a consequent rapid reduction in activity. Regeneration by burning off the coke is possible at temperatures of about 600 ^{o}C for older types of catalyst, up to 675-700 ^{o}C for newer catalysts. Many successive regenerations are possible, provided that care is taken not to overheat the catalyst and prolonged contact with steam at high temperature is avoided.

Reactor systems: The foregoing discussion shows that the reactions of catalytic cracking can be represented by the network of Figure XIII.3. The cracking reactions are endothermic (about 200 kJ/kg feedstock cracked) and the oxidative

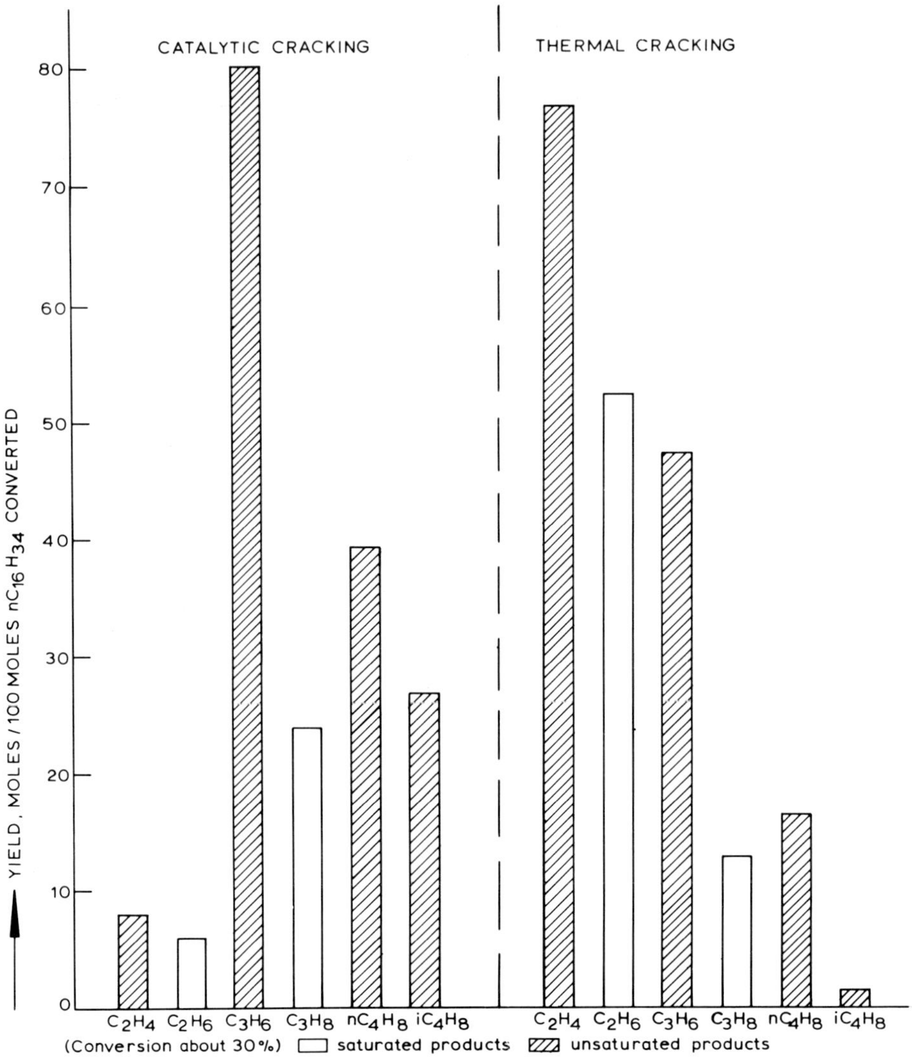

Figure XIII.2. Product distribution in thermal and catalytic cracking of n-hexadecane

regeneration is highly exothermic. A process system based on this network should not only be as selective for gasoline components as possible but also have the following desirable properties:

- continuous processing, in spite of the catalyst regeneration required;
- sufficient heat input in the cracking step to evaporate the feed and supply the heat of reaction;
- good heat removal during regeneration to prevent overheating of the catalyst with consequent activity loss;

- the cracked products and the regeneration products should be collected separately;
- any CO left in the regeneration gases should be used as fuel to recover energy from it;
- minimum environmental pollution, e.g. by SO_2, NO_x and CO;
- low energy consumption, i.e. optimal heat recovery;
- contact between catalyst and steam should be kept to a minimum in view of its deactivating influence.

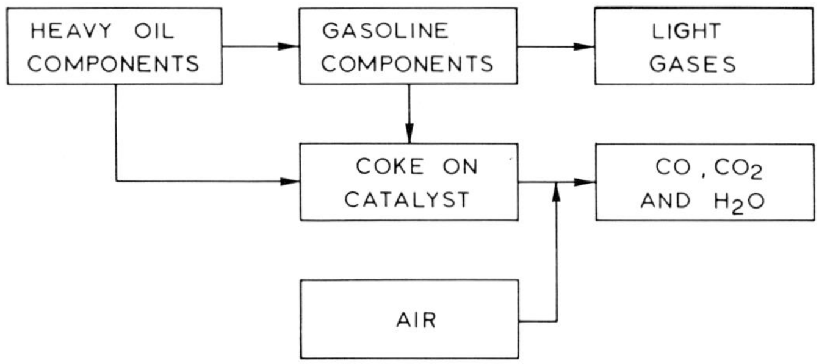

Figure XIII.3. Catalytic cracking (schematic reaction network)

The oldest type of plants (the Houdry process) contained fixed-bed catalytic reactors. These were used in cyclic regenerative operation, i.e. a reactor was operated in a cycle consisting of cracking, inert flush, regeneration and inert flush, followed by similar cycles. A number of parallel reactors was installed to ensure that the equipment downstreams of the reactor section could be operated continuously. Major disadvantages of such units were:

- high investment costs for the reactor section, the more so since high-quality materials had to be used because of the alternately reducing and oxidizing atmospheres and the temperature fluctuations to which reactors were subjected;
- control and safety required much attention;
- heat removal during regeneration and heat supply during cracking was accomplished with molten salt mixtures flowing through cooling/heating tubes in the catalyst beds;
- maintenance and operating costs were high.

Much better performance is shown by the moving-bed reactors of the Thermofor®

Catalytic Cracking process (Figure XIII.4.). In this case the catalyst moves through the reactor by gravity, entering in cocurrent flow with the feed. Steam is added to the transfer lines upstream of the reactor and at the outlet to isolate it from the oxidizing regimes in the catalyst feed line and the regenerators. The steam at the reactor outlet also serves to strip off any remaining volatile oil components from the catalyst. Coke-containing catalyst then moves downwards to the regenerator where reactivation occurs by burning with air. In earlier units (see figure) it was necessary to cool the catalyst to prevent sintering. With present-day catalysts, much less coke is formed and cooling is no longer necessary. The hot regenerated catalyst is transferred to the reactor inlet by means of an air lift. It contains sufficient heat to supply the heats of reaction and feed evaporation. Cracked gasoline having research octane numbers up to about 95 is recovered in a distillation section.

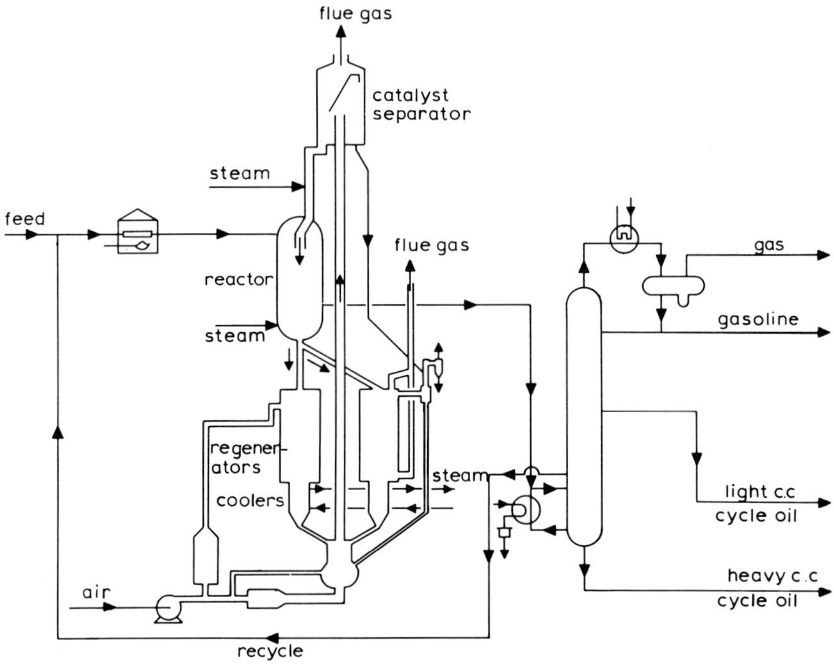

Figure XIII.4. Thermofor catalytic cracking process (air lift)

Fluid-bed reactors and regenerators are, however, applied much more widely than moving-bed reactors. Early designs of the reactor section consisted of a reactor, a regenerator and a stripper (Figure XIII.5.a.). It is also possible to combine reactor and stripper (Figure XIII.5.b.). Catalyst particles of 20-200 μm are used in the fluidized state, in which the solid/gas mixture resembles to some extent a liquid in that it can be transported by applying a small pressure gradient in the system. Moreover, a fluidized bed has a rather sharp upper boundary, like a liquid. Very rapid heat and mass transfer between solid and gas is possible. Since the solid is mixed very well, the temperature over the fluid bed is uniform. Of course, the strength of the catalyst particles should be high to minimize attrition losses.

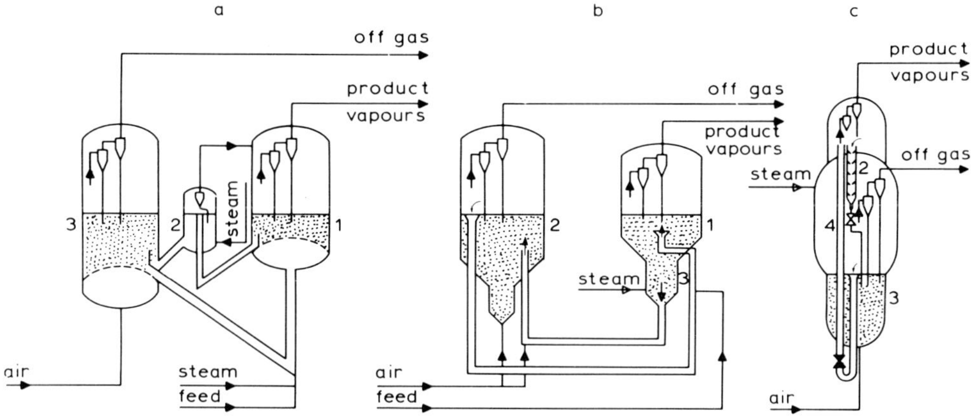

Figure XIII.5. Examples of reactor system for catalytic cracking
a. early designs with separate stripper
b. systems with combined reactor-stripper vessel
c. system with internal riser reactor
1. reactor
2. stripper (section)
3. regenerator
4. riser reactor

Preheated feed is introduced together with some steam in the line leading to the reactor; in the dilute catalyst/oil mixture the feed evaporates and cracking starts. Usually, the catalyst/oil weight ratio is between 8 and 12. The product stream leaves the reactor at the top after having been separated from entrained catalyst particles by cyclones. The catalyst moves from reactor to stripper, where some steam is used to strip off adsorbed hydrocarbons. Coke is then burned off in the regenerator and the hot regenerated catalyst returned to the reactor. With older types of catalyst some deactivation occurred, in spite of the excellent temperature control in the regenerator. Such activity losses by sintering are much smaller with present-day zeolite containing cracking catalysts supported by amorphous silica-alumina and deactivation by deposition of metals (V, Ni) is now becoming a limiting factor. The activity level of the catalyst inventory is kept at the desired level by drawing off a small amount of spent catalyst and replacing it by fresh catalyst.

Heat transfer in the reactor is direct. The hot catalyst from the regenerator contains just so much heat that the feed can be evaporated, heated to the reaction temperature and the heat of reaction is supplied. Also any heat loss to the surroundings is compensated. In this way substantial savings are obtained because the unit is heat-balanced and cooling surfaces need not be built into the reactors. However, fluid reactor beds have the disadvantage that some mixing of the gas phase occurs. Since the process can be represented by a network containing a successive reaction system (Figure XIII.3.), the selectivity to gasoline, the intermediate product, is reduced to some extent. This disadvantage is much smaller in the most recent reactor systems in which most of the cracking occurs in the riser, i.e. the transfer line leading to the reactor. The cocurrent plug flow regime in the riser leads to maximum selectivity for a given oil/catalyst combination. The fluidized bed in the reactor is now as shallow as possible to avoid further cracking of gasoline components. This became feasible only when very much more active catalysts containing synthetic zeolites were developed. An example of a riser reactor system in which the riser runs through the regenerator bed to the small reactor located on top of the system is shown in Figure XIII.5.c. With these systems, gasoline having clear research octane numbers of 92-95 can be obtained.

Catalysts: Activated mineral clays were applied as catalysts in the first cracking units. These soon proved to have too low an activity and were replaced by synthetic amorphous aluminosilicates prepared by controlled coprecipitation of SiO_2 and Al_2O_3. The acidity of these materials can be understood from Figure XIII.6. Amorphous silica is built up from Si atoms linked to each other via oxygen bridges, each oxygen atom being shared by 2 Si atoms. At the surface excess negative charges are present because surface oxygen atoms are connected to one Si atom only. This is compensated by surface coverage with OH groups having but a low acidity. If one now replaces a Si atom by Al (see figure), the resulting structure is no longer electrically neutral because Al is trivalent. The excess negative charge is balanced by an extra proton giving Brønsted acidic properties. Lewis acids also present at the surface (see Figure XIII.6.) can abstract a hydride ion to form a carbonium ion. Both types of structure contribute to the cracking activity.

The first generation of silica/alumina catalysts contained up to 15% Al_2O_3.

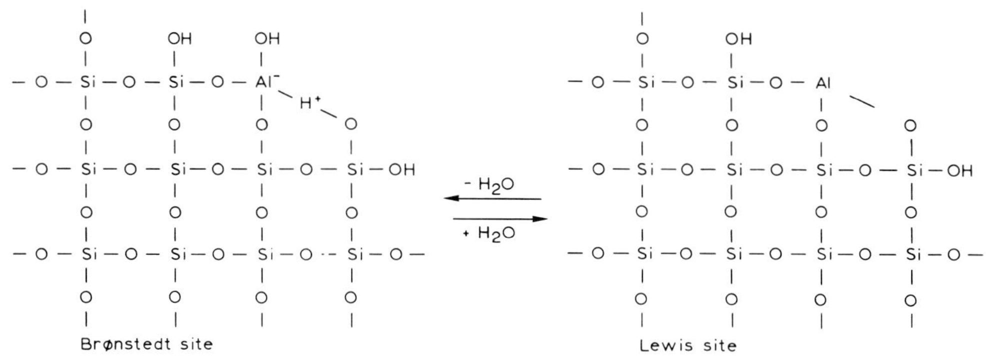

Figure XIII.6. Structure of silica/alumina cracking catalysts

Later, improved catalysts with 20-30% Al_2O_3 were adopted because of their higher activity and greater stability. In general, catalysts for fluid-bed cracking should have the following properties:

- abrasion resistance;
- resistance against deactivation by steam and by regeneration products;
- resistance against sintering;
- an adequate particle size distribution (20-200 μm) for good fluidization;
- a high porosity and a high surface area for minimum mass transport retardation and high activity. Fresh silica/aluminas have surface areas of about 600 m^2/g.

Present-day plants use still more active catalysts containing synthetic zeolites, crystalline aluminosilicates in which the excess negative charge of the AlO_4^- groups is neutralized by the presence of alkali and/or earth alkali metal ions. Most of these can be replaced by protons to obtain strong acids; if all the metal ions are replaced the crystal lattice disintegrates. Even more acidic catalysts are produced when alkali or alkaline earth ions are exchanged for rare earth ions. These catalysts are so active that they are applied as an active ingredient imbedded in a matrix of amorphous silica/alumina. The most recent zeolite catalysts used in riser reactor systems promote cracking in a few seconds.

The introduction of these highly active catalysts is an extreme example of how profoundly new developments in catalysis can affect process structure and performance. Major advantages over amorphous SiO_2/Al_2O_3 catalysts are:

- much better activity, selectivity and lower coke make;
- higher regeneration temperatures leading to more complete burn-off and giving a flue gas containing very little CO;

- better resistance against sintering and much improved steam stability;
- higher gasoline yields and octane numbers;
- higher yields of propene and butenes, potential alkylation feedstocks;
- riser cracking is feasible; with old type catalysts very long reactors would have been required;
- lower rate of circulation of catalyst and lower catalyst consumption.

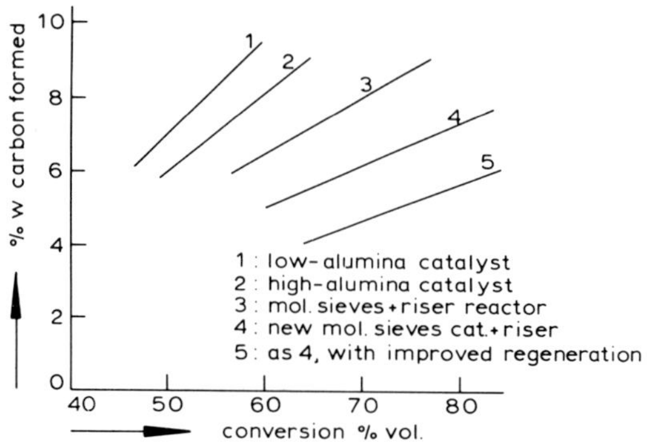

Figure XIII.7. Improvements in catalytic cracking

These improvements not only lead to more and better products but also to reduced investment and less environmental pollution. For instance, older units frequently contain a CO boiler to raise steam by combustion of CO present in the regenerator off-gas. With new catalysts it is possible to regenerate so completely that such a boiler is no longer needed. Moreover, the flue gases of CO boilers contain more NO_x because of the higher combustion temperature and also more SO_2, because more sulfur-containing coke must be burnt than in the regenerators of riser crackers. Thus, substantial savings can be obtained. The extra energy obtained by complete combustion of CO in the regenerator can now be used for heat-balancing the cracking process. Figure XIII.7. shows how conversion and coke make have changed over the years. Impressive economic gains have been obtained as a result of process and catalyst improvements.

Feedstocks: The content of metals like V, Ni and Fe in the feed should be low because these accelerate dehydrogenation and coke formation. This is the reason why distillates are almost always used as feed. Temporary poisoning is caused by basic nitrogen compounds and the content of very heavy aromatics (asphaltenes) should be as low as possible to minimize coke formation. It is possible to increase the yield of gasoline having higher octane number and lower sulfur content by catalytic pre-hydrogenation of the feed over catalysts like Co/Mo-sulfides on alumina. This also reduces the content of nitrogen compounds and leads to savings in the cost of after-treatment of the gasoline. The components having roughly the same boiling range as the feedstock are separated by distillation. Part of this cycle oil can be returned to the feed stream to increase gasoline yield, and the remainder is used as a fuel oil component or as hydrocracking feedstock.

Catalytic reforming

General: As the trend towards higher engine severities continued, products made by catalytic cracking and thermal reforming could no longer meet the more stringent octane number specifications, the more so because their sensitivity was too high on account of the high olefin content. Therefore, processes were developed to make high-octane gasoline components of low olefin content. Among the processes developed was catalytic reforming over noble metal catalysts on a carrier, a process which has grown to be the second largest oil conversion process of to-day. Its products consist of lightly branched paraffins and aromatics.

The feedstock for catalytic reforming is naphtha having a boiling range of about 60-190 ^{o}C. It is pretreated to remove catalyst poisons like sulfur and nitrogen compounds. The catalyst, Pt with or without activators on an acidic alumina carrier, is deactivated by sulfur and nitrogen compounds if the feed does not meet strict specifications.

Catalyst and reaction mechanism: The above bifunctional catalysts (compositions in Table XIII.6.) promote several reactions:

- the metal function accelerates hydrogenation and dehydrogenation, cyclization and hydrocracking reactions;
- the acidic function causes isomerization reactions and acid-catalyzed cracking reactions.

If nothing but naphtha is passed over the catalyst at the usual reactor inlet temperatures of up to 525 ^{o}C, catalyst activity would soon be lost because of excessive carbon lay-down. The activity can be stabilized by adding hydrogen to the feed and operating at 1.0-5.0 MPa total pressure. About 5-10 moles H_2

Table XIII.7. Catalytic reforming reactions

Reaction	Equilibrium data at 475 °C		
1. *Dehydrogenation of cyclohexanes*	% dehydrogenated at:		
		P_{H_2} = 3.5 MPa	P_{H_2} = 2.5 MPa
	C_6	73	82
	C_7	76	88
	C_8	78	94
2. *Isomerization of cyclopentanes to cyclohexanes*	% isomerized:		
	C_6	8	
	C_7	66	
	C_8	90	
3. *Dehydroisomerization of cyclopentanes*	% dehydroisomerized at:		
		P_{H_2} = 3.5 MPa	P_{H_2} = 2.5 MPa
	C_6	27	52
	C_7	83	94
	C_8	93	98
4. *Hydrocracking of (cyclo) alkanes*	complete conversion into saturated fragments		
5. *Dehydrocyclization of n-alkanes*	% dehydrocyclized at:		
		P_{H_2} = 3.5 MPa	P_{H_2} = 2.5 MPa
	C_6	3	12
	C_7	30	62
	C_8	57	85
6. *Hydrodealkylation of aromatics*	complete conversion into benzene and alkane		
7. *Isomerization of alkanes*	average number of branches per mole		
	C_4	0.34	C_7 1.31
	C_5	0.75	C_8 1.23
	C_6	1.05	

per mole feed is used. The liquid hourly space velocity is 1-2 vol naphtha per vol catalyst per hour.

Table XIII.6. Catalytic reforming bimetallic catalyst

Composition, % wt		
	Pt	0.3 - 0.8
	Re	0.2 - 0.5
	Cl	0.6 - 1.0
	Carrier	Al_2O_3

The acid function of the catalyst weakens during processing because of loss of chlorine. It is necessary to optimize acidity by passing a feed containing traces of an organic chloride over the catalyst. The water content of the feed should be kept within certain limits depending on the catalyst used. If it is too high, the carrier surface is hydrated and acidity lowered, leading to lower isomerization activity.

The reactions proceeding during catalytic reforming are shown in Table XIII.7., along with some data on equilibrium compositions. Further information is tabulated in Table XIII.8.

It follows that those reactions in which aromatics are formed contribute most to the improvement of product octane number. The dehydrocyclization is particularly effective through the conversion of very poor alkanes to excellent aromatics. The isomerization of alkanes is not very important for octane improvement in view of the low average degree of branching in the product at the high reaction temperature applied here. Hydrocracking contributes to octane number improvement by removal of high-molecular weight alkanes. However, cracking of naphthenes, precursors of aromatics, is a disadvantage because it lowers the yield. Dealkylation of aromatics boiling in the gasoline range is not a desirable reaction: it does not improve product octane number and decreases the yield to some extent.

Another conclusion drawn from Table XIII.7. is that the hydrogen partial pressure should not be too high because this reduces the equilibrium yield of aromatics. For this reason, the first generations of catalytic reformers operated at a moderate hydrogen partial pressure which was low enough to give sufficient aromatics but high enough to prevent catalyst activity decline by carbon deposition. Still lower pressures are desirable, however, mainly because less hydrocracking occurs. Therefore, the development of the process has gone in the general direction of lower hydrogen partial pressures, combined with catalyst and/or process system improvements to reduce carbon formation and consequent catalyst activity decline.

Table XIII.8. Rates and heat effects of catalytic reforming reactions at 475 ^{o}C

	reaction rate	heat of reaction (kJ/mole)
1. Dehydrogenation of cyclohexanes to aromatics	very high	+200 to +230
2. Isomerization of cyclopentanes to cyclohexanes	high	0 to +20
3. Dehydroisomerization of cyclopentanes to aromatics	moderately high	+200 to +230
4. Alkane isomerization	high	negligible
5. Dehydrocyclization of alkanes to aromatics	low (C_6) to moderately high (C_8)	+220 to +250
6. Hydrocracking of alkanes	low to moderately high; increases with mol.weight	-10 to -25
7. Hydrodealkylation of aromatics	low	-25

Another point worth noting is that the equilibrium yield of benzene is lower than that of toluene and the xylenes. Additional reasons are that cyclization of n-hexane requires linkage of two methyl groups, i.e. the breaking of two primary CH-bonds, which is apt to lead to slow kinetics. Furthermore, the only benzene precursor usually present in naphthas in fairly large concentrations is methylcyclopentane: unsubstituted aromatics and naphthenes like cyclohexane occur in rather smaller amounts.

Process form: The strongly endothermic nature of the aromatization reactions and the high temperatures have led to the use of a number of adiabatic reactors in series (2-5, usually 3) with interstage reheating in furnaces. The oldest form of the process used such a high pressure that long catalyst life was obtained without in situ regeneration. At present, semi-regenerative, fully regenerative or continuous process configurations are applied.

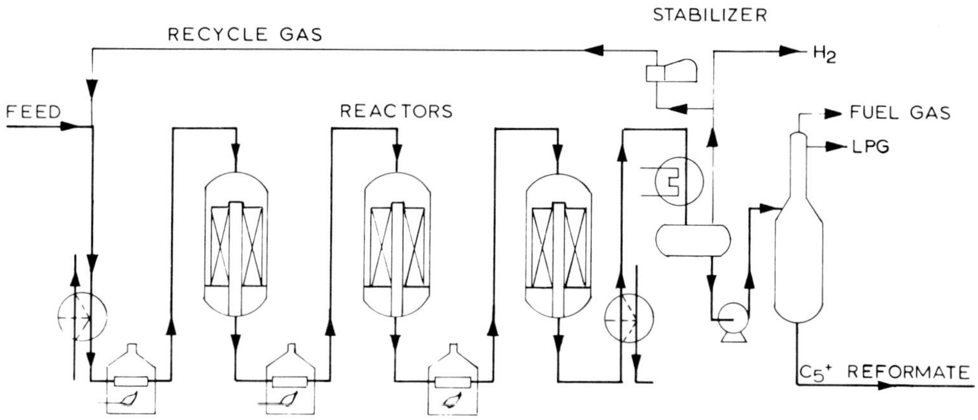

Figure XIII.8. Semi-regenerative catalytic reformer (radial flow reactors)

In the semi-regenerative system (Figure XIII.8.) naphtha feed is mixed with hydrogen, passed through a furnace and fed to the first reactor. The flow is radial towards the centre of the reactor. In the first stage a very rapid temperature decrease is observed (Figure XIII.9.) because of the very endothermic dehydrogenation reactions and dehydroisomerization reactions. Since the naphthene concentration as well as the rate of these reactions are both high, the first reactor is usually small. After reheat, further dehydroaromatization reactions occur in reactor 2; there is also some contribution from the exothermic hydrocracking reactions because the temperature does not decrease so rapidly. The temperature change is smallest in reactor 3, usually the largest of the set; it is possible to observe no temperature change, or even a small increase because of hydrocracking. If a highly naphthenic feed must be processed some refiners prefer to run the first reactor at a lower inlet temperature to reduce losses of valuable naphthenes by hydrocracking.

After a few months' processing, carbon is burnt off to restore catalyst activity. This must be done very carefully to avoid catalyst deactivation by overheating. Air/inert gas mixtures are usually applied: the oxygen concentration should be 0.5-2 vol%. The chlorine content of the catalyst should be restored in a second step of the regeneration procedure.

Fully regenerative catalytic reforming is needed when higher octane numbers or better yields at a given octane level must be reached. The hydrogen partial pressure must now be so low and the temperature so high that catalyst

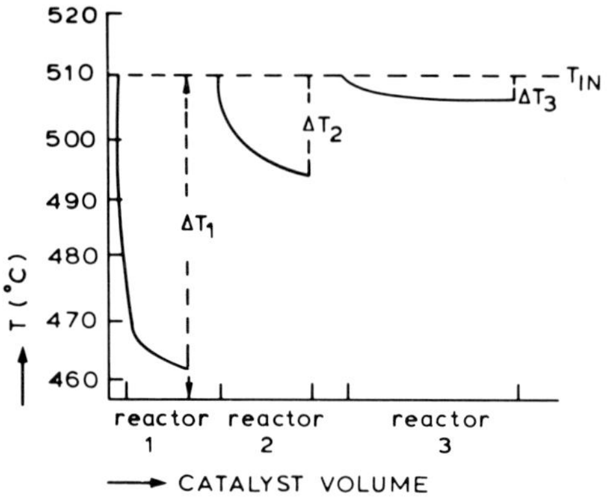

Figure XIII.9. Temperature profiles in catalytic reformer reactors

deactivation becomes serious. A fourth reactor is added here. One of the reactors is periodically taken out of reforming service and regenerated by coke burning and re-chlorination. It is then put back into operation and the next reactor is regenerated, etc. The regeneration frequency depends on the operational severity required; one regeneration per 48-120 h is quite feasible.
To decide on the economic feasibility of a change-over to fully regenerative operation, it is necessary to weigh the higher operating, investment and catalyst consumption costs against the increased processing flexibility obtained and the benefits of obtaining higher product octane numbers at a given yield (or alternatively, a higher gasoline yield at a given octane rating). It is, however, uncertain whether fully (cyclic) regenerative reforming will remain competitive with the continuous process discussed below.

In continuous catalytic reforming a moving bed of catalyst is used (Figure XIII.10.). The regenerated catalyst is fed to the top of the reactor system and slowly moves down whilst coke is deposited. Here again, interstage heating is applied. The catalyst is transported to a separate regenerator where coke is burnt off. After heat exchange and cooling the product is fed to a low-pressure separator where gas/liquid separation occurs. The gas is compressed and cooled further to remove light hydrocarbons (isopentane/n-pentane) as completely as possible. The liquid phase from the low-pressure separator is added to the compressed gas to serve as an absorbent for the light hydrocarbons, i.e. to increase the amount of C_5 hydrocarbons removed from the gas phase.

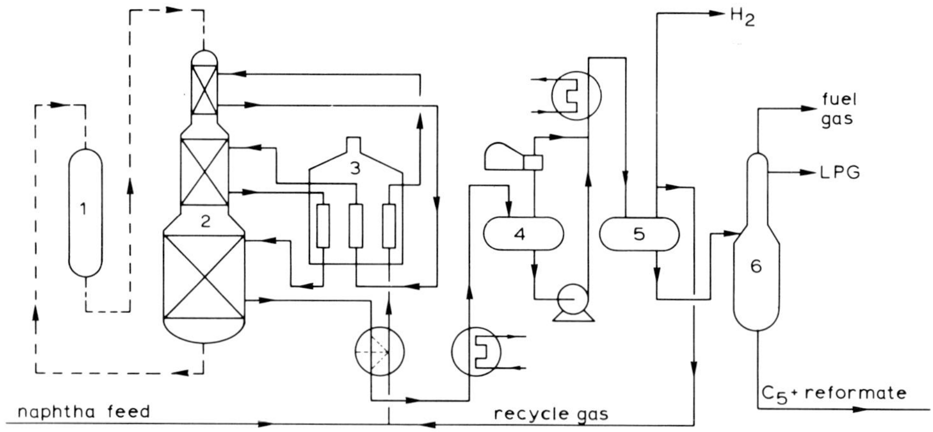

Figure XIII.10. Continuous catalytic reformer

1. regenerator

2. reactors

3. furnace

4. low-pressure separator

5. high-pressure separator

6. stabilizer

This is done to prevent these low-boiling materials from entering the recycle hydrogen stream.

Effect of process conditions: Figure XIII.11. shows example of the effects of various process conditions on gasoline yield, product quality and catalyst life. The predicted effects of pressure increases are indeed observed: at lower pressure less product is obtained, except from naphtha (1) which is very rich in naphthenes. Higher reactor inlet temperatures at otherwise unchanged conditions produce a better octane number, but at the expense of a lower yield. These correlations were determined using catalyst A, Table VIII.6., correcting for possible loss of activity during the experiment. At higher operating severity the amount of hydrogen produced is also higher. This is important economically because there is a tendency to demand fuels of lower sulfur content. In most refineries the quantity of by-product hydrogen is still sufficient to cover the needs of hydrotreating operations. If, however, hydrocracking (see below) and residual oil desulfurization must be introduced on a large scale, separate hydrogen plants must be added.

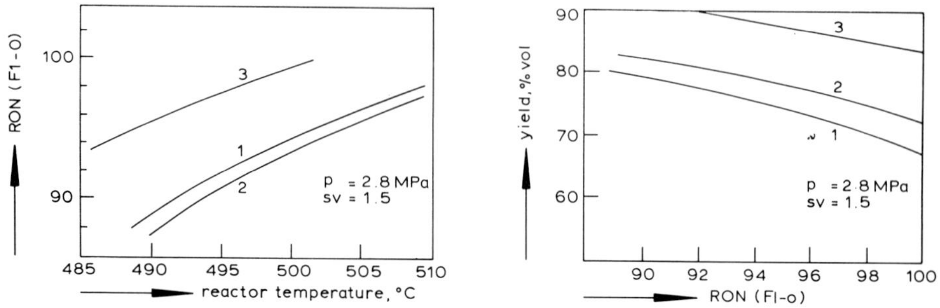

NAPHTHA No:	Composition, vol % paraffins	naphthenes	aromatics
1	69	20	11
2	54	42	4
3	29	61	10

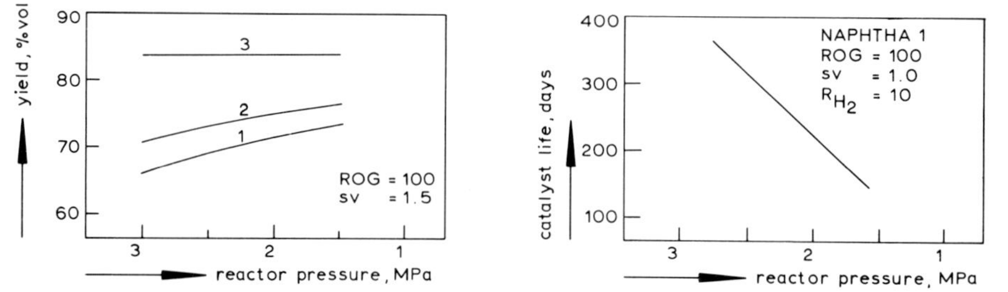

Figure XIII.11. Relation between reforming conditions and process performance

Hydrocracking

Bifunctional catalysts containing a hydrogenation/dehydrogenation function and an acidic function can also be used in a cracking process at temperatures of approximately 400 ^{o}C and high hydrogen partial pressures (10-15 MPa). A high gasoline yield is obtained and the product, being free of olefins, has a low sensitivity and a high octane number. Moreover, the process can be applied to feedstocks which are difficult to process by catalytic cracking, such as catalytically cracked cycle oils, gas oils from coking operations, and the like. Apart from the possibility of achieving a better balance between production and demand of gasoline and heavy fuel oils, the yield of middle distillates, e.g. jet fuels of high quality, can be increased. Another application of hydrocracking is in the production of lubricating oil base stocks with very attractive viscosity/temperature behavior from desasphalted residues. In short, hydrocracking is a much more versatile tool for the refiner than either catalytic reforming or catalytic cracking.

Catalysts used or claimed contain a fairly strong acidic function, usually fluorinated alumina or silica/alumina with or without a rare earth-modified synthetic zeolite. For processing of sulfur-containing stocks a hydrogenation component of the sulfide type, such as Co/Mo-, Ni/Mo- or W-Ni-sulfides is used. Otherwise, noble metals such as Pt are applied. If the latter is used to process high-sulfur feeds, the cracking process can be carried out in two stages, the first of which contains a non-sulfur sensitive catalyst and the second a sensitive catalyst. This necessitates interstage removal of H_2S and NH_3 from the product of the first stage because these contaminants would poison the hydrogenation and acidic sites of the second-stage catalyst, respectively. There are, however, claims for new catalysts that can be used without interstage separation of H_2S and NH_3.

A block diagram of a two-stage hydrocracker is shown in Figure XIII.12. It follows that many products can be obtained at the expense of a rather costly high-pressure unit and considerable amounts of H_2, viz. up to 4-5%w on feed.

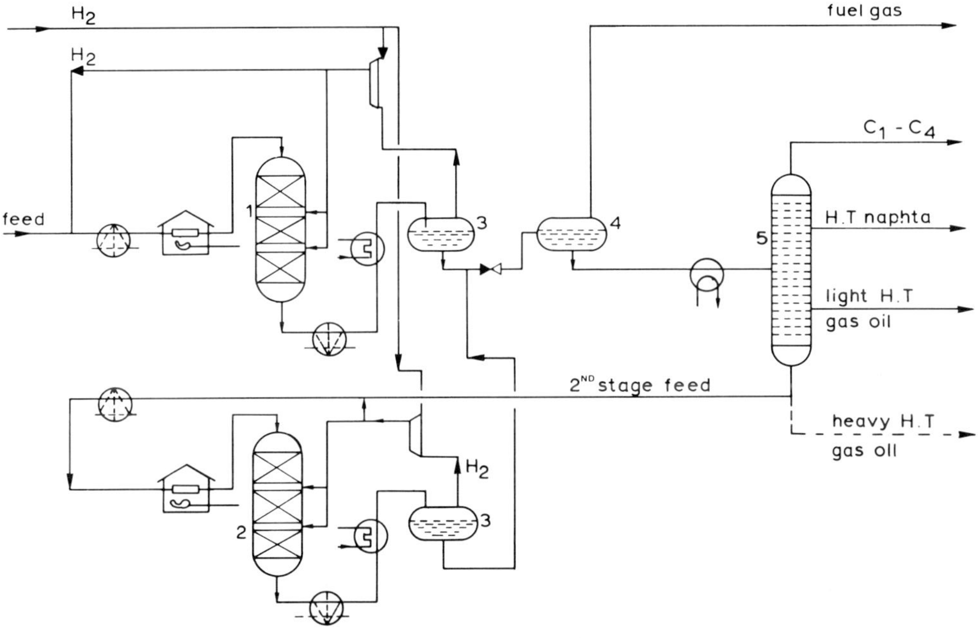

Figure XIII.12. Two-stage hydrocracker

1. *first-stage reactor*
2. *second-stage reactor*
3. *H.P. separators*
4. *L.P. separators*
5. *fractionator*

By-product hydrogen from catalytic reforming is usually insufficient and a separate hydrogen plant must be built, which adds to the investment costs. This is one of the reasons why the introduction of hydrocracking operations has been relatively slow.

The feed stream to the hydrocracker should contain some sulfur to keep Co/Mo catalysts of stage I in the sulfided form. If low-sulfur feeds must be processed, it may be necessary to introduce H_2S by injection of elemental sulfur which is subsequently hydrogenated.

The hydrocracking reactions are fairly strongly exothermic. Therefore, the reactor usually contains a number of catalyst beds, with cold hydrogen injection between the beds to absorb the heat of reaction. The result is a high pressure drop across the reactor, even at the relatively low space velocities (1.0-1.5 vol.feed/vol.catalyst/h) and high pressures, i.e. low linear gas velocities, used here. Wider beds can hardly be applied for constructional reasons. It is even necessary to use a number of reactors in parallel in high-capacity plants.

With regard to gasoline quality, the C_5/C_6-fraction is found to have very good octane numbers. The C_4-fraction contains much isobutane, which can be used to advantage in alkylation (see later) to make highly branched, i.e. high-quality, paraffins. The iso/normal ratio of the butanes and the C_5/C_6-fractions is much higher than corresponds to thermodynamic equilibrium at the reaction temperatures applied. There must obviously be kinetic reasons for this. The naphtha is a good gasoline component and it can even be further upgraded by catalytic reforming, if the necessity arises.

Hydrocracking may become more important as legal requirements to reduce the lead content of gasolines are adopted more widely, or if the gasoline/fuel oil production ratio becomes higher. This point will be touched upon in the discussion of future trends at the end of this chapter.

Isomerization

The octane numbers of the light (C_5/C_6) fractions of motor gasoline are often rather low. Although some improvement is possible by adding butanes, this can only be done to a limited amount because the Reid vapor pressure should not exceed the specified limit. A better method is to raise the front end octane number by isomerizing the C_5/C_6 components of straight-run distillates. The first processes developed for this purpose employed strong acid catalysts ($AlCl_3.HCl + SbCl_3$ or $HSbF_6$) for liquid-phase isomerization. These processes are rather expensive, however, because of their corrosivity, high reagent cost and the occurrence of unwanted side reactions. In these respects, a better type of process is gas-phase hydroisomerization, i.e. isomerization over a bifunctional catalyst (Pt/acidic function) in the presence of excess hydrogen. A mechanism of hydroisomerization is shown in Figure XIII.13.

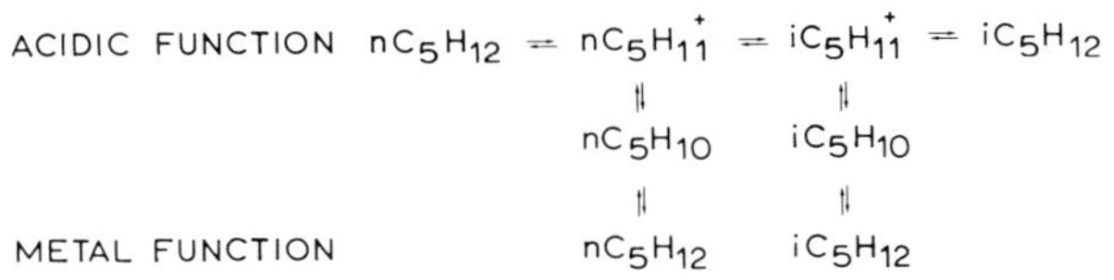

Figure XIII.13. Mechanism of hydroisomerization

Although the acid function is responsible for the isomerization reaction, Pt is indispensable because it keeps the concentrations of alkenes and carbonium ions low. The catalysts used contain either chlorinated alumina (BP, UOP) or an acidic zeolite as the carrier (Shell). As high an isomerization activity as possible is desirable because equilibrium should be established at the lowest possible temperature. As the temperature is lower, the average degree of branching is higher, and also the octane number. This is shown in Figure XIII.14.

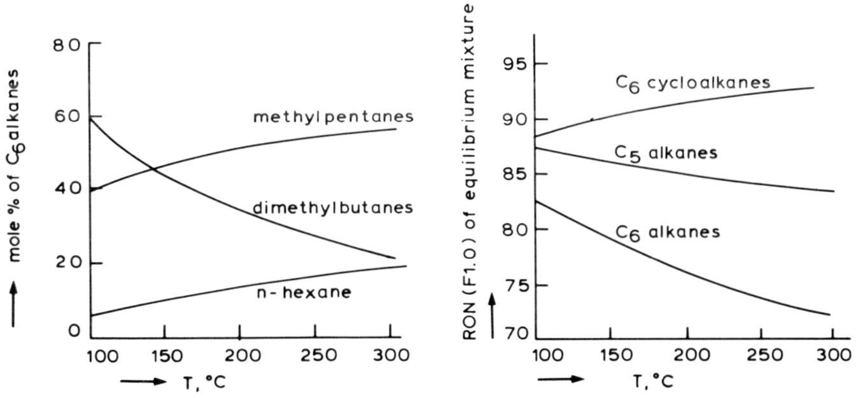

Figure XIII.14. Effect of temperature on composition and research octane number of equilibrium mixtures

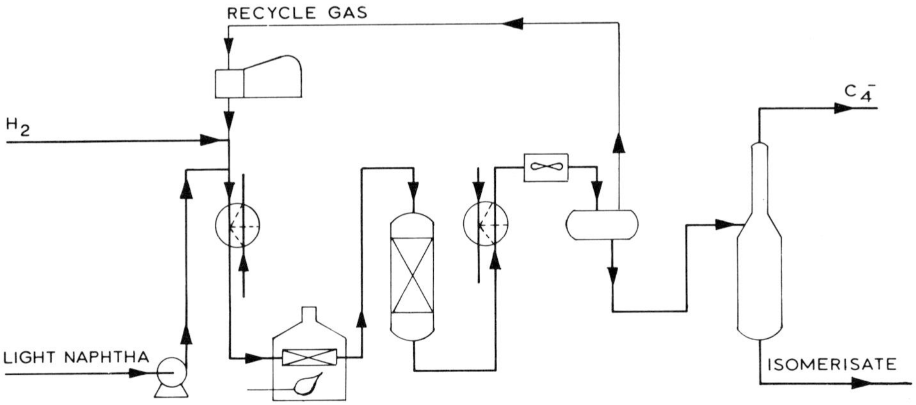

Figure XIII.15. Shell Hysomer Process

A flowchart of the Shell Hysomer Process is shown in Figure XIII.15. Light naphtha is mixed with 2-3 H_2/mole feed, preheated and then passed over a single fixed-bed reactor. The heat of reaction is quite small and adiabatic operation is easily accomplished. The hydrocarbons are separated from the gas phase and then stabilized, i.e. light components removed by distillation. The hydrogen is returned via a recycle compressor. The process operates at about 260 ^{o}C and 2.5 MPa total pressure. With liquid hourly space velocities of about 1.0 vol. feed per vol.catalyst per hour octane improvements of up to 10 units are obtained. The yield is almost quantitative.

Even better octane quality can be reached when the isomerization is combined with a separation. This is shown in Figure XIII.16., where the isopentane and pentane are separated by distillation; the latter is recycled to the hydro-isomerization reactor. A still greater octane number improvement is possible by completely removing the normal alkanes from the product stream by adsorption in molecular sieves, type 5A, which retain n-paraffins but hardly any iso-paraffins. Union Carbide's Isosiv process employs several adsorbent beds operated in a cyclic regenerative manner. Adsorption is carried out at 300 ^{o}C and 0.4-0.5 MPa; desorption by pressure reduction and applying vacuum (pressure swing desorption). Each cycle lasts 5-10 minutes. The raffinate contains 1-2% n-alkanes and the extract up to 90%. Similar separations are possible with heavier hydrocarbons (up to C_{14}). Energy consumption is fairly high. The combination of the Hysomer and Isosiv processes, known as the TIP (Total Isomerization Process) process constitutes one of the most successful methods for making high octane light gasoline components. The UOP Molex process (Figure XIII.17.) applies a desorbent to recover the adsorbed n-alkanes. It contains a single adsorption

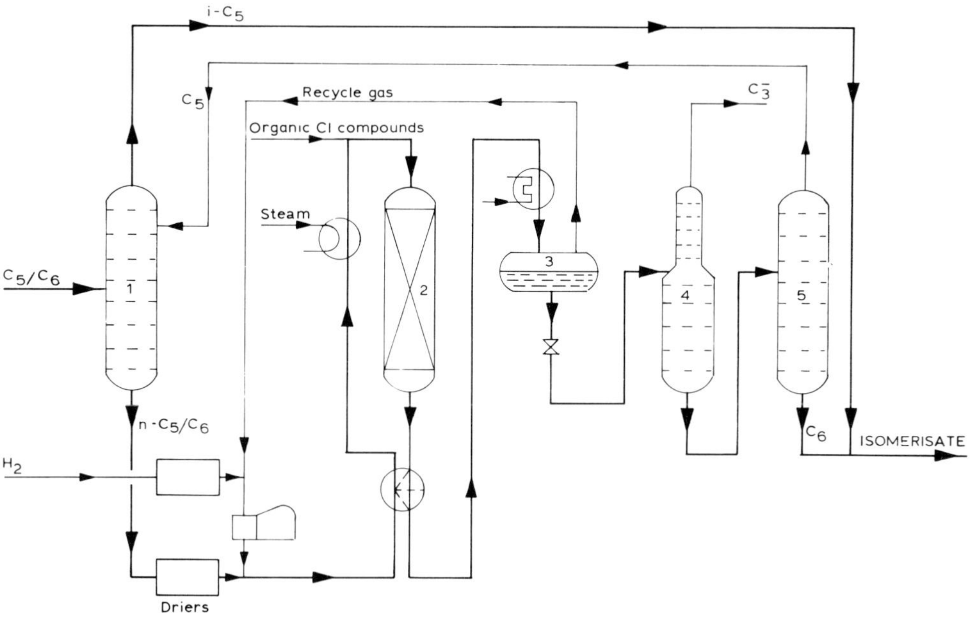

Figure XIII.16. UOP Penex process with nC_5 recycle

1. deisopentanizer

2. reactor

3. separator

4. stabilizer

5. C_5/C_6 splitter

column and simulates a continuous countercurrent operation by moving the feed points and product streams along the column by means of a rotary valve. In the figure, positions 3, 6, 9 and 12 are being used. As soon as the separation quality diminishes, the valve rotates to positions 4, 7, 10 and 1. The raffinate and the extract are separated from the desorbent by distillation. Purities are similar to those obtained in the Isosiv process.

The above isomerization processes and their combinations with adsorptive separations are being accepted rather slowly by the industry, mainly because of the high processing costs.

Alkylation and polymerization

The large amounts of olefins produced in catalytic cracking can be used to

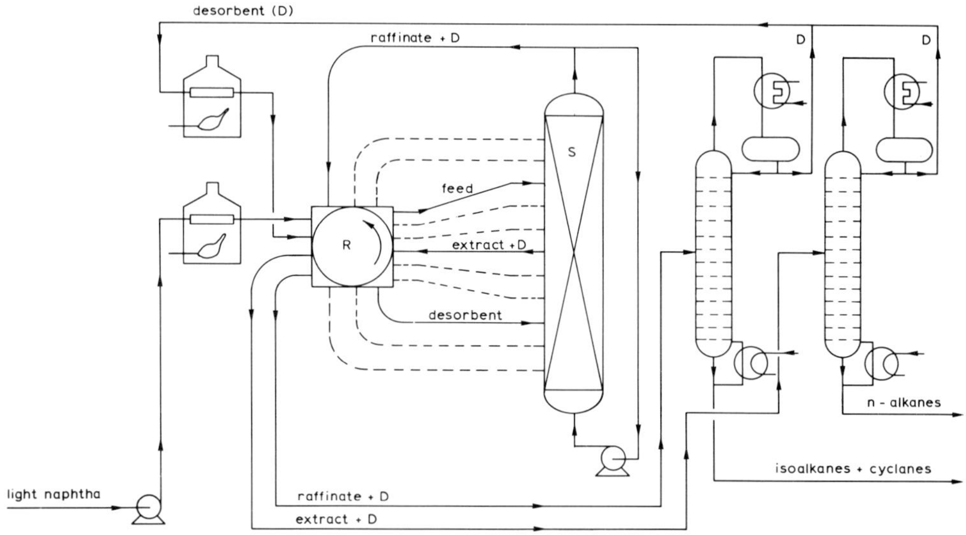

Figure XIII.17. UOP Molex process for separating normal alkanes from isoalkanes Rotary valve shifts position of feed and product lines along molecular sieve columns

make high-quality gasoline by alkylation. In the petroleum industry, the term alkylation is used to indicate the process of addition of an olefin to a saturated, branched-chain alkane, usually isobutane, but in a few cases also isopentane. The resulting branched isoalkanes have high research octane numbers, viz. between 90 and 100, and a very low sensitivity.

The alkylation reactions are catalyzed by strong acids in the liquid phase according to a chain mechanism in which carbonium ions add to olefins to give a higher molecular weight carbonium ion. Termination occurs through hydrogen abstraction from an isoalkane (see Figure XIII.18.). Since this is a relatively slow step in the chain, it is necessary to work with a large excess of isoalkane (4-12 moles/mole olefin) in the feed. Otherwise, higher molecular weight oligomers of the olefins would be produced by continued addition to olefins followed by dissociation of the heavy carbonium ions.

The latter reactions have also been applied in the petroleum industry to make strongly branched olefinic gasoline components and propylene tetramer used for alkylating benzene to produce a raw material for the manufacture of synthetic detergents. This process, essentially a gas-phase oligomerization over H_3PO_4/ silica, is called polymerization in the industry. It has lost much ground because

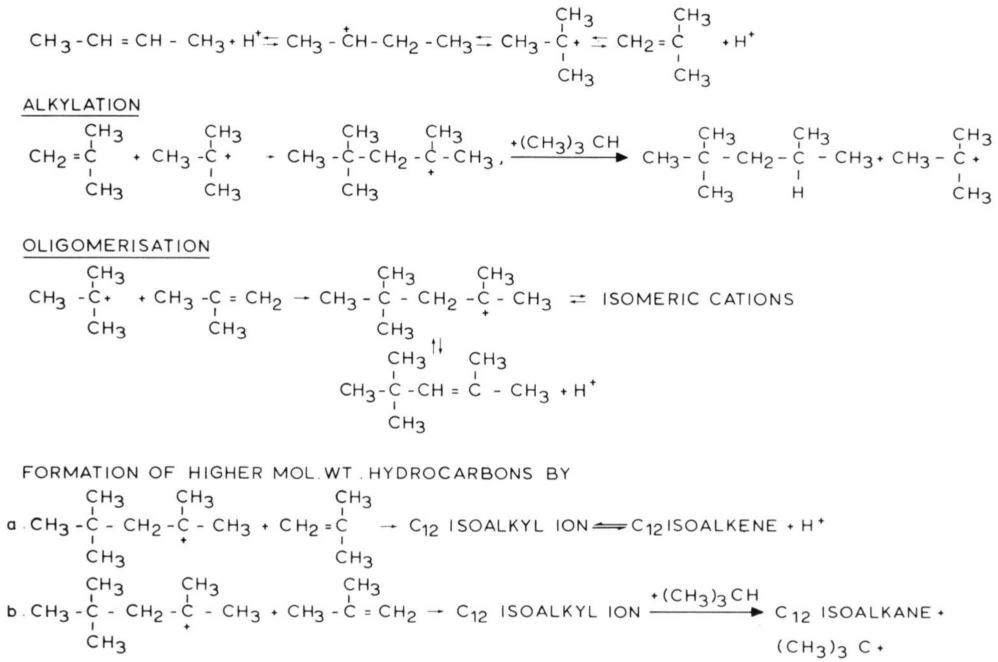

Figure XIII.18. Reactions in alkylation process
Formation of isoalkanes is favored by a high ratio isobutane/alkane in the feed

of the high cost of olefins, the relatively high sensitivity of the product as a gasoline component and also because many countries have banned detergents based on dodecylbenzene. The reason is that these detergents containing strongly branched side chains cannot be broken down by microorganisms and cause serious environmental pollution.

Catalysts used in alkylation are HF and sulfuric acid. The former is active at 20-40 ^{o}C whereas the latter requires lower temperatures (0-20 ^{o}C) to avoid oxidative attack on the hydrocarbons by which tars and SO_2 are formed. The contact time in HF alkylation is 10-20 min and in sulfuric acid alkylation up to 30 min. The ratio of hydrocarbon to acid volume fed to the reactor is about 1. Very intense agitation is needed to ensure adequate mixing of the two phases because of the large difference in density.

The olefins used in alkylation are usually propylene and butenes but ethylene and pentenes are sometimes included in the feed. Separation from the corresponding alkanes is not necessary, either because these do not react or because they are desirable (branched-chain) feed components. Likewise, butane is not separated from isobutane. It is much easier first to alkylate and then to separate by distillation because the differences in boiling point are much greater in the alkylate product stream.

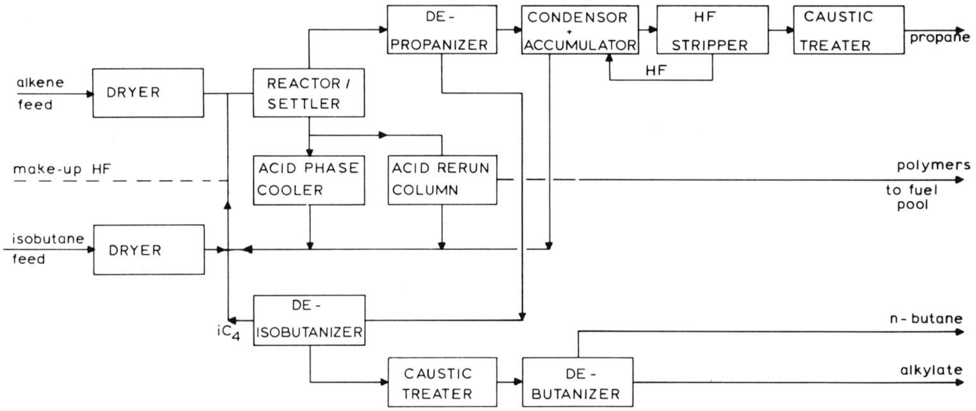

Figure XIII.19. Block diagram of HF alkylation

Figure XIII.19. shows a block diagram of HF alkylation. The two feed streams are dried to remove water which would weaken the acid and cause low reaction rates. The hydrocarbons are then thoroughly mixed with cold acid and sent to the reactor. The pressure is high enough to keep the entire mixture in the liquid phase. The reaction products are allowed to settle and the hydrocarbon phase is passed overhead. The acid stream is cooled to remove the heat of reaction and then recycled after having been mixed with fresh feed. Most of the acid is cooled and recycled but a small side stream of acid is withdrawn and fed to the acid rerun column which separates it into fluorine-containing tars + water and purified HF. The HF is returned to the reactor system. The product stream from the reactor consists of propane, n-butane, alkylate (C_5 and higher) and traces of tar. It is separated into components by distillation; this is often done by a single column from which propane is obtained as the top product. A side stream containing the isobutane and some propane is recycled to the alkylation reaction section. The separation also produces n-butane and, of course, alkylate. These product streams are washed with aqueous lime or alkali to remove traces of HF. Any effluent from the plant should be carefully treated to prevent pollution by organic fluorine compounds.

A disadvantage of HF alkylation is that special and costly precautions and intensive training for protection of operating personnel are required. Also, equipment where aqueous HF-containing streams is processed should be made of Monel or Monel-clad steel. Carbon steel is suitable for equipment containing dry HF streams.

Sulfuric acid alkylation is also used on a relatively large scale. The choice between the two processes depends on many factors such as feedstock prices and purity, capital costs, sulfuric acid price (the losses of H_2SO_4 are much greater than for HF), licensing costs and possibilities, environmental and safety requirements, and many others.

Future developments

Accurate predictions of future trends in gasoline manufacture are difficult to make. A first and fairly obvious reason is that any change in the demand pattern of a main oil product like LPG, middle distillates, naphtha and residual fuels is bound to affect the overall production pattern of petroleum refineries, also of gasoline manufacture. Another cause of change is the new environmental legislation enforced or being introduced in many areas where gasoline consumption is high. The new regulations are already affecting the quality criteria of gasolines and, more slowly, the design of internal combustion engines. This is particularly so for the limits that are now being set for emissions of CO, hydrocarbons and nitrogen oxides, and to a lesser extent for sulfur oxide emissions. Furthermore, there is a mounting pressure to eliminate lead compounds from gasolines for environmental reasons. If the gasoline engine is not modified for operation on low-lead or lead-free gasoline, the requisite high octane quality must be reached by introducing more selective conversion or synthesis processes.

The above picture is further complicated by the sharp increase in crude oil prices since the oil embargo of 1973/1974 and by the increasing awareness that the world's oil reserves are limited. Thus, the new legislation is not only important in connection with cleaning up the environment, but it also has far-reaching economic consequences, as is outlined below.

Emission control and the gasoline engine

The amounts of CO, hydrocarbons and nitrogen oxides emitted from car exhausts at cruising conditions are small. A limited part of the fuel/air charge near relatively cool surfaces, e.g. in the laminar layer at the cylinder wall, does not react and appears in the exhaust gas. Much larger amounts of unburned hydrocarbons and organic oxygen compounds are emitted during deceleration. The vacuum in the inlet system is then so deep that early flash evaporation of the gasoline causes the formation of a rather rich fuel/air mixture. Even at complete oxygen consumption part of the hydrocarbons are not burned at such conditions. Another reason for the incomplete hydrocarbon oxidation under deceleration is that the pressure in the combustion chamber is somewhat lower than in the inlet manifold. Some exhaust gas then flows back through the inlet valve and mixes with the fresh fuel/air charge, causing incomplete combustion.

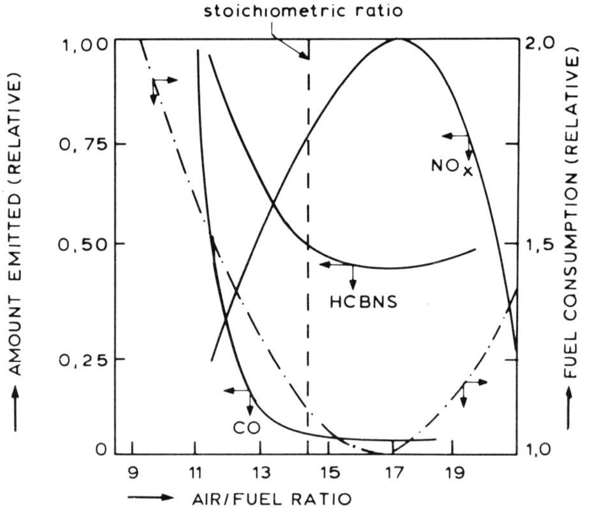

Figure XIII.20. Effect of air/fuel ratio on emissions and fuel consumption

The amounts of CO and hydrocarbons emitted in the exhaust gas depend on the air/fuel ratio, as follows from Figure XIII.20. It is concluded that these emissions can be reduced substantially by using rather lean mixtures: the amounts of CO and hydrocarbons are smallest when the engine operates just above the stoichiometric air/fuel ratio. However, NO_x formation is then aggravated (see figure) because the temperature peak during the engine cycle is then maximal. At higher air dosages it is lowered because of the cooling effect of the excess air. Nitrogen oxide emission is directly linked with the peak temperature: above 1500 K the formation of NO_x via the Zeldovitch reactions

$$O_2 \rightleftarrows 2\,O\cdot$$
$$O\cdot + N_2 \rightleftarrows NO + N\cdot$$
$$N\cdot + O_2 \rightleftarrows NO + O\cdot$$

is governed entirely by equilibrium. Below 1500 K NO formation is limited by kinetics. Figure XIII.20. shows that changes in mixture strength alone are insufficient to reduce all three emissions simultaneously.

Still, this is what is required in the new legislation enacted mainly in Japan and the USA, in particular in California where smog formation caused by exhaust emissions is particularly objectionable in the Los Angeles area. Table XIII.9. lists the US Federal requirements as they have developed since about 1970.

Table XIII.9. Emissions by gasoline engines in the USA (in g/km)

Pollutant:	Emissions	Federal requirements		
	< 1970	1973	1975A	1975B
carbon monoxide	52	17	2.0	2.0
hydrocarbons	5.2	1.8	0.25	0.25
nitrogen oxides	--	1.9	1.2	0.24

The limits for 1973 and 1975 apply to new cars built in or after the years indicated. It appears that almost a quarter of the fuel fed to engines of pre-1970 cars leaves the exhaust unburned or only partly oxidized. Emissions can be reduced by careful tuning which has the advantage of much lower fuel consumption but reduces power output. Further reduction is possible by a number of means (table XIII.10.). Exhaust gas recirculation reduces the peak temperature in the engine cycle and, therefore, NO_x emission, without leading to increased CO and hydrocarbon emissions. Another means of reducing engine temperatures is by lowering the compression ratio. This causes a further reduction of NO_x emissions but has the disadvantage of leading to lower engine efficiency. These changes made it possible to meet the 1973 requirements, at the expense of a rather large increase in fuel consumption (see Table XIII.11.).

Table XIII.10. Means for reducing emissions by gasoline engines

NO_x emission can be reduced by	CO and hydrocarbon emission can be reduced by
- using less air than stoichiometric	- using excess air
- exhaust gas recycle	- optimizing spark timing
- catalytic NO_x conversion	- improving carburation
- improving engine construction	- improving engine construction
	- thermal or catalytic exhaust gas oxidation

Table XIII.11. Effects of emission control on gasoline consumption

Consumption (average), 1970	100
Consumption (optimized), 1970	85-90
Consumption with 1973 requirements	114
Consumption with 1975A requirements	107
Consumption with 1975B requirements	108-112

The limits set for 1975 are too stringent to meet by the above, fairly simple means, and catalytic exhaust gas purification became necessary. The 1975A standards were met by using the system of Figure XIII.21.a. The engine operates at an air/fuel ratio just above stoichiometric (15:1) and exhaust gas recirculation is employed for NO_x control. CO and hydrocarbons are removed to the required limit by catalytic oxidation over a Pt/Pd-catalyst using secondary air. It is now necessary to use unleaded gasolines because lead compounds would soon poison the noble metal catalyst. The lower limit of NO_x emission specified in 1975B cannot be met in this manner; another drawback of this system is that fuel consumption increases because of the lower compression ratio needed with lead-free gasoline. Alternatively, the octane number of the gasoline must be brought to a higher level than the base gasoline previously used which was upgraded by lead addition. The cost of manufacturing unleaded gasoline is much higher than that of leaded gasoline of the same octane quality, mainly because of the lower yield on crude.

The more exacting limits set in specification 1975B would require the use of a dual catalytic reactor system for exhaust gas clean-up (Figure XIII.21.b.). It contains a noble metal catalyst for NO_x removal; however, a sufficiently stable catalyst for this purpose is not yet available. At any rate, it will be necessary to operate the engine with a low air/fuel ratio because NO_x removal should be carried out under reducing conditions. Catalysts for the selective reduction of NO in the presence of oxygen have not yet been developed. If low air/fuel ratios are used, the specific fuel consumption goes up still further; moreover, the price of the car rises substantially because two reactors containing expensive catalysts must now be applied. This is possible with large-capacity engines, but with small engines such as are used in Europe and Japan the 'driveability' of the car is adversely affected.

A further problem is that the available reserves of platinum are limited and are unlikely to last much longer than 20 years. Moreover, proven reserves are

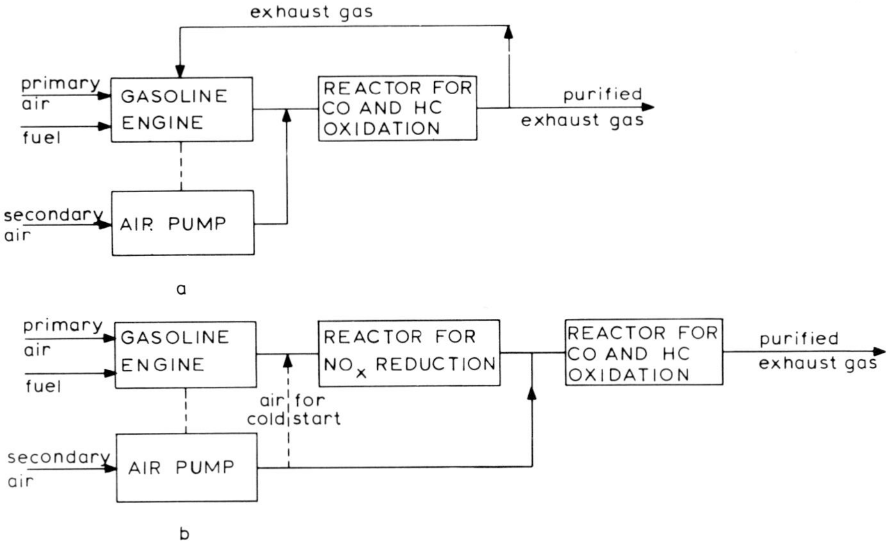

Figure XIII.21. Exhaust gas clean-up systems

a. single reactor; proportional exhaust gas recycle

b. dual reactor; primary air/fuel ratio below stoichiometric

Cold start: NO_x low; HC/CO very high. Excess air also to reactor

available only in Canada, South Africa and the USSR. Thus, this solution to the problem (an application of catalysts which, quantity-wise, will surpass all other catalysts combined if enforced on all cars in Europe, the USA and Japan), is rather unattractive. It is, therefore, probable that engine manufacturers will continue to develop new types of engines operating at high air/fuel ratios and low average temperature, whilst ensuring complete combustion.

Sulfur in gasolines contributes little to the total emission of SO_x. However, if noble metal oxidation catalysts are applied for exhaust gas purification, further oxidation of SO_2 to SO_3 is a distinct possibility because Pt catalyses this reaction. Even if the conversions attained are low, the SO_3 produced may be hydrated to give noticeable and possibly harmful concentrations of sulfuric acid above or near motorways when traffic is heavy. This problem is still being investigated.

Consequences for oil refining

According to the above, emissions can be reduced by switching to lead-free gasolines. The costs are considerable, as is shown in Table XIII.12. Very much

Table XIII.12. Costs of reducing lead content of West-European gasolines (prices of 1976)

Lead content specified (g/l)	0.6	0.4	0.15
RON of base gasoline needed	92.0	93.9	96.3
Investments (in Dfl. 10^9)	--	1.1	2.8
Extra crude oil needed (10^6 tons per year) at:			
- unchanged sensitivity	--	< 1.5	< 4.5
- lower sensitivity	--	1.0	3.0

capital for investment in new equipment or for debottlenecking existing units is required and more crude oil must be processed. The economic damage can remain limited if the intermediate lead level is adopted and if a higher sensitivity can be tolerated. Nevertheless, the new specifications imply that much effort by the industry is needed to meet them. The literature does not contain many clear strategies by which this can be done. Moreover, there are other simultaneous changes affecting the oil industry such as lower specifications for maximum sulfur content of distillate and residual fuels and a trend to change to coal in electric power generation. Therefore, only rather general predictions for the future are possible:

- in large markets, refineries containing several of the major process units, (Figure XIII.22.), such as catalytic cracking and reforming and hydrocracking plus some additional processes like alkylation and isomerization, are more economical than simpler refineries. Thus, hydroskimming refineries (Figure XIII.23.) producing a limited product package by distillation, reforming and hydrotreating with the hydrogen by-product from reforming do not appear to be the optimum choice for e.g. Western Europe;
- combination of fuels production with the manufacture of raw materials for the chemical industry (ethylene, propylene, butadiene, aromatics) is to be preferred to fuels production only;
- as gasoline sensitivity must be lower, the route via aliphatics (isomerization and alkylation) should receive more attention relative to the 'aromatics approach' in which catalytic reforming is the main gasoline producer;
- if the use of coal increases at the expense of residual fuel oil, more residue processing capacity must be installed (coking, visbreaking) to produce middle distillates. The same applies to the hydrocracking capacity needed to convert middle distillates into jet fuels, naphtha and gasoline components. Apart from higher investments more refinery fuel is then needed, i.e. refining becomes more energy-expensive.

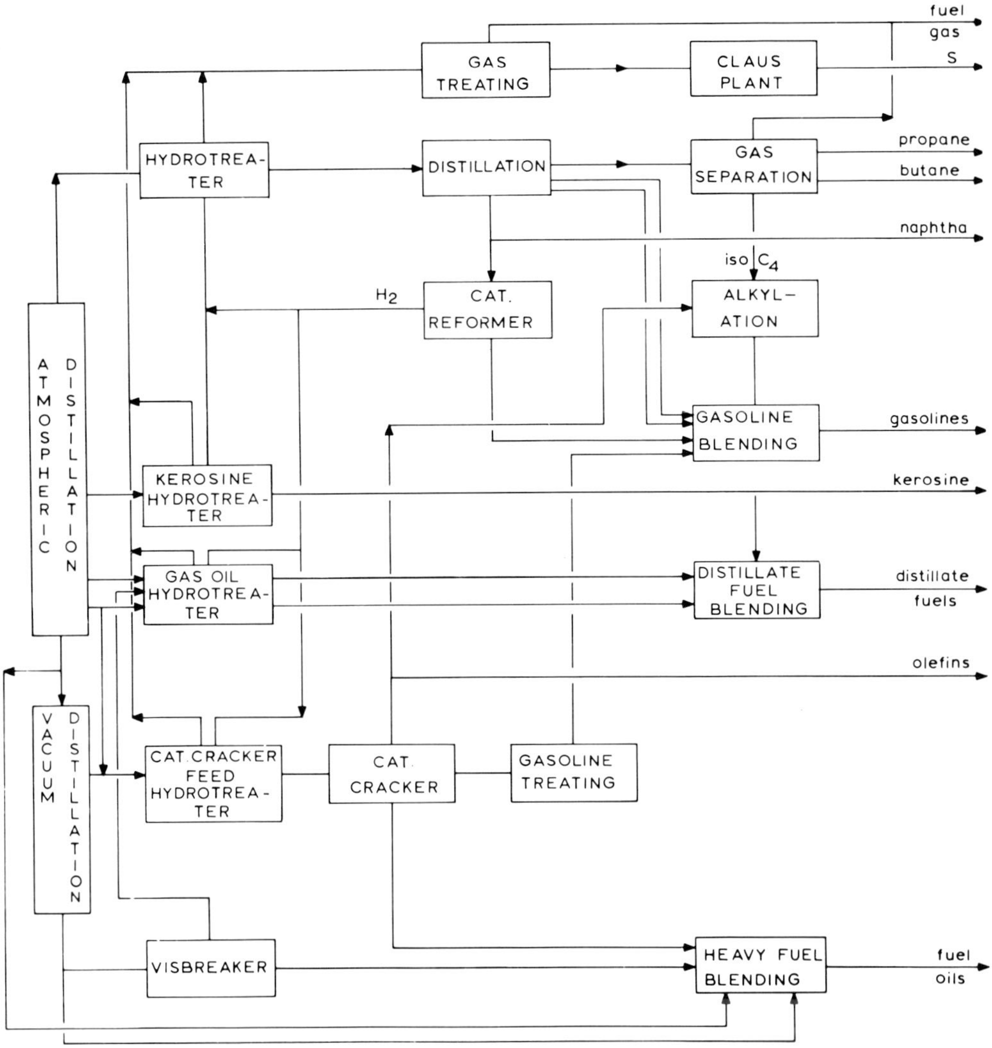

Figure XIII.22. Block diagram of a fuels refinery

The above changes in refinery processing required by new environmental specifications all lead to increased crude oil consumption. This trend can to some extent be counteracted by replacing older refinery equipment by newer and more economic units. For instance, introduction of continuous catalytic reformers containing better catalysts improves the yield of reformate by up to 5% compared with semi-regenerative reforming at the same severity level. Similar benefits can be derived from the conversion of older fluid catalytic crackers to riser crackers. It is expected that much of the construction activity in Western Europe

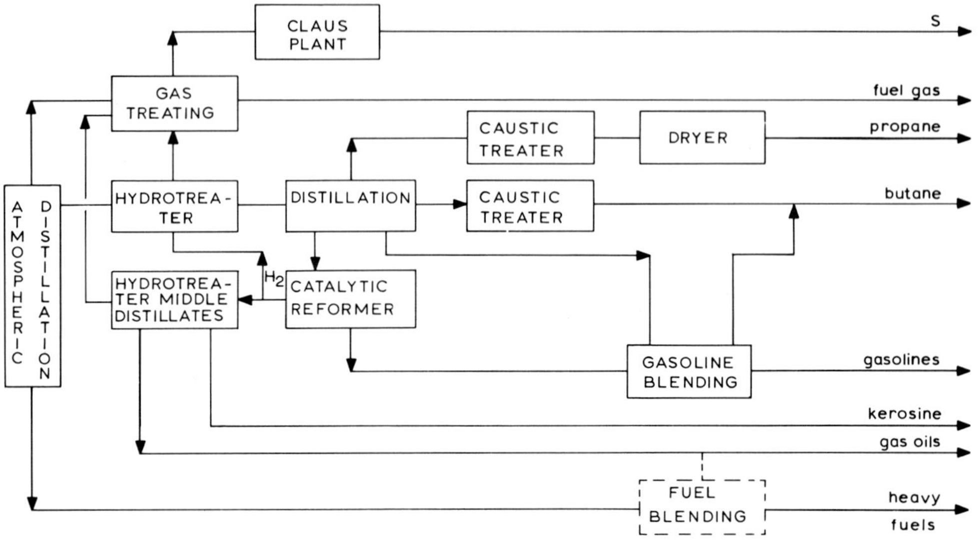

Figure XIII.23. Block diagram of a hydroskimming refinery

will be confined to such revamping of older units. New refineries are unlikely to be built there in the near future because of the present capacity surplus. The trend to add capacity in the oil-producing countries will no doubt continue in the next decade.

REFERENCES

General

1. Hobson, G.D., 'Modern Petroleum Technology', Appl. Science Publ. Ltd., Barking, U.K.

2. Bland, Editors W.F., R.L. Davidson, 'Petroleum Processing Handbook', McGraw-Hill Book Co., New-York 1967.

3. Kobe, Editors K.A., J.J. McKetta Jr., 'Advances in Petroleum Chemistry and Refining', Interscience Publ. Inc., New-York 1958; Americal Society for Testing Materials, Book of ASTM Standards (Petroleum and Lubricants), Philadelphia.

Special

1. Erickson, R.A., G.F. Asselin, Chem.Eng.Progr. 61 (3), 53-58 (1965). Figure XIII.14. is reproduced from this article with permission.

2. Kouwenhoven, H.W., W.C. van Zijll Langhout, Oil Gas Int. 11 (11), 64-68. Figure XIII.15. is reproduced from this article with permission.

3. Broughton, D.B., Chem.Eng.Progr. 64 (8), 60-65 (1968). Figure XIII.17. is reproduced from this article with permission.

4. Dabelstein, W.E.A., E.K. Walter, E.G. Zeschmann, P. Hofbauer, B.W.H. Wiedemann, Proc. World Petr. Congr., 9th, Vol. 6, 159-219 (1975). Table XIII.12. is reproduced from this article with permission.

5. Refining Process Handbook,Hydrocarbon Process, 51 (9), 131 (1972). Figure XIII.4. is reproduced from this article with permission.

6. Refining Process Handbook, Hydrocarbon Process, 49 (9), 167-274, (1970). Figure XIII.8. is reproduced from this article with permission.

7. Refining Process Handbook, Hydrocarbon Process, 53 (9), 103-221 (1974). Figure XIII.16. is reproduced from this article with permission.

8. Refining Process Handbook, Hydrocarbon Process, 57 (9), 97-224 (1978). Figure XIII.10. is reproduced from this article with permission.

XIV. ENERGY AND THE ENVIRONMENT

Introduction

The transformation of fossil fuels into electrical energy, usually via thermal energy, is a sector of the process industry in which rapid developments are occurring and in which chemical engineering takes a prominent part. Moreover, much work is under way to convert solid fuels, which are relatively difficult to handle, into fuels qualitatively superior in terms of ease of transportation, distribution and application. Further incentives for the extensive development efforts are the impending shortage of oil and natural gas as well as their high and still rising cost. Another major cause of much of the current activity is that energy transformation processes often adversely affect the environment, particularly in respect of emission of nitrogen and sulfur oxides and/or dust.

Some of the potential solutions to the above energy problems are discussed in this chapter. After a brief review of the motives for the new developments, the desulfurization of fuels and flue gases and new methods of combustion of coal are reviewed. The gasification of solid and heavy liquid fuels, the transformation of coal into high-grade liquid or gaseous fuels as well as combinations with electricity generation are then discussed, with emphasis on chemical engineering aspects. The chapter ends with a brief look into the future.

Fuel reserves and consumption

Since World War II the consumption of primary energy has risen continually until the early '70s, as follows from Figure XIV.1., but marked changes occurred in the distribution over the various sources of energy. The consumption of petroleum products increased markedly at the expense of coal consumption for reasons outlined in Chapter III. Other sources such as geothermal, hydrothermal and wind energy remained of secondary importance. The growth of nuclear energy was and still is much below expectations because of lack of public acceptance caused by concern over safety and other risks.

This picture changed very suddenly after the oil embargo of 1973/1974, when the growth of energy consumption became much smaller or disappeared temporarily.

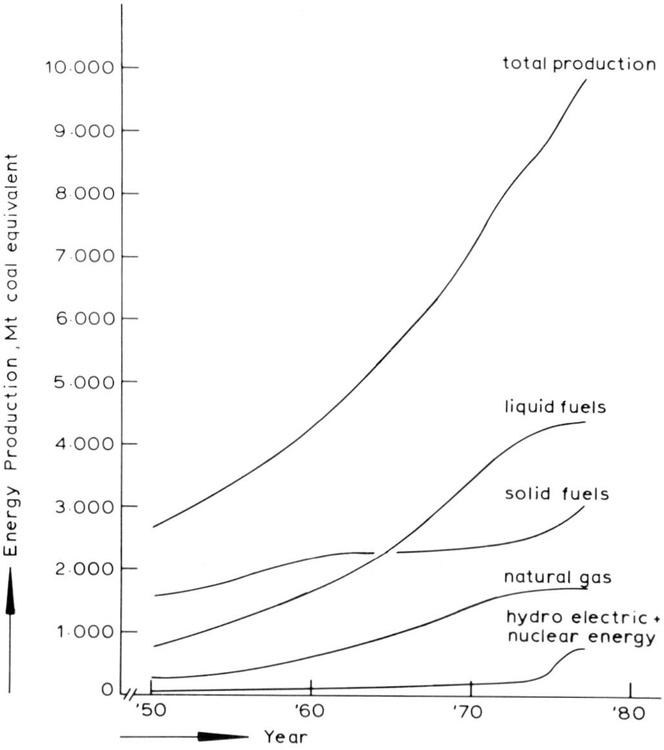

Figure XIV.1. World energy production

Now, a few years after the event, consumption has returned to its former growth rate in the USA and other industrial nations, in spite of extensive programs intended to convince the public that the continued increase in energy consumption is a threat to the economic and political stability of the industrialized world. As a result, it is expected that during the last decade of the 20th century local shortages of oil and natural gas will occur with increasing frequency, mainly because of the decline in easily accessible reserves. Moreover, the fact that the Organization of Petroleum Exporting Countries (OPEC) has become a major power in world economics, dictating and enforcing price increases of crude oil, is among the main causes of the present uncertain state of the economies of most industrialized countries. It is also one of the incentives for the intensive development programmes now in progress.

Since world reserves of coal are much greater than those of petroleum (Table XIV.1.) methods for improvement of combustion in terms of higher efficiency and reduced environmental effects are being sought. Another, but less extensive program, is concerned with the exploitation of the very large oil shale and tar sand formations located chiefly in Canada and the USA. As with coal, formidable technical, logistic and environmental problems must be overcome, and a major contribution to world energy is still decades away. Eventually, energy sources other than fossil fuels will have to be utilized: the table shows that the problem can be solved only through the development of nuclear fission energy or similarly difficult technology.

Table XIV.1. Fossil fuel reserves (in 10^9 metric tons)

	Coal	Lignite	Oil
Recoverable reserves *	430	161	80
Reserves in place **	1,077	340	142
Additional reserves ***	7,063	2,291	131
Total reserves	8,570	2,792	353

* Known reserves recoverable at the present technical and economic conditions.

** Known reserves recoverable at present or in future technical and economic conditions

*** Sum of known reserves not recoverable economically and technically and probable reserves in unexplored areas.

Environmental aspects

Conversion of fossil fuels into thermal or electrical energy is usually associated with environmental side effects. Most important are the emission of ash, dust, and sulfur and nitrogen oxides. Another consequence is thermal pollution caused by hot water and to a lesser extent by hot air, used as coolants. The discussion below is confined to sulfur and nitrogen oxides because the most important methods of combating environmental pollution can be illustrated in this way. The problem of sulfur dioxide emission is that locally high concentrations are injurious to health, cause damage to vegetation and lead to corrosion. It is possible to reduce SO_2 emissions substantially. Table XIV.2. gives a survey of methods for reducing the SO_2 content of the combustion products, some of which are discussed later in this chapter. In some cases it is sufficient to disperse the SO_2 by venting the sulfur-containing waste gases through very tall stacks. Although this method is frequently criticized because the amount emitted is not

Table XIV.2. Methods of reducing SO_x emissions in power generation

Process type	Examples
Fuel desulfurization followed by combustion	1. Desulfurization of natural gas (liquid-phase)*
	2. Desulfurization of liquid fuels (catalytic hydrogenation)
	3. Desulfurization of coal - by solvent refining with a donor solvent - by catalytic hydrogenation
Simultaneous combustion and desulfurization	1. Injection of limestone or dolomite during (oil) combustion
	2. Fluidized-bed combustion in the presence of lime or dolomite (oil and coal)
Gasification by partial oxidation followed by desulfurization and complete combustion	1. With O_2 (oil), followed by (liquid-phase) H_2S removal and combustion
	2. Of coal: partial combustion in a fluid bed in the presence of CaO, followed by complete combustion
	3. Of coal: oxidative gasification followed by desulfurization and combustion
Desulfurization after combustion	1. Wet scrubbing of flue gases with SO_2 removal agent (regenerative or non-regenerative)
	2. Dry flue gas desulfurization processes (regenerative)

* see Chapter VI.

reduced, it should be kept in mind that sulfur dioxide emissions caused by human activities are of the same order of magnitude as natural (volcanic) emissions. Similarly, it can be concluded that partial removal of SO_2 from flue gases often suffices to prevent the occurrence of high local ground concentrations. This is important because reduction of the emissions from e.g. 100 to 30% may cost less than a further reduction from 30 to 10%. Costs are high in terms of investment and operating costs, and appreciable amounts of energy are sometimes required to prevent this type of pollution.

The present position with regard to nitrogen oxides is much less satisfactory. Unlike sulfur oxide emission, the presence of NO_x in flue gases is not associated with fixed nitrogen in the fuel but caused by high temperature oxidation of nitrogen. As outlined in the discussion on NO_x emission by internal combustion engines (see Chapter XIII), the endothermic equilibrium

$$N_2 + O_2 \rightarrow 2\ NO \quad (\Delta H_s^o = +91.4\ kJ/mole\ NO)$$

is established relatively rapidly above about 2000 K. Below that temperature the rate of oxidation is so low that it governs the amount of NO_x in flue gases. Thus, the combustion temperature should be kept low to reduce NO_x formation (see Table XIV.3.).

Table XIV.3. NO_x content of flue gases from fuel oil (at stoichiometric combustion)

Temperature, ^{o}C	vol % NO_x
1500	0.2
1700	0.5
1900	1.4
2100	1.8

Another method is to cool the combustion gases relatively slowly to allow time for nearly complete NO_x decomposition. Even with the more costly equipment then needed, it is impossible to obtain flue gases with less than 200-500 ppmv NO_x. NO_x removal processes are in a much less advanced state of development than those for SO_2 removal, principally because NO_x cannot (yet) be reduced selectively at acceptable costs in the presence of the excess oxygen in flue gases. The only known method is to reduce NO_x with ammonia on a copper-containing catalyst. This method is applied in a modified flue gas desulfurization process using a Cu-containing solid. A few other known flue gas desulfurization processes can be adapted to give partial nitrogen oxide removal.

Desulfurization of fuels

Catalytic desulfurization of petroleum fractions

The catalytic desulfurization of petroleum fractions is practised very widely in modern refineries to lower product sulfur content, improve oil processing and upgrade products (see Chapter XIII.). Apart from sulfur removal reactions (Table XIV.4.), denitrogenation and hydrocracking may also occur, particularly at higher operating severities. Metal complexes, if present in the feed, are partly decomposed and the inorganic constituents deposited on the catalyst, leading to activity decline.

The conditions of treatment depend on the type of feedstock. Vapor phase processing is possible with light naphthas at relatively low temperatures (around 300 °C) and pressures (0.5-2.0 MPa). Higher-boiling distillates are processed in three-phase (G/L/S, trickle-phase) reactors, temperature and pressure generally being higher and space velocity lower (T = 300-375 °C, p = 1.0-3.5 MPa and LHSV = 10-1) because heavier fractions contain more refractory aromatic sulfur compounds. The most common catalyst is Co/Mo-sulfide on alumina ; a more acidic carrier can be used, but only at the expense of increased hydrocracking activity. Another known combination of active ingredients is W/Ni-sulfide. Like the Co-Mo catalysts, this sulfur resistant catalyst was originally discovered in Germany during World War II in work related to coal hydrogenation.

Table XIV.4. Reactions during hydrodesulfurization.

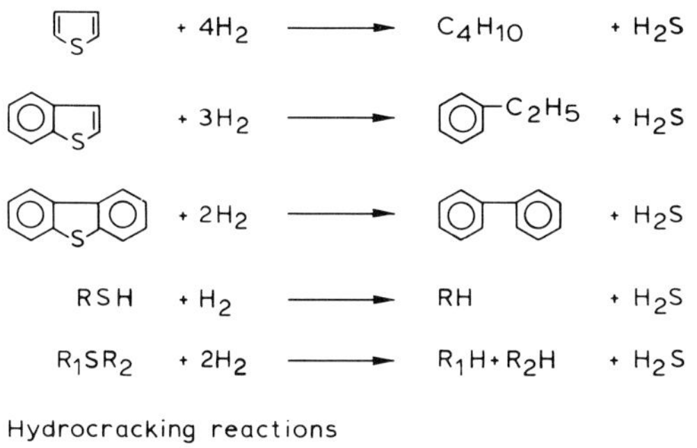

Figure XIV.2. shows a simplified flowchart of a gas oil hydrotreater. The reactor is a trickle-phase reactor, a fixed bed of catalyst through which liquid and gas are passed co-currently, preheated oil and gas entering at the top. Since the reactions are exothermic a temperature rise occurs over the reactor. The reactor effluent is cooled, gas and liquid are separated and the liquid product is stripped for light ends removal. The vented gas is processed to remove sour gases, hydrocarbons and other contaminants, or used as fuel after sour gas removal.

In recent years, hydrodesulfurization of residues has received considerable attention because new and more stringent specifications for the maximum S content of fuel oil were enforced in many areas, notably in the USA and Japan. These lower sulfur contents could no longer be obtained by conventional methods, such as blending residues from low-sulfur crude oils with desulfurized distillates. Residue desulfurization is more difficult than distillate hydrotreating because of the presence of metal (V, Ni) complexes, potent catalyst poisons. Moreover, residues often contain asphaltenes, very heavy aromatic compounds

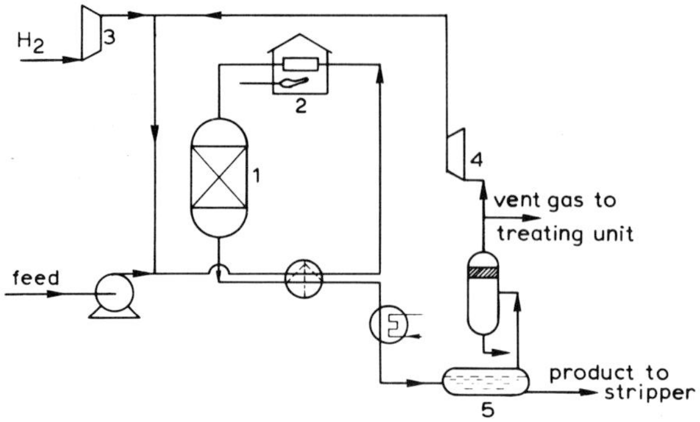

Figure XIV.2. Trickle phase hydrodesulfurizer

1. reactor

2. preheater

3. fresh gas compressor

4. recycle compressor

5. high pressure separator

dissolved colloidally in the oil, which cause coke formation if special precautions are not taken. These include: improved catalysts and higher hydrogen partial pressures (4.0-10.0 MPa), often combined with lower space velocities. The hydrogen partial pressure used affects process costs in two ways:

- under more severe conditions, more hydrogen reacts with the oil because unsaturated fragments and aromatics are hydrogenated. This is clearly seen in Figure XIV.3. Since the hydrogen consumption is a major cost item in residue desulfurization, this is a disadvantage;
- the total amount of fuel to be hydrogenated as well as the higher hydrogen take-up in many cases cause by-product hydrogen from catalytic reforming to fall short of requirements. Thus, a separate hydrogen plant must be built (added investment costs) and operated (more fuel lost in the hydroprocess + hydrogen manufacture) without benefits other than the lower sulfur content.

New catalysts with improved activity and stability were developed to reach the sulfur content of 0.1 %w which some authorities planned to enforce in the mid-1970s. Such catalysts were made by adapting the active ingredient as well as changing the support. Reduced carbon deposition at a given hydrogen partial pressure was obtained. This result can be exploited in two ways:

1. longer catalyst life is possible than with previous generation catalysts at otherwise unchanged conditions; this results in a cost reduction because

longer run periods between catalyst changes are possible when using fixed-bed reactors. Also, catalyst consumption is lower;

2. at the same catalyst life, a lower hydrogen partial pressure can be used, other conditions being kept the same. This saves on hydrogen cost (see Figure XIV.3.).

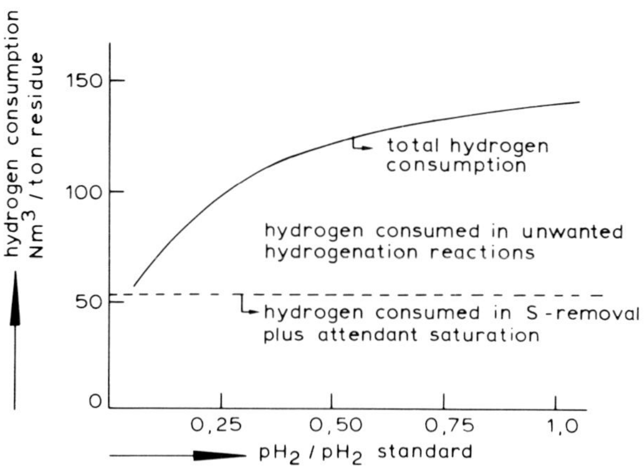

Figure XIV.3. Hydrogen consumption at 76% desulfurization of a Kuwait long residue

Another possibility is to apply a moving-bed reactor. Shell have developed such a reactor in combination with a special catalyst which is very selective towards metals removal and can absorb a heavy load of inorganic reaction products. Combination of a demetallation reactor with a second reactor in which desulfurization is the main reaction type then gives optimal catalyst utilization.

Technically, it is possible to lower the sulfur content to about 0.1 %w when applying the most modern catalysts. However, although it was the intention in the USA to prescribe this limit as the allowable maximum, it does not seem likely at the moment that sulfur contents below 0.3 %w will be required. The main reason for this is that the reduction from 0.3 to 0.1 %w is very costly energy-wise, which is not attractive in view of the rising fuel cost. Even a reduction to 0.3 %w goes at the expense of a sizeable fraction of the fuel because by-product hydrogen from catalytic reforming is not sufficient quantity-wise and some fuel must be sacrificed to make hydrogen. A study by Concawe, a joint study group of major oil companies which is located in the Netherlands, has shown that the cost of fuel desulfurization rises sharply with the need to lower the sulfur content of the product (Table XIV.5.A.); similarly, deep

Table XIV.5. Cost of controlling SO_2 emissions from oil refineries

range considered	: from hydroskimming to extensive fuels refinery
units added	: hydrodesulfurizer(s), amine treating of gases and Claus unit
capacity	: 1 000 000 ton/year crude throughput
products	: LPG, regular and premium gasoline, hydrotreated kerosenes, gas oil pool 0.3 %w S, residual fuel oil %S between 1 and 3
prices	: in US $ valued per 1975-01-01. location Western Europe

	Percent desulfurization of effluent gases		
	95	97	99*
investment for desulfurization 10^6 US $ (ton/hour SO_2 reduction)	0.5 - 0.8	0.6 - 1.0	9.5 - 12.7
extra operating cost ($/ton SO_2 removed)	13 - 19	15 - 20	530 - 590
incremental fuel requirement (%)** for increasing % deS from 95 to 99	--	--	0.6 - 2.5

* a Claus tail gas treating unit must be added

** on crude intake

Table XIV.5.A. Gas oil hydrodesulfurization capacity required and costs of desulfurizing to different S content (location: Western Europe, 1980, 5 Mt/year)

Sulfur content of feed, %w	S in product, %w	Capactiy needed, 10^6 ton/year	Energy consumed* % of product
1.05	0.64	0.5	1.2
1.05	0.22	1.2	2.9
1.05	0.12	1.7	4.2
0.75	0.37	0.5	1.2
0.75	0.23	0.75	2.2
0.75	0.12	1.24	3.2

* includes direct and indirect energy requirement and the energy for hydrogen manufacture.

sulfur removal from refinery off-gases is much more expensive than moderate desulfurization, in respect of investments as well as operating costs. This is a clear example of a compromise between the conflicting requirements of energy economics and environmental protection. In this connection, application of the newly developed desulfurization catalysts referred to above is worthwhile: one can either desulfurize more deeply at the same hydrogen consumption, or desulfurize less deeply with appreciable savings in hydrogen requirements.

An effect of the postponement of more exacting sulfur specifications of fuels is that the number of new projects for residue desulfurization units has decreased markedly, i.e. by about 60%, immediately after it was announced that 0.3 %w instead of 0.1 %w S would be the new maximum specified for the near future

Desulfurization of coal

Desulfurization of coal is being studied in several countries to produce either a highly viscous or solid reaction product containing relatively little sulfur, or a range of products with very low sulfur content, from which fuel oils, middle distillates, gasoline and LPG can be obtained. Some of the methods under investigation are briefly discussed below.

Solvent refining of coal is a process in which ground coal (< 2 mm) is mixed with a donor solvent such as partly or completely hydrogenated naphthalene derivatives and hydrogen-containing gas. At reaction temperatures of up to 450 °C and 14 MPa pressure about 70-85% of the sulfur is converted into gaseous reaction products, mainly H_2S and COS. Pyritic sulfur is removed almost completely whereas about one third of the organically bound sulfur is left in the solvent-refined coal. The treated product contains ½-1 %w S; it melts at about 150 °C into a viscous product that can be handled like heavy fuel oils. The solvent is recovered and is catalytically hydrogenated prior to recycle. A small amount of light hydrocarbons is a by-product; the quantity of hydrocarbon gases obtained (about 1-2 %w on coal feed) is usually sufficient to produce the hydrogen required for the treatment. Before hydrogen can be produced from this gas, desulfurization is necessary (see block diagram Figure XIV.4.). Advantages of this desulfurization method are that hydrogen consumption is low and that catalysts and auxiliary chemicals are not required. However, the sulfur content of the product may have to be reduced further by catalytic hydrogenation (see below) in areas where strict sulfur specifications are set. Moreover, special equipment is needed for storage and combustion of the high-melting fuel.

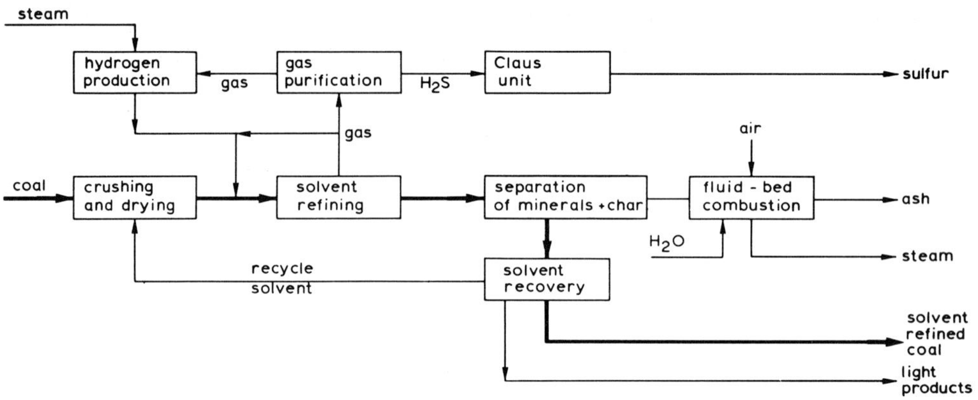

Figure XIV.4. Solvent refining of coal

Direct catalytic hydrogenation of coal at high temperatures (above 500 °C) and at pressures of 30.0-70.0 MPa was carried out on an industrial scale in Germany just before and during World War II to obtain a broad spectrum of hydrocarbon products of low sulfur content. The maximum production capacity in the period 1940-1945 amounted to 4 x 10^6 tons coal per year. Coal, brown coal or coal tar were used as feedstocks. A simplified block diagram is shown in Figure XIV.5.

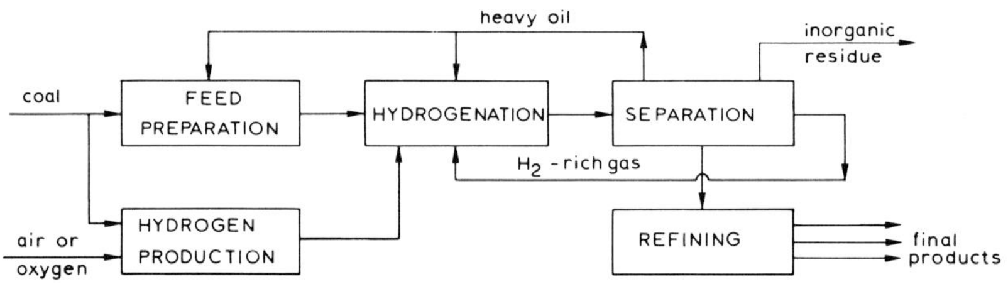

Figure XIV.5. Block diagram of coal hydrogenation

Although a small research effort was devoted to this subject around 1960, attempts to improve this process were resumed on a much larger scale soon after the oil embargo of 1973, in particular in the USA. Several pilot plants are now in operation; as far as is known supported Co-Mo-sulfides are used as catalysts. Improvements in catalyst activity have led to the use of lower pressures, i.e. 5.0-20.0 MPa.

The products are rich in aromatics and contain little or no sulfur. As the amount of hydrogen introduced increases, which is brought about by longer residence times, higher pressures and higher catalyst/coal ratios, the proportion of lighter products increases (Figure XIV.6.). In principle it is possible to process the liquid products into e.g. lubricating oils, but applications other than direct combustion have so far not been considered. The main incentive for this type of work is the wish to become economically and strategically less dependent on the OPEC countries.

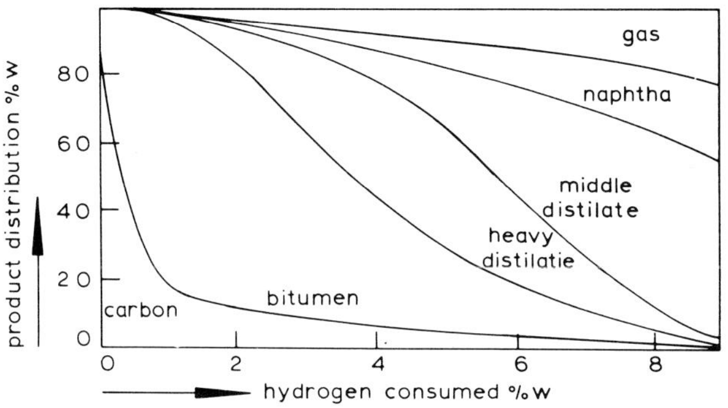

Figure XIV.6. Product distribution in coal hydrogenation

It is also possible to synthesize high-molecular weight hydrocarbons from coal by gasifying it (see later in this chapter) to obtain a synthesis gas consisting of H_2/CO mixtures, and then converting this gas into liquid and, if desired, solid hydrocarbons. A block diagram of this Fischer-Tropsch route is shown in Figure XIV.7. If a fixed-bed reactor containing an Fe/Co catalyst is applied (225-240 ^{o}C; 2.5 MPa) products consisting of LPG, gasoline, middle distillates, oil and wax are produced. An alternative method employs a fluid-bed reactor with Fe catalyst at about 325 ^{o}C and 2.0-2.5 MPa; in this case, part of the higher-boiling fractions contain sizeable proportions of alcohols. Like direct coal hydrogenation, this route for converting solid into liquid fuels was developed in Germany before World War II.

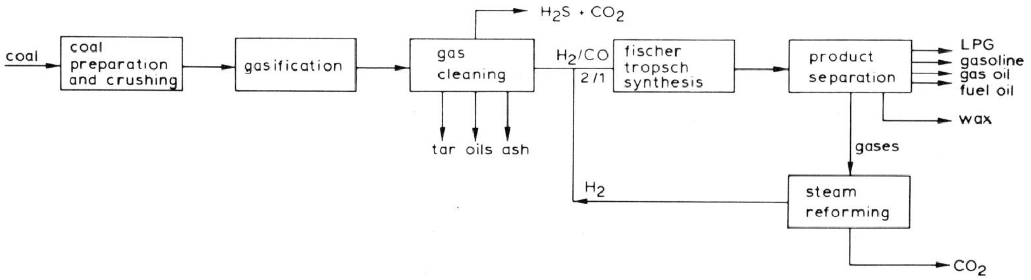

Figure XIV.7. Block diagram of Fischer-Tropsch process

Simultaneous combustion and desulfurization

In the mid-1960s, when coal prices dropped because of increased competition by oil and natural gas, work was started in the USA and Great Britain (National Coal Board) to develop combustion of coal in a fluidized bed. This work was intensified when it was recognized that combustion could be combined with desulfurization by adding a sulphur dioxide *acceptor* to the fluidized bed, such as limestone ($CaCO_3$) or dolomite ($CaCO_3.MgCO_3$). Both solids dissociate into oxide(s) and CO_2 at the optimum combustion temperature, 850-1000 ^oC, and react with SO_2 and oxygen to give $CaSO_4$. Magnesium sulfate is not stable at these conditions. An additional advantage is that much less NO_x is formed because of the low combustion temperature. In the early 1970s, demonstration plants of 30-50 MW capacity were built and operated successfully. Up to 90% desulfurization was obtained. The process can use coal as well as heavy fuel oil as feed.

However, a large excess of acceptor is necessary to effect high degrees of desulfurization, as follows from Figure XIV.8. Thus, between 8 and 10 tons of ash-containing spent acceptor is formed per ton S in the fuel. In other words: if coal containing 1 %w S is used and the degree of desulfurization is 90%, more than 250 tons of spent limestone is produced annually per MW installed capacity. Moreover, the spent material exhibits strongly alkaline reaction and the $CaSO_4$ tends to harden by hydration. Another disadvantage is that mining as well as disposal of the spent acceptor is environmentally unattractive.

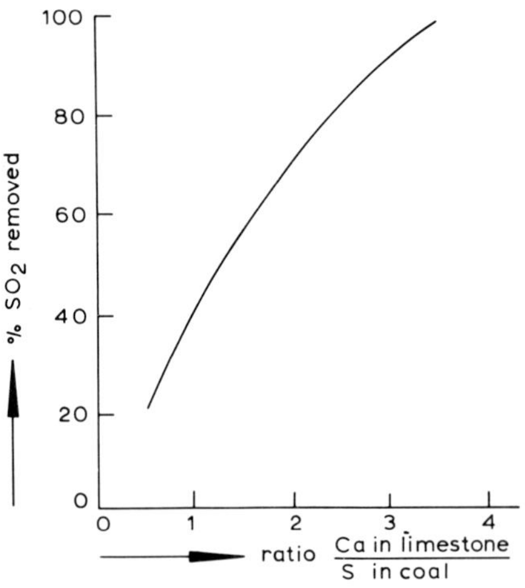

Figure XIV.8. Desulfurization efficiency during fluid-bed combustion of coal in the presence of limestone

Some improvement can be obtained by working at higher pressure, e.g. at about 1.0 MPa. Combustion efficiency and acceptor utilization are somewhat higher, but dolomite must be applied in pressurized combustion instead of limestone because the latter is insufficiently reactive. Nevertheless, these improvements do not yet make the process fully satisfactory. An additional problem is that complete solids removal is imperative.

Attempts are now being made to apply the spent acceptor as a cheap source of gypsum in the building industry. However, it does not seem likely that this outlet can absorb the very large quantities expected from widespread use of desulfurizing combustion.

In principle, it also is possible to regenerate the sulfate by reduction, but high temperatures and much energy are required. By comparison, methods in which partial oxidation is followed by (1) removal of the H_2S formed at the prevailing reducing conditions and (2) a second combustion step (see below) are expected to be superior to pressurized fluidized bed combustion + sulfate regeneration. Similar objections can be raised against desulfurization by injection of an aqueous dolomite suspension into the boiler downstream of the combustion zone. This method, which is known to be applicable on a large scale in power stations, can at best be considered as an interim solution.

Gasification

Residual fuel oils can be gasified with oxygen or by partial oxidation at high temperature. Since this type of process has been discussed in Chapter V., attention will be focused here on the gasification of coal with oxygen, air, steam and/or hydrogen. Depending on the intended application, the following products are obtained:

Producer gas : the product of gasification of coal with air and little or no steam, has a low heating value (0.5×10^4 kJ/Nm^3) because of its high nitrogen content. It is also referred to as low-BTU gas.

Syngas : produced by gasifying coal with steam and oxygen, is used in many applications, such as for the synthesis of ammonia and methanol, in industrial boilers or power stations, and for making liquid fuels by the Fischer-Tropsch route (see above). Its low heating value is about 1.1×10^4 kJ/Nm^3.

SNG : substitute natural gas with a low heating value of about 3.5×10^4 kJ/Nm^3 is obtained by converting syngas into methane.

After a brief survey of gasification reactions and general process aspects, some examples of gasification processes are discussed below. In all cases attention is given to the problem of sulfur removal.

The reactions between coal and gasification agents are very complex. Rates and product composition depend not only on the agent used, but also on the type of coal, amount and composition of ash, coal porosity as developed during volatilization and gasification, heating rate, pressure and several other process variables. In general, the following main stages can be distinguished:

1. As the coal is heated to the reaction temperature, moisture is stripped off. Reaction sets in at about 400 ^{o}C; in the first low-temperature region (400 to about 800 ^{o}C) loosely bound carbon in side chains and relatively small aromatic nuclei are converted into gaseous products. Rates of reaction are very high with all gasification agents (O_2, H_2O and H_2). If hydrogen is used, much methane is formed along with naphtha and tar.
2. The particulate coal left after low-temperature gasification stage (1) is less reactive towards hydrogen but it still reacts very rapidly with steam and oxygen between 800 and 1000 ^{o}C. Higher molecular weight hydrocarbons are now entirely converted into CO, CH_4 and CO_2.
3. After stage (2) is completed, *char* is left, i.e. residual coke which is quite unreactive towards H_2 and requires long residence times and high temperatures for complete conversion with steam and oxygen. In some processes char is not gasified but burned to supply process heat.

The reactions occurring can be represented schematically in the following way:

$$C + H_2O \rightarrow CO + H_2 \qquad (\Delta H^o_s = +121.1 \text{ kJ/mole C}) \qquad \text{(a)}$$

$$C + 2\,H_2 \rightarrow CH_4 \qquad (\Delta H^o_s = -\ 85.2 \text{ kJ/mole C}) \qquad \text{(b)}$$

$$C + CO_2 \rightarrow 2\,CO \qquad (\Delta H^o_s = +161.5 \text{ kJ/mole C}) \qquad \text{(c)}$$

The heats of reaction show that gasification with steam alone requires a source of heat. This can be done from an external source, e.g. a nuclear reactor; it is, however, much more common to make the process autothermal. This is done by adding so much oxygen to the steam that the reaction

$$C + O_2 \rightarrow CO_2 \qquad (\Delta H^o_s = -393.5 \text{ kJ/mole C}) \qquad \text{(d)}$$

supplies the heat absorbed by reactions (a) and (c). An alternative is to combine steam gasification with heat production in a nuclear reactor.

Another possibility is to generate heat by the reaction

$$CaO + CO_2 \rightarrow CaCO_3 \qquad (\Delta H^o_s = -177 \text{ kJ/mole}).$$

This is done in Consolidated Coal's CO_2 acceptor gasification process. For practical purposes, gasification reactions (a) through (d) can be considered irreversible. The composition of the product gas can be calculated from feed composition, temperature and pressure by taking the following equilibria into account:

$$CO + H_2O \rightleftarrows CO_2 + H_2 \qquad (\Delta H_s^o = -41.0 \text{ kJ/mole CO}) \qquad (e)$$

$$CO + 3\ H_2 \rightleftarrows CH_4 + H_2O \qquad (\Delta H_s^o = -206.6 \text{ kJ/mole CO}) \qquad (f)$$

The above reactions and equilibria show that it is profitable to gasify under pressure. More methane is then formed according to equilibrium (f), giving a product with a higher heating value. Moreover, since methane is formed by exothermic reactions, less coal needs to be burned by reaction (d). Other advantages are that gas purification is easier and that smaller gasification reactors can be used because reaction rates are somewhat higher at higher pressures. The product gas is contaminated by ash, char, particles, H_2S, COS, NH_3, HCN and in many cases higher-molecular weight components (tars). It is necessary to purify the gas before it can be applied.

Producer gas by partial combustion with air

It was shown in Chapter V. that fuel oils can be gasified by partial combustion with air or oxygen. If air is used, producer gas is obtained having a high nitrogen content. This does not significantly decrease the flame temperature during combustion, but transport over long distances is uneconomic and the gas should be used at or near the gasification plant site. Among the uses is the generation of electric power in combined cycle systems containing steam turbine-driven electric generators as well as generators powered by gas turbines. Such systems can be based on fuel oil or coal; a desulfurization operation must be applied to minimize emission of SO_x. Desulfurization can be effected during or after gasification, as is shown below.

Simultaneous desulfurization and gasification occurs in the system of Figure XIV.9. Heavy fuel oil is gasified in a fluid-bed reactor containing CaO (Exxon). The oxide reacts with H_2S to CaS and water; a low-sulfur combustible gas is obtained which can either be sent to a gas turbine or used to generate steam. A stream of solid particles is continuously withdrawn from the reactor and introduced into a fluid-bed regenerator, where air oxidizes the acceptor to form CaO and SO_2. The SO_2 can be converted into sulfur or sulfuric acid; regenerated CaO is returned to the partial combustion reactor. A small amount of CaO is withdrawn from the system to prevent build-up of contaminants (e.g.

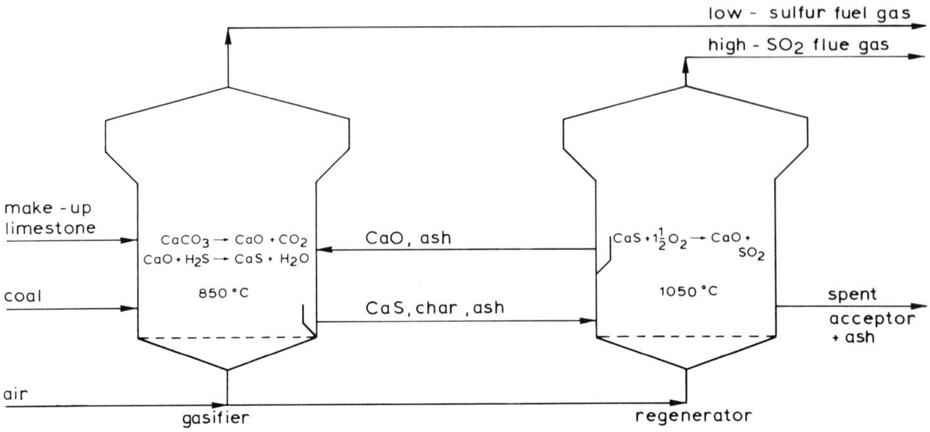

Figure XIV.9. Partial combustion + desulfurization

inorganic matter from the fuel); make-up $CaCO_3$ is added, which dissociates into CO_2 and CaO at gasification conditions. Apart from reducing pollution by SO_x, this system also minimizes emission of NO_x because much less nitrogen oxides are formed when burning the producer gas in a boiler or a gas turbine. Reasons are the relatively low combustion temperature and the effects of improved mixing leading to shorter flame lengths, i.e. shorter residence times.
It is also possible to use coal as feed to fluid-bed gasifiers of the type described above. However, coal usually contains so much ash that sizeable proportions of solid acceptor must be bled off, giving rise to high limestone consumption. On the other hand, the ash contributes to the desulfurization.
It is not yet clear whether simultaneous combustion and desulfurization is a better method of processing coal than gasification by partial oxidation followed by desulfurization of the fuel gas, or even by flue gas desulfurization after full combustion.

Partial oxidation followed by desulfurization is practised on an industrial scale: both Shell and Texaco have developed processes based on fuel oil. An example of use as part of a combined cycle power plant is shown in Figure XIV.10. Fuel oil is gasified with air in a firebrick-lined reactor and the gases cooled in a waste heat boiler generating steam at 8.0-9.0 MPa and about 500 ^{o}C.
The producer gas is further cooled by heat exchange with boiler feed water, washed with water to remove soot, ash, NH_3 and HCN and then desulfurized by countercurrent washing with an alkaline liquid. The gas is sent to a gas turbine driving both the air compressor for the gasification reactor and an electric

generator. The exhaust gases of this turbine are cooled in a steam superheater to recover heat from them. The steam is used to drive a steam turbine/electric generator combination. Condensate is recycled to the boiler.

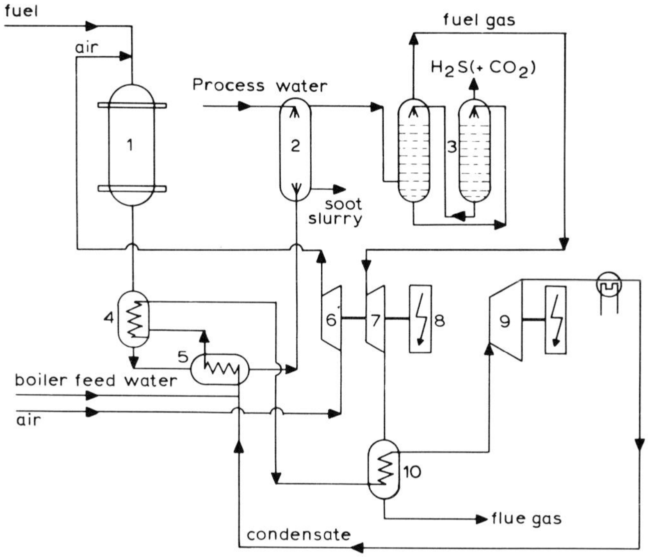

Figure XIV.10. Example of a combined cycle system

1. *reactor*
2. *soot removal*
3. *gas purification*
4. *boiler*
5. *preheater*
6. *air compressor*
7. *gas turbine*
8. *generator*
9. *steam turbine*
10. *heat exchanger*

This combined cycle system has similar advantages as the fluid-bed gasifier/ desulfurizer discussed above. To be useful for electric power plants, however, rapid changes in load should be possible to follow electricity demand. Moreover, a combined cycle system should be flexible enough to be brought on stream to 90% of its full capacity in about one hour, if it is to be used for supplying peak loads. Computer simulations of the non-stationary behavior of oil-fired combined cycle systems have shown this to be possible.

If gas turbine inlet temperatures of about 950 ^{o}C are used, efficiency is about equally high as for conventional oil-fired power stations. Thus, the reduced environmental pollution is obtained without loss in efficiency. If the inlet temperature could be raised still further, savings in operating cost, mostly by reduced fuel consumption, would be obtainable. Capital cost then is marginally lower than for conventional stations. If this is to become feasible, materials with better high-temperature properties should be developed for gas turbines.

Coal gasification by partial combustion followed by a separate desulfurization step is also a likely future component of combined cycle systems. Although partial oxidation at substantially atmospheric pressure could be applied, it is profitable to use pressure gasification because most desulfurization processes perform much better at higher pressure. The first application of atmospheric fluidized-bed reactors was the Winkler gas generator used in Germany during World War II to produce synthesis gas. Efforts are now under way to develop fluid-bed pressure gasifiers to generate producer gas from coal. Since fuel gases obtained in this way contain tar components, NH_3, HCN, COS as well as H_2S, extensive gas purification is needed.

In principle all types of gasifier (see below) can be built into a combined cycle plant. However, hydrogasification seems less attractive for this purpose because of the tar content of the product. The same applies to the Lurgi gasifier, a moving bed reactor through which gas and fuel pass countercurrently. Cocurrent gasifiers operating at high temperatures (e.g. Koppers Totzek reactor) are preferred in this respect because the cocurrent flow and the very high temperature cause the gas to be low in tar. However, oxygen must be used as the gasification agent to obtain the required high reactor temperature.

Synthesis gas production

Partial oxidation with oxygen of coal leads to synthesis gas that can be applied for making ammonia, methanol or to liquid fuels by the Fischer-Tropsch route (Figure XIV.7.).Fluidized bed gasifiers are suitable for this purpose. The Koppers Totzek gasifier (Figure XIV.11.) developed in Germany is an attractive alternative because it operates on any type of coal that can be reduced to powder with an average particle diameter of about 1 mm. In this entrained bed reactor oxygen and coal particles flow cocurrently and gasification occurs

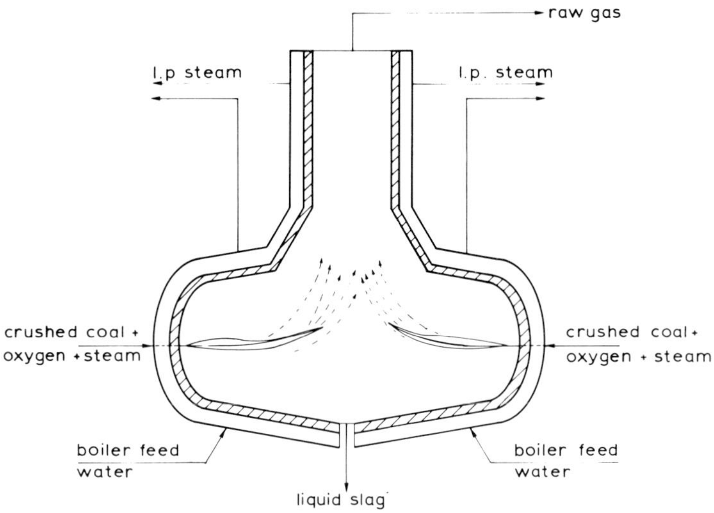

Figure XIV.11. Koppers-Totzek gasifier

in a fraction of a second. Molten ash is solidified by collecting it in water at the bottom of the gasifier. Part of the inorganics are carried overhead in the form of fly ash droplets which should be given time to solidify before separation is possible. Since very high temperatures (over 2000 K) are needed for such fast reaction the gasifier cannot run on air and oxygen plant must be

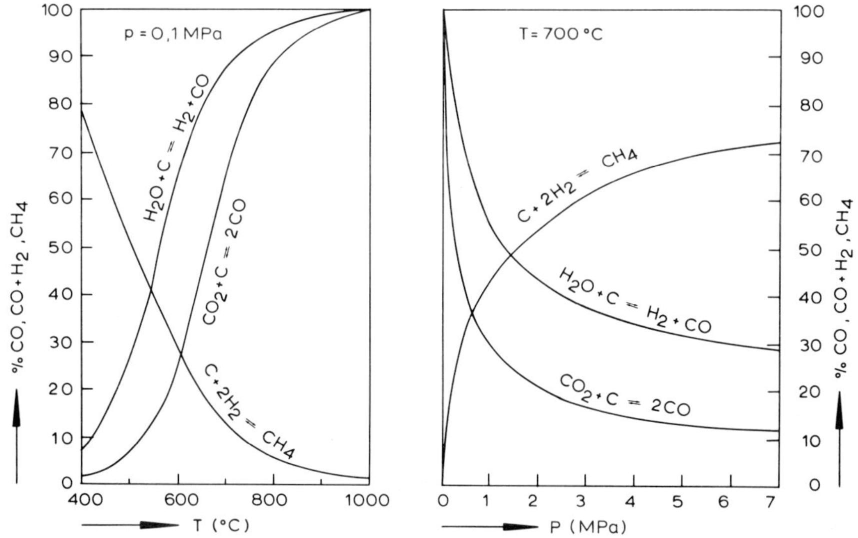

Figure XIV.12. Equilibrium composition of product gas

added. So far, the Koppers-Totzek gasifier has been used industrially at atmospheric pressure, where it performs satisfactorily. Efforts are new being made to raise gasifier pressure in order to improve thermal efficiency and downstream gas purification operations. At the high temperatures prevailing in the reactor little methane is formed (see Figure XIV.12.), although the methane content is somewhat higher at higher pressure it does not contribute much to raising the heating value of the gas. An advantage of the Koppers-Totzek gasifier is that the tar content of the product gas is quite low.

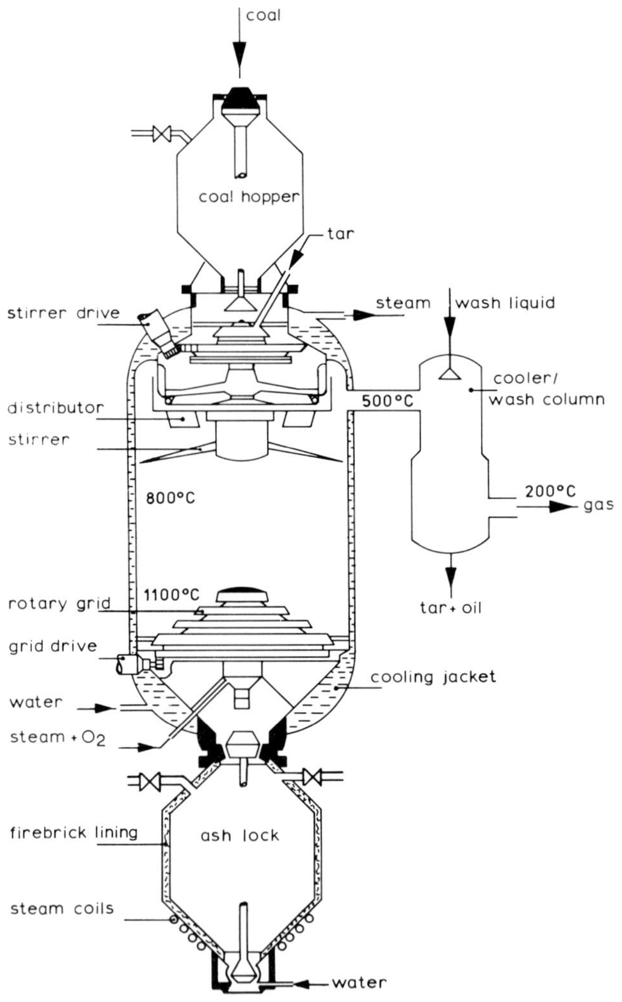

Figure XIV.13. Lurgi pressure gasifier

Another well-known process is offered by Lurgi (Figure XIV.13.). Oxygen and steam enter at the bottom of the gasifier and coal at the top. A combustion zone is generated in the lower half of the reactor having a temperature of about 1000 °C. The hot combustion gases generate CO and H_2 in a reducing zone above the combustion zone (reactions c. and a. of page 290); tar, light naphtha components and a fair amount of methane are formed in the top zone of the reactor. The result is a medium-BTU gas with higher heating value than the gas obtained with the Koppers-Totzek reactor. Enhanced formation of methane occurs at higher pressures; this is one of the reasons why a pressurized version of the Lurgi gasifier was developed. The tar must be removed from the gas prior to desulfurization; it can be returned to the reactor for further conversion into combustible gas. Liquid ash is removed at the bottom.

Drawbacks of the Lurgi gasifier are that coal lumps with a limited range of particle sizes are needed and that coal with a high tendency to caking cannot be processed in it unless the feed is pretreated by a superficial oxidation. Moreover, like the Koppers-Totzek reactor, the capacity per reactor is still somewhat lower than desirable for application in large combined cycle systems or SNG plants.

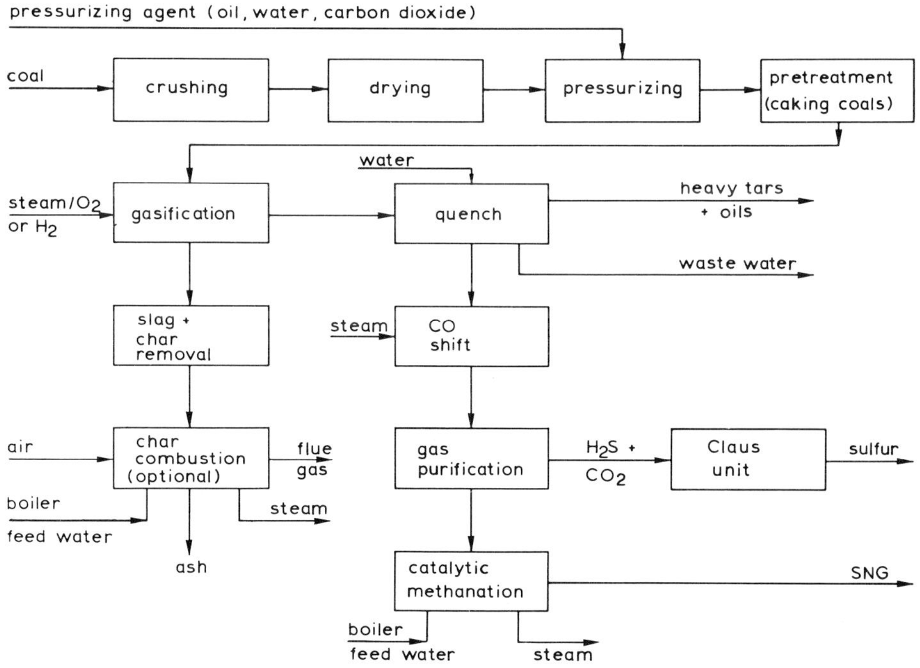

Figure XIV.14. Generalized block diagram of SNG process

Substitute natural gas

In contrast to producer gas and synthesis gas, SNG has such a high heating value per unit volume that transport through pipelines and mixing with natural gas is feasible. To achieve this, coal must be converted into a product gas containing high percentages of methane. The general form of processes for SNG is shown in block diagram Figure XIV.14. The pretreatment may include drying, size adjustment and, in case caking coals must be processed in gasifiers sensitive to the mechanical properties of the feed, preoxidation. When operating at high pressure to promote methane formation, the feed must be brought to the reactor pressure by screw extrusion or pressurizing with a driving gas. After gasification fly-ash is removed, e.g. by cyclones, and the gas freed of heavy tars, NH_3, HCN, water and other impurities. The CO_x/H_2 ratio is then adjusted by removing CO_2, if necessary preceded by CO shift. The final operation is catalytic methanation to CH_4 and water.

The most desirable overall conversion to methane is:

$$2\,CO + 2\,H_2O \rightleftarrows CH_4 + CO_2 \qquad (\Delta H_s^o = -5 \text{ kJ/mole C}) \qquad \text{(g)}$$

In theory, no reactants other than water would be needed if the rates of reaction were sufficiently high. This is so only at high pressures and temperatures; thus, energy should be supplied to the system. As the temperature increases, however, the equilibrium

$$C + 2\,H_2 \rightleftarrows CH_4 \qquad (\Delta H_s^o = -85 \text{ kJ/mole C}) \qquad \text{(h)}$$

becomes less favorable, as follows from Figure XIV.12. It does help to raise the pressure, but because acceptable rates are possible only at temperatures far in excess of 1000 ^oC, gas with a high enough methane content is not formed. If this is to be done, gasification at moderate temperatures (below about 600-800 ^oC) is necessary, requiring complex reactors or separate generation of H_2. An example of producing hydrogen in a separate process is the use of the iron/steam reaction

$$Fe + H_2O \rightleftarrows FeO + H_2 \qquad \text{(i)}$$

The oxide is reduced with primary synthesis gas containing CO:

$$FeO + CO \rightleftarrows Fe + CO_2 \qquad \text{(j)}$$

The CO-containing gas can be produced by partial oxidation of coal or *char*, i.e. the unreactive part of the coal left after hydrogasification and the first phase of steam/oxygen gasification (see page 290). Other proposed processes such as the Hy-gas process under development by the US Institute of Gas Technology (IGT) contain complex high-pressure reactors in which countercurrent flow of coal and gases is effected (see Figure XIV.15.). A coal slurry in hydrocarbon liquids produced in the process is fed to the top section of the reactor, where the process liquid and water evaporate by the action of the product gases passing through. The solid then moves downwards and enters mixing zone <u>4</u> from where the particles are carried upwards by the hydrogen-

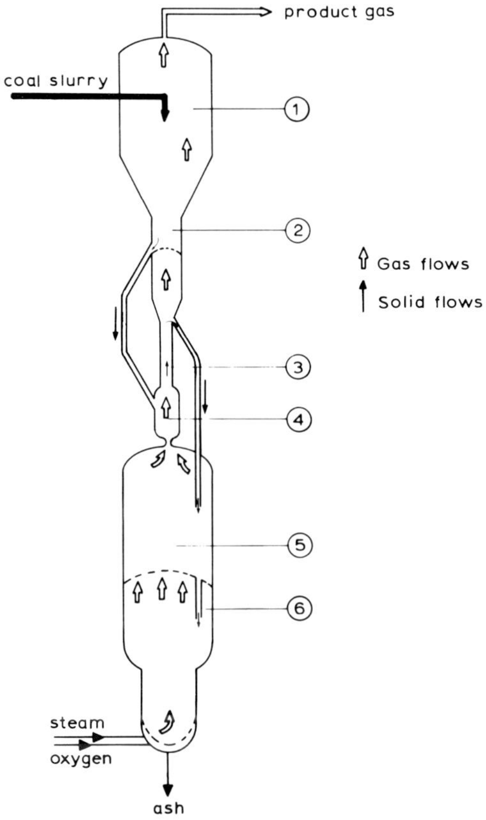

Figure XIV.15. IGT reactor

1. slurry dryer

2. evaporator

3. riser - first hydrogasifier

4. G/S mixer

5. fluid bed - second hydrogasifier

6. oxy-gasifier

containing gas through riser reactor 3 where the main hydrogasification reactions occur. The fluid bed above the riser acts as a solid/gas separator; the solid moves downwards to fluid-bed zone 5 where hydrogasification is completed. The char left from this operation is used to make CO/H_2-containing gas by gasification with steam and oxygen (zone 6). Ash leaves at the bottom. Although in principle this type of reactor appears more attractive than e.g. Lurgi gasifiers, many engineering problems must be overcome to commercialize it since the various gasification stages all occur in one reactor and because changes in reactor conditions are bound to affect all stages. Moreover, the reactor is very large: the pilot plant reactor is about 40 m high.

Apart from gasification, methanation also presents problems. The methanation reactions

$$CO + 3\,H_2 \rightleftarrows CH_4 + H_2O \text{ and } CO_2 + 4\,H_2 \rightleftarrows CH_4 + 2\,H_2O$$

are very exothermic. The best way of removing heat from the methanators is still being sought. Possibilities are a multitubular reactor containing catalyst pellets or tubes with a catalyst coating, multi-bed reactors with interstage cooling by injecting cold gas between the stages and liquid-phase methanation. The problem is to maintain catalyst activity over a wide temperature range; with cold gas recycle, an unduly high pressure drop must be avoided. A complicating factor is carbon deposition on the catalyst, which occurs at high temperatures and can be avoided by adding sufficient steam. These problems are less severe in cases where hydrogasification is applied because the methanator feed then contains fair amounts of methane causing the heat evolved to be smaller.

It is beyond the scope of this chapter to describe other novel reactor designs for SNG manufacture. The reader is referred to several excellent reviews on this subject.

Flue gas desulfurization

Most gasification processes described above are either fairly new or still in the development stage. Since fuel desulfurization prior to combustion is sometimes impossible and often quite costly, there is a need for methods of desulfurizing flue gases. If sufficiently effective and cheap flue gas treatments could be developed, these might well be more economic than combinations of gasification and desulfurization or fuel desulfurization alone. This is particularly so for installations like power stations operating on high-sulfur fuel oil or coal. Moreover, in several countries new legislation requires addition of flue gas desulfurizers to existing plants, in the form of 'retrofit' units. The conditions at which flue gas desulfurizing equipment must operate are somewhat exotic:

- very large quantities of flue gas must be treated: about 10^6 Nm^3 per 100 MW installed capacity;
- the flue gas contains low concentrations of SO_2 (0.1-0.3 vol%), a few percent oxygen, CO, CO_2 and water vapor;
- ash particles are usually present in flue gas; with coal feed the ash concentration can be as high as 20 g/Nm^3;
- the treated gas should not only be low in SO_2 (80-90% desulfurization is usually needed) but its temperature should be high enough (120-140 ^oC) to provide buoyancy and to prevent the occurrence of a condensation plume at the top of the stack. Condensation in the stack should be avoided for reasons of corrosion.

As a result, nearly all stack gas desulfurization processes developed so far are based on a chemical concentration step involving the oxidation S(IV) - S(VI) as the driving force. Major requirements of these processes, in addition to low capital and operating costs, are:

1. the sulfur should be recovered in a saleable form or in a form in which it does not create secondary pollution (see the example of limestone or dolomite injection discussed earlier). Sulfur is the preferred product because of ease of storage and world-wide demand;
2. energy losses (equivalent to increased fuel consumption) should be kept to a minimum. Among the energy losses pressure drop in the reactors used and fuel requirements for reheating the flue gas after treatment at low temperature are important;
3. a candidate flue gas desulfurization process should perform well under varying loads of stack gas at varying sulfur content of the gas;
4. the process should be easy to operate, also for supervisory personnel of electric power stations unused to chemical processes;
5. the process should be very reliable because interruptions in power generation cannot be tolerated.

A survey of some flue gas desulfurization methods is shown in Figure XIV.16.

Wet SO_2 removal processes

In most wet flue gas desulfurization processes the gas is cooled in a first washing step that also removes fly-ash. A second wash is then carried out to reduce the SO_2 content to the desired low level; usually 90% desulfurization is the target.

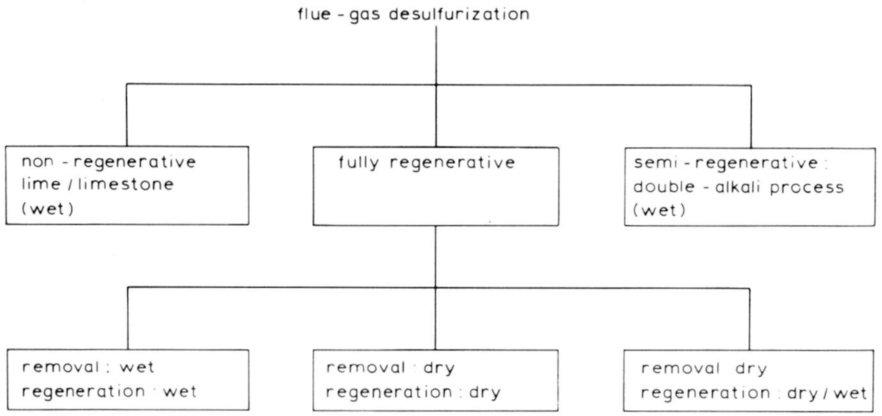

Figure XIV.16. Flue-gas desulfurization processes; nature and type of operation (commercially or semi-commercially applied processes).

Scrubbing with river water was the first process to be commercialized (Battersea Power Station near London, 1927), but it was discontinued because of high cost, river pollution and the occasional occurrence of objectionable odors. Scrubbing with seawater is still being considered, but this process introduced sea pollution with ecological consequences that cannot be adequately weighed; it is obviously suitable only for large power plants situated on the coast. More generally useful are processes in which the second washing step is done with an alkaline liquid, e.g. aqueous solutions or slurries containing alkali carbonates (usually Na_2CO_3), ammonia, lime, lime + limestone, lime + alkaline fly-ash or dolomite. With few exceptions, such processes are non-regenerative and yield a by-product which is difficult to dispose of. Sludges from single-stage washing operations contain ash components like heavy metals and even, in some cases, radioactive materials from the ash along with unconverted alkaline materials. This is the main reason why two-stage processes are now common. Products from two-stage washes range from gypsum, which can sometimes be purified to give wall-board grade gypsum, to sodium sulfate and ammonium sulfate, a low-grade fertilizer for which markets can hardly be found. Several of the wet SO_2 removal processes listed in Table XIV.6. have now been commercialized.

Table XIV.6. Examples of wet flue gas desulfurization processes

Non-regenerative processes

	absorbent	product	status early 1978	first unit in
Limestone/gypsum	$CaCO_3$	$CaSO_4$	46	1933
Lime/gypsum	CaO	$CaSO_4$	37	1964
Chiyoda Thoroughbred	H_2SO_4/Fe^{+++}	$CaSO_4$	14	1972
Fläkt-Hydro	sea water	--	1	1975

Regenerative processes

	active agent in			
	absorption	regene-ration	status early 1978*	first unit in
Wellman Lord	Na_2SO_3	heat	26	1970
Double alkali	NaOH	$CaCO_3$	30	1973
MgO scrubbing	MgO	heat	6	1971

* Number of commercial plants in North America, Western Europe and Japan.

Some flue gas desulfurization processes can be modified to give partial removal of NO_x as well. An example is the Chiyoda Thoroughbred process, which is based on oxidation of HSO_3^- with O_2 in an aqueous medium catalyzed by Fe ions. If a small amount of ozone is added to the stack gas more than 50% of the NO_x is converted into nitrate. This method appears to be rather costly because it requires a fair amount of electrical energy for ozone formation.

Disadvantages of wet SO_2 scrubbing are:
- the washing temperature is low (50-80 ^oC) and reheating to 125-130 ^oC is required to provide buoyancy and prevent the formation of a white condensation plume at the top of the stack;
- the pressure drop over the scrubbers is rather high and is a significant cost factor in view of the large volumes of gas to be processed.

Although wet scrubbing is not very favorable energy-wise, these processes are now well-established, giving reliable service once the teething troubles have been overcome. Another advantage over the dry methods discussed below is that the investment costs are lower.

Dry flue gas desulfurization methods

Dry SO_2 removal processes are in a much less advanced stage of development, although the first attempts to design dry flue gas treating methods are almost as old as the wet processes. In most cases some form of carbon was applied, usually to remove SO_2 from smelter off-gases. These early processes are not very suitable for cleaning power station flue gases because of the relatively low SO_2 concentration in the feed and the very large gas volumes to be processed.

A promising candidate process in which specially activated carbon is applied was developed by Bergbau-Forschung (see Figure XIV.17.). When flue gas containing SO_2 and oxygen is passed over this type of carbon sulfuric acid is formed on the surface according to

$$2\ SO_2 + O_2 + 2\ H_2O \rightarrow 2\ H_2SO_4$$

Regeneration is accomplished by heating to about 500 ^{o}C:

$$H_2SO_4 + C \rightarrow SO_2 + CO + H_2O$$

The first step, adsorption, is done in a moving bed of relatively large lumps of carbon to minimize the pressure drop. Solid and gas flow co-currently through the reactor. Fly-ash not removed previously by electrostatic precipitation is trapped by the coke and separated at the bottom of the moving bed reactor before regeneration in a second moving bed reactor. Hot sand is applied as a

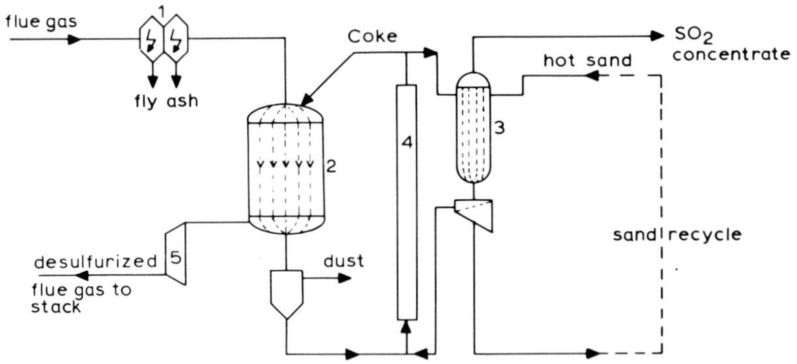

Figure XIV.17. Principle of Bergbau-Forschung process

1. *electrostatic precipitator*
2. *moving bed reactor*
3. *moving bed regenerator*
4. *coke lift*
5. *blower*

heat carrier because otherwise the rate of regeneration would be too low, requiring too large a reactor. Sand and carbon are separated before the former is recycled to the adsorption step. The SO_2 set free in the regenerator is sent to a Claus unit for conversion into sulfur.

The carbon should have the following properties:
- high capacity for SO_2 at adsorption conditions (120-180 ^{o}C);
- high affinity for SO_2 but low reactivity towards oxygen;
- adequate mechanical strength to withstand abrasion and high pressure at the bottom of the adsorption reactor; this strength should be maintained up to high degrees of carbon conversion;
- low pressure drop, i.e. relatively large particles should be used.

In general, these requirements can be met, although careful preparation of the carbon and optimum processing conditions are needed to maintain the strength of the carbon. The process has advantages over wet scrubbing in that the pressure drop is lower and because there is no need to reheat the flue gas. Energy consumption is, therefore, more favorable than for wet processing, but investment costs are higher. The process is operating on a demonstration plant scale in West Germany.

A second dry process with interesting possibilities was developed by Shell and tested for desulfurization of flue gases from both oil- and coal-fired boilers. In this process, CuO/Al_2O_3 acts as the sorbent for SO_2:

$$CuO + SO_2 + \tfrac{1}{2} O_2 \rightarrow CuSO_4 \qquad (a)$$

Regeneration is by reaction with H_2 or crude gas produced by e.g. steam reforming of naphtha or natural gas:

$$CuSO_4 + 2\ H_2 \rightarrow Cu + SO_2 + 2\ H_2O \qquad (b)$$

Since flue gas usually contains a little oxygen CuO is formed during subsequent sulfation period:

$$Cu + \tfrac{1}{2} O_2 \rightarrow CuO \qquad (c)$$

Thus, a *cyclic regenerative process* is possible in which regeneration and sulfanation periods are separated by an inerts flush. The attractive feature of this cycle is that both regeneration (reaction b) and sulfation/oxidation (reactions a + c) proceed smoothly at about 400 ^{o}C, obviating the need to heating or cooling of the solid *acceptor*. This makes it possible to apply a fixed-bed reactor, thus avoiding the attrition and wear problems that have plagued earlier processes like the Bureau of Mines' Alkalized Alumina process, which was abandoned because of excessive acceptor losses. Shell developed a special type of reactor for the process having a low pressure drop, the *parallel passage reactor*. In this reactor, the flue gas flows through channels between shallow beds of solid acceptor contained in baskets built of wire screen (see Figure XIV.18.). The rate of acceptance of SO_2 is a function of the rate of mass transfer through the wire screen. Two or more reactors of this type are needed to ensure continuous flue gas processing.

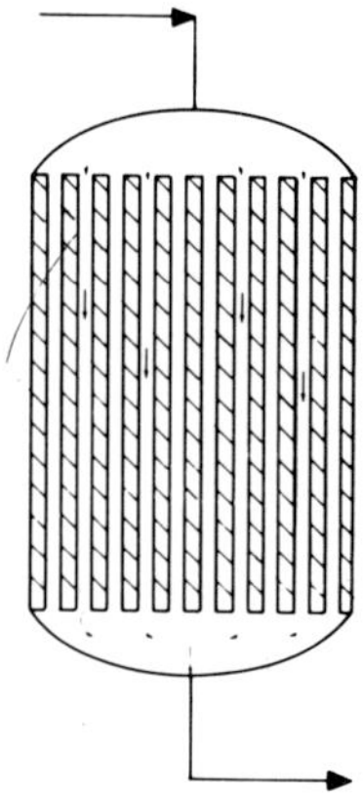

Figure XIV.18. Principle of parallel passage reactor

A simplified flowchart of the SFGD process is shown in Figure XIV.19. Two reactors are used in swing operation. The regeneration off-gas rich in SO_2 and free of oxygen is fed to an absorber after having been cooled by a water quench. The SO_2 is set free in the stripper and subsequently converted into sulfur. This absorber/stripper arrangement is needed to smooth the SO_2 concentrations fluctuations inherent to this non-stationary process. The net result is that SO_2 is not only concentrated by a factor of 100 or more, but also that it is separated from the oxygen in the flue gas.

The conversion to Cu during regeneration is nearly quantitative only if mixtures of steam and hydrogen are used. Otherwise, SO_2 is reduced to S or H_2S by H_2 and reacts with $CuSO_4$, CuO and/or Cu to give CuS. This is an unwanted side reaction, because the CuS formed by it is reoxidized to sulfate or CuO and SO_2 by oxygen in the flue gas during the subsequent sulfation:

$CuS + 2\ O_2 \rightarrow CuSO_4$ or $CuS + 1\frac{1}{2}\ O_2 \rightarrow CuO + SO_2$

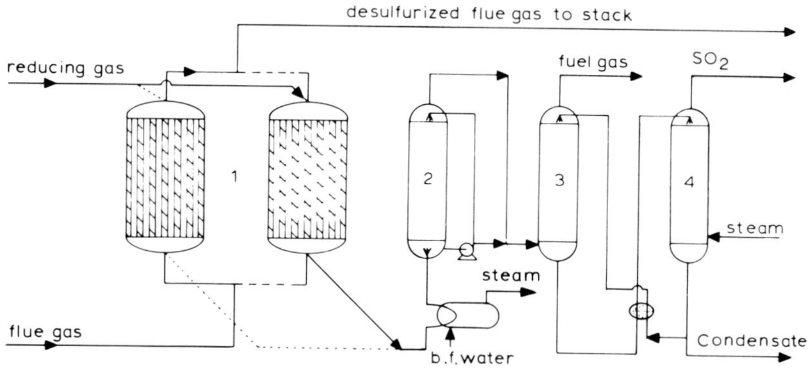

Figure XIV.19. SFGD process

1. reactors

2. quench column

3. absorber

4. stripper

Both reactions are even more exothermic than the oxidation of Cu to CuO (reaction c, above) and cause sintering with consequent capacity and activity loss of the acceptor. The steam probably acts by suppressing catalysis by Al_2O_3 of the reduction of SO_2 by H_2 to H_2S or sulfur. Disadvantages of the SFGD process are the high costs of investment and hydrogen: if side reactions are neglected the net reaction is the formation of water from hydrogen and oxygen. However, since most of the heat evolved is given up to the flue gas it can be recovered by heat exchange and the net energy consumption is low. Advantages are that the process is very easy to operate and automate, and that it can be modified to remove NO_x as well. This is done by injection of NH_3 in the flue gas; Cu ions catalyze the selective reduction of NO_x by NH_3 in the presence of oxygen.

Future prospects

The current search for new and improved methods of using fossil fuels other than oil and natural gas, combined with the attempts to find better ways of preventing environmental pollution by sulfur and nitrogen oxides and ash constituents, together represent perhaps the most massive development the world has ever seen. Its outcome is probably decisive for the prosperity of industrialized as well als developing countries over the next few decades. In particular, studies on coal processing and use are now given much attention, the most significant trends being:

- improvement of the capacity and thermal efficiency of combustion and gasification processes; in this connection there is a clear trend towards higher gasification pressures and the use of hydrogasification;
- methods of synthesizing building blocks for more complex chemicals from primary gasification products obtained from coal are being sought; examples are the manufacture of ethylene, propylene, butadiene and aromatics from coal-based methanol and, more attractively, directly from synthesis gas.

However, if coal is to be used on a much larger scale the methods of producing it must be considerably improved. It is expected that much work must be done to recover coal or shale and tar sand oil from underground deposits by methods other than mining.

As for methods of SO_2 pollution abatement, it appears probable that most efforts will be focussed on desulfurization of coal or crude gas obtained from coal. Although in some instances flue gas desulfurization may still be economically feasible, flue gas treatment is expected to be an interim solution rather than a final one. Methods without the danger of secondary pollution, e.g. by solids, are potentially the most attractive. This means that sulfur should be the final product. The prevention of pollution by nitrogen oxides requires further development of improved methods of combustion.

REFERENCES

General

1. Cheremisinoff, P.N., R.A. Young, 'Pollution Engineering Practice Handbook', Ann Arbor Science Phubl. Inc., Ann Arbor, Michigan, 1975.

2. Pfeiffer, J.B., 'Sulfur removal and recovery from industrial processes', Adv. Chem. Ser., Nr. 139, Am. Chem. Soc., Washington D.C. 1975.

3. Meyre, B., 'Sulfur, Energy and Environment', Elsevier Scient. Publ. Co., Amsterdam 1977.

4. Yaverbaum, L., 'Fluidized-bed Combustion of Coal and Waste Materials', Noyes Data Co., Park Ridge, N.J. 1977.

5. Wen, C.Y., S. Tone, Chem. Reac. Eng. Rev. ASC Symposium Ser. 72, 56-109 (1978).

6. Proc. 2nd and 3rd Int. Conf. on the Control of Gaseous Sulphur and Nitrogen Compound Emission, University of Salford, U.K. 1976, 1979.

Special

1. World Energy Supplies 1950-1974, United Nations, New-York 1976.
   Figure XIV.1. is reproduced from this publication with permission.

2. Montagna, A.A., W.C. Sun, J.A. Frayer, Proc. World Petr. Congr., 9th Vol. 6, 65-75 (1975).
   Figure XIV.3. is reproduced from this article with permission.

3. Montfoort, A.G., Chem. Weekbl. 70, S5-S12 (1974).
   Figure XIV.6. is reproduced from this article with permission.

4. Ullmanns Encyklopädie der technischen Chemie, 4th Ed., Vol. 14, 374 (1978).
   Figure XIV.12. is reproduced from this book with permission.

5. Figure XIV.13. is published with permission from Lurgi Kohle und Mineralöltechnik, Frankfurt/Main, W.-Germany.